CYBER-PHYSICAL SYSTEMS

Applications, Challenges, and Research Directions

Research Notes on Computing and Communication Sciences

CYBER-PHYSICAL SYSTEMS

Applications, Challenges, and Research Directions

Rajanikanth Aluvalu, PhD
Samarjeet Borah, PhD
Lalit Garg, PhD
Editors

APPLE ACADEMIC PRESS

First edition published 2026

Apple Academic Press Inc.
1265 Goldenrod Circle, NE,
Palm Bay, FL 32905 USA

760 Laurentian Drive, Unit 19,
Burlington, ON L7N 0A4, Canada

CRC Press
2385 NW Executive Center Drive,
Suite 320, Boca Raton FL 33431

4 Park Square, Milton Park,
Abingdon, Oxon, OX14 4RN, UK

Apple Academic Press exclusively co-publishes with CRC Press, an imprint of Taylor & Francis Group, LLC

Library and Archives Canada Cataloguing in Publication

Title: Cyber-physical systems : applications, challenges, and research directions / Rajanikanth Aluvalu, PhD, Samarjeet Borah, PhD, Lalit Garg, PhD, editors.
Names: Aluvalu, Rajanikanth, editor. | Borah, Samarjeet, editor | Garg, Lalit, 1977- editor
Description: First edition. | Series statement: Reseach notes on computing and communication sciences | Includes bibliographical references and index.
Identifiers: Canadiana (print) 2025020665X | Canadiana (ebook) 20250206668 | ISBN 9781779643414 (hardcover) | ISBN 9781779643421 (ebook)
Subjects: LCSH: Cooperating objects (Computer systems)—Industrial applications. | LCSH: Robotics.
Classification: LCC TK7895.E42 C93 2026 | DDC 006.2/2—dc23

Library of Congress Cataloging-in-Publication Data

CIP data on file with US Library of Congress

ISBN: 978-1-77964-341-4 (hbk)
ISBN: 978-1-77964-342-1 (ebk)

DOI: 10.1201/9781779643421

RESEARCH NOTES ON COMPUTING AND COMMUNICATION SCIENCES

ABOUT THE SERIES

Editor-in-Chief:

Dr. Samarjeet Borah
Department of Computer Applications,
Sikkim Manipal Institute of Technology,
Sikkim Manipal University (SMU), Majhitar, East Sikkim, India

Brief Description of the Series:

Computing can be defined as the practice in which computer technology is used to do a goal-oriented assignment. It covers design and development of hardware and software systems for various purposes. Computing devices are becoming an integral part of life now-a-days, including desktops, laptops, hand-held devices, smartphones, smart home appliances, etc. The evolution of the Internet of Things (IoT) has further enriched the same. The domain is ever growing and opening up many new endeavors, including cloud computing, social computing, ubiquitous computing, parallel computing, grid computing, etc.

In parallel with computing, another field has emerged that deals with the interconnection of devices. It is communication, and without which, the modern world cannot be thought of. It works with a basic purpose of transferring information from one place or person to another. This technology has a great influence in modern-day society. It influences business and society by making the interchange of ideas and facts more efficient. Communication technologies include the Internet, multimedia, e-mail, telephone, and other sound-based and video-based communication means.

This book series consists of both edited volumes as well as selected papers from various conferences. Volumes of the series contain the latest research findings in the field of communication engineering, computer science and engineering, and informatics. Therefore, the books cater to the needs of researchers and readers of a broader spectrum.

Coverage and Approach:

The series

- Covers a broad spectrum of research domains
- Presents on market-demanded product-based research works
- Discusses the latest developments in the field

The book series broadly considers contributions from the following fields:

- Artificial Intelligence and Expert Systems
- Big Data Analytics
- Broadband Convergence System and Integration Technologies
- Cellular and Mobile Communication
- Cloud Computing Technologies
- Computational Biology and Bioinformatics
- Computer and Information Security
- Computer Architecture
- Computer Graphics and Video Processing
- Control Systems
- Database Management Systems
- Data Mining
- Design Automation
- Digital Signal Processing
- GSM Communication
- High Performance Computing
- Human-Computer Interaction
- IoT and Blockchains
- Machine Learning

- Natural Language Processing
- Next Generation Communication Technologies
- Operating Systems and Networking
- Pervasive Computing and Cyber-Physical Systems
- Robotics and Automation
- Signal Processing
- Smart Internet of Everything
- SOC and System Platform Design Technologies
- Social Network Analysis
- Soft Computing

Types of Volumes:

This series presents recent developments in the domains of computing and communications. It will include mostly the current works and research findings, going on in various research labs, universities and institutions and may lead to development of market demanded products. It reports substantive results on a wide range of computational approaches applied to a wide range of problems. The series provides volumes having works with empirical studies, theoretical analysis or comparison to psychological phenomena. The series includes the following types of volumes:

- Conference Proceedings
- Authored Volumes
- Edited Volumes

Volumes from the series must be suitable as reference books for researchers, academicians, students, and industry professionals.

To propose suggestions for this book series, please contact the book series editor-in-chief. Book manuscripts should be minimum 250–500 pages per volume (11 point Times Roman in MS-Word with 1.5 line spacing).

Books and chapters in the series are included in Google Scholar and selectively in Scopus and possibly other related abstracting/indexing services.

Books in the *Research Notes on Computing and Communication Sciences* series

- **Applied Soft Computing: Techniques and Applications**
  Editors: Samarjeet Borah and Ranjit Panigrahi
- **Intelligent System Algorithms and Applications in Science and Technology**
  Editors: Sunil Pathak, Pramod Kumar Bhatt, Sanjay Kumar Singh, Ashutosh Tripathi, and Pankaj Kumar Pandey
- **Computing and Communications Engineering in Real-Time Application Development**
  Editors: B. K. Mishra, Samarjeet Borah, and Hemant Kasturiwale
- **Cyber-Physical Systems: Applications, Challenges, and Research Directions**
  Editors: Rajanikanth Aluvalu, Samarjeet Borah, and Lalit Garg
- **Applied Soft Computing Techniques: Theoretical Principles and Practical Applications**
  Editors: Samarjeet Borah, Ratna Raja Kumar Jambi, Sharifah Sakinah Syed Ahmed, and Mahendra Prabhakar Deore
- **Intelligent Computing and Analytics in Image Processing**
  Editors: Samarjeet Borah, Jambi Ratna Raja Kumar, Laxmi Bewoor, Rahul Joshi, and Mahendra Prabhakar Deore

ABOUT THE EDITORS

Rajanikanth Aluvalu, PhD
Director, Symbiosis Institute of Technology, Hyderabad, India

Rajanikanth Aluvalu, PhD, is an accomplished professor currently working as Director at the Symbiosis Institute of Technology, Hyderabad, India. He has a wealth of over 21 years of experience and contributions in the field of computer science engineering and academia. His extensive tenure as a professor, coupled with his leadership roles in reputed academic institutions and professional organizations, reflects his commitment to education and research. He holds a PhD in Computer Science and Engineering, specializing in cloud computing. He also holds the IGIP International Engineering Educator certification. He practices outcome-based education (OBE) and believes in interactive learning through the use of technology and innovations in pedagogical practices. He is serving as a member of the boards of studies for various institutions. He has a strong understanding of the NBA and NAAC accreditation processes. Prof. Aluvalu demonstrates a strong foundation in both theoretical knowledge and practical expertise. His significant publication record comprises over 150 research articles in various peer-reviewed journals and conferences. He has edited seven books for Springer, De Gruyter, CRC, IGI Global, and Apple Academic Press. He has also edited special issues for *Soft Computing* (Springer), *International Journal of Data Analysis Techniques and Strategies* (Inderscience), and SPCE journals. He is the recipient of a best faculty advisor award from the IEEE Hyderabad Section as well as the IUCEE (Indo Universal Collaboration for Engineering Education) Faculty Fellow Award (2018). He is a life member of the International Society for Technology in Education and a member of the Association for Computing Machinery and Machine Intelligence Research Labs (MIR Labs). He served as a resource person for various technical programs, faculty development programs, etc. He has organized various international conferences and delivered keynote addresses.

Samarjeet Borah, PhD
Professor and Head, Department of Computer Applications, SMIT, Sikkim Manipal University (SMU), Sikkim, India

Samarjeet Borah, PhD, is currently working as Professor and Head in the Department of Computer Applications, SMIT, Sikkim Manipal University (SMU), Sikkim, India. Dr. Borah has carried out various funded projects from AICTE (GoI), DST-CSRI (GoI) etc. He has organized various workshops and conferences at national and international levels. Dr. Borah is involved with various book volumes and journals of repute from Springer, IEEE, Inderscience, and IGI Global as Editor or Guest Editor. He is the editor-in-chief of the book and proceedings series *Research Notes on Computing and Communication Sciences*, published by Apple Academic Press, USA. His areas of research are data mining, data science, and machine learning.

Lalit Garg, PhD
Associate Professor, Department of Computer Information Systems, University of Malta

Lalit Garg, PhD, is an Associate Professor in the Department of Computer Information Systems at the University of Malta, and an honorary Lecturer at the University of Liverpool, UK. He has also been a researcher at the Nanyang Technological University, Singapore, and Ulster University, UK. He has supervised over 200 master's dissertations and DBA and PhD theses and has published over 150 publications in refereed journals and conferences, and he has also edited five books and holds 22 patents. His research interests are in business intelligence, machine learning, data science, deep learning, cloud computing, mobile computing, etc. He participates in many EU and locally funded projects, including a one-million-euro Erasmus+ Capacity-Building project in Higher Education (CBHE), titled Training for Medical education via innovative eTechnology. MediTec. He also received Malta Council of Science and Technology's Space Research funds. The University of Malta has awarded him the 2021–22 Research Excellence Award for exploring novel intelligent computing methods for healthcare requirements forecasting, allocation, and management (NICE-Healthcare).

CONTENTS

CONTRIBUTORS

Ananya Aggarwal
BML Munjal University, Gurugram, Haryana, India

Anubhav Agrawal
BML Munjal University, Gurugram, Haryana, India

Kuruba Ankitha
Department of Electronics and Communication Engineering, G. Pullaiah College of Engineering and Technology Kurnool, India

Pradeep Kumar Arya
Symbiosis Institute of Technology, Pune, Maharashtra, India

Jyoti Bali
KLE Technological University, Hubballi, India

N. P. G. Bhavani
Department of Electronics and Communication Engineering, Saveetha Institute of Medical and Technical Sciences, Tamil Nadu, Chennai, India

Pratham Bhoir
Department of Computer Engineering, Dwarkadas J. Sanghvi College of Engineering, Mumbai, India

Yaka Bulo
Department of Electronics and Communication, NIT Arunachal Pradesh, Papum Pare, Arunachal Pradesh, India

Kishor Kumar Reddy C.
Department of Computer Science and Engineering, Stanley College of Engineering and Technology for Women, Hyderabad, India

Parimita Dash
School of Law, Kalinga Institute of Industrial Technology (a Deemed to be University), Bhubaneswar, Odisha, India

Shatarupa Dash
Department of Computer Science and Engineering, ITER, Siksha 'O' Anusandhan University, Bhubaneswar, India

B. Dasharath
Department of ECE, G.Pullaiah College of Engineering and Technology, Kurnool, Andhra Pradesh, India

Urmi Dedhia
Department of Computer Engineering, Dwarkadas J. Sanghvi College of Engineering, Mumbai, India

Uday Kiran Dhane
Department of ECE, G. Pullaiah College of Engineering and Technology, Kurnool, India

T. Dheeraj
Department of ECE, G.Pullaiah College of Engineering and Technology, Kurnool, Andhra Pradesh, India

Srinath Doss
Botho University, Botswana

Arunkumar Giriyapur
KLE Technological University, Hubballi, India

Smita Gupta
Department of Electronics and Communication, NIT Arunachal Pradesh, Papum Pare, Arunachal Pradesh, India

Marlia M. Hanafiah
Department of Earth Sciences and Environment, Faculty of Science and Technology, Universiti Kebangsaan Malaysia (UKM), Bangi, Selangor, Malaysia
Centre for Tropical Climate Change System, Institute of Climate Change, Universiti Kebangsaan Malaysia (UKM), Bangi, Selangor, Malaysia

B. Hemanth
Department of ECE, G. Pullaiah College of Engineering and Technology, Kurnool, Andhra Pradesh, India

Deepika Roselind J.
Department of Computer Science and Engineering, St. Joseph's Institute of Technology, Chennai, India

Ranjana Jadhav
Bharati Vidyapeeth (Deemed to be) University, Institute of Management and Entrepreneurship Development, Erandwane, Pune, India

K. Jagadeep
Department of ECE, G. Pullaiah College of Engineering and Technology, Kurnool, Andhra Pradesh, India

M. Jyothirmai
Department of ECE, Jawaharlal Nehru Technological University Anantapur, Ananthapuramu, Andhra Pradesh, India
Ravindra College of Engineering for Women, Kurnool, Andhra Pradesh, India

Ramesh Karnati
Department of Computer Science and Engineering, Vardhaman College of Engineering, Hyderabad, India

N. Kavitha
Department of Electronics and Communication Engineering, Saveetha Institute of Medical and Technical Sciences, Tamil Nadu, Chennai, India

Manohar Kedar
SNJB's Late Sau. K. B. Jain College of Engineering, Neminagar, Chandwad, Nashik, Maharashtra, India

Junaid Md. Khan
Department of Electronics and Communication, NIT Arunachal Pradesh, Papum Pare, Arunachal Pradesh, India

C. B. Kolanur
KLE Technological University, Hubballi, India

Gautam Kumar
Department of Electronics and Communication, NIT Arunachal Pradesh, Papum Pare, Arunachal Pradesh, India

Sandeep Kumar
Meluha International School, Hyderabad, India

Shubhasri Kundu
School of Electrical Engineering, Kalinga Institute of Industrial Technology (a Deemed to be University), Bhubaneswar, India

Sai Ramesh L.
School of Computer Science Engineering, VIT, Chennai, India

Manoj Kumar Maharana
School of Electrical Engineering, Kalinga Institute of Industrial Technology (a Deemed to be University), Bhubaneswar, Odisha, India

Areena Mahek
School of Computing and Digital Media, Depaul University, Chicago, IL, USA

Kush Maniar
Department of Computer Engineering, Dwarkadas J. Sanghvi College of Engineering, Mumbai, India

Rajani Meshram
Jayakar Knowledge Resource Centre, Savitribai Phule Pune University, Pune, Maharashtra, India

Nilakshi Maruti Mule
Department of Computer Engineering, Smt. Kashibai Navale College of Engineering, Savitribai Phule Pune University, Pune, Maharashtra, India

Sikhinam Nagamani
Department of Computer Science Engineering, Indian Institute of Information Technology, Tiruchirappalli, Tamil Nadu, India

Yogesh Nagargoje
CSMSS CHH. Shahu College of Engineering, Kanchanwadi, Paithan Road, Chhatrapati Sambhajinagar, Aurangabad, India

Sindhu Nair
Department of Computer Engineering, Dwarkadas J. Sanghvi College of Engineering, Mumbai, India

C. V. Lakshmi Narayana
CSE Department, AITS, Rajampet, Andhra Pradesh, India

Dipti Durgesh Patil
Department of Information Technology, MKSSS's Cummins College of Engineering for Women, Karve Nagar, Savitribai Phule Pune University, Pune, Maharashtra, India

Amarendra Pattnaik
School of Law, Kalinga Institute of Industrial Technology (a Deemed to be University), Bhubaneswar, Odisha, India.

Anil Bhanudas Pawar
Art's Science and Commerce College, Kolhar TAL, Rahata, Ahmednagar, India

Mallellu Sai Prashanth
Symbiosis Institute of Technology, Hyderabad Campus, Symbiosis International University, Pune, India

Arjyadhara Pradhan
School of Electrical Engineering, KIIT, Bhubaneswar, India

M. Siva Prasad
Department of ECE, G. Pullaiah College of Engineering and Technology, Kurnool, India

Linkan Priyadarsini
School of Electrical Engineering, Kalinga Institute of Industrial Technology (a Deemed to be University), Bhubaneswar, Odisha, India

Anisha P. R.
Department of Computer Science and Engineering, Stanley College of Engineering and Technology for Women, Hyderabad, Telangana, India

Varun Racharla
CSE Department, AITS, Rajampet, Andhra Pradesh, India

N. Ramamurthy
Department of Electronics and Communication Engineering, G. Pullaiah College of Engineering and Technology Kurnool, India

S. China Ramu
Department of CSE, Chaitanya Bharathi Institute of Technology (CBIT), Hyderabad, India

B. Uday Kiran Reddy
Department of ECE, G. Pullaiah College of Engineering and Technology, Kurnool, India

G. Vrishank Reddy
Meluha International School, Hyderabad, India

H. Venkateshwara Reddy
Department of Computer Science and Engineering, Vardhaman College of Engineering, Hyderabad, India

N. Renugadevi
Department of Computer Science Engineering, Indian Institute of Information Technology, Tiruchirappalli, Tamil Nadu, India

Swati Sucharita Roy
Department of Computer Science and Engineering, ITER, Siksha 'O' Anusandhan University, Bhubaneswar, India

Kanimozhi S.
School of Computer Science Engineering, VIT, Chennai, India

R. V. S. Srikasyap Sagar
Meluha International School, Hyderabad, India

Bharat Jyoti Ranjan Sahu
Department of Computer Science and Engineering, ITER, Siksha 'O' Anusandhan University, Bhubaneswar, India

Narapureddy Sandhya
CSE Department, AITS, Rajampet, Andhra Pradesh, India

Jangiti Uma Sankar
CSE Department, AITS, Rajampet, Andhra Pradesh, India

K. R. Senthilkumar
Sri Krishna Arts and Science College, Coimbatore, India

Gadige Shirisha
Department of Electronics and Communication Engineering, G. Pullaiah College of Engineering and Technology Kurnool, India

Ranbir Singh
BML Munjal University, Gurugram, Haryana, India

Galisetty Sravani
CSE Department, AITS, Rajampet, Andhra Pradesh, India

Nandivargam Swapna
Department of Electronics and Communication Engineering, G. Pullaiah College of Engineering and Technology Kurnool, India

Shilpa Tanvashi
KLE Technological University, Hubballi, India

K. Sai Teja
Department of ECE, G. Pullaiah College of Engineering and Technology, Kurnool, India

T. Tirupal
Department of ECE, G. Pullaiah College of Engineering and Technology, Kurnool, India

Surya Narayan Tripathy
School of Electrical Engineering, Kalinga Institute of Industrial Technology (Deemed to be) University, Bhubaneswar, India

Ravi Uyyala
Department of CSE, Chaitanya Bharathi Institute of Technology (CBIT), Hyderabad, India

Ramana Murthy B. V.
Department of Computer Science and Engineering, Stanley College of Engineering and Technology for Women, Hyderabad, India

Lokita Varma
Department of Computer Engineering, Dwarkadas J. Sanghvi College of Engineering, Mumbai, India

Viswanadham Vamsi
CSE Department, AITS, Rajampet, Andhra Pradesh, India

Muni Sekhar Velpuru
Department of Information Technology, Vardhaman College of Engineering, Hyderabad, India

PREFACE

Cyber-physical systems (CPS) cleverly merge computing algorithms with physical operations, thereby enabling the state-of-the-art response system. CPS integrates the virtual and physical dimensions and enhance efficiency, automation, and interconnectedness in various sectors such as healthcare, transportation, manufacturing, etc. This amalgamation plays a major role in shaping the potential of using intelligent technologies in interconnected systems. Hereby keeping in view the same, we present a new book *Cyber-Physical Systems: Applications, Challenges, and Research Directions*, which consists of 23 chapters.

The assortment of research in this volume explores various facets of robotics and engineering. One study focuses on the cutting-edge methodologies to ensure the safe navigation of industrial robots. Another integrates robotics and audio technology to enhance audio performance through noise cancellation. The third delves into the reduction of total harmonic distortion in nine-level inverters, employing a holistic approach that combines theoretical analysis, practical solutions, and data visualization techniques. These scenarios demonstrate a wide range of innovative developments at the intersection of robotics, automation, and engineering.

This thorough compilation of research areas involves various technological dimensions followed by their practical implementations. Emphasis on the legal aspects is also important in the case of autonomous vehicles, and a thorough examination of the complexities linked to inadvertent liability is required. Selective weighting histogram shifting can be used to address image-processing-related issues in CPS. Analyzing antenna design for 5G communication involves a thorough examination within the context of multiple input multiple output (MIMO) technology. Attention-based coupled frameworks can be applied to accurately outline roads and detect potholes, thereby aiding in the maintenance of road infrastructure. The integration of CPS in elections aims to enhance electoral integrity through decentralized and secure voting methods. Additional studies delve into media insights, recognition of player actions in sports, virtual simulation of a piano, vehicle-to-grid technology, intelligent helmets, and applications of IoT in various domains, including automation of libraries, nowcasting of precipitation, management of traffic, and analysis of breath for health predictions. This collection serves to emphasize the extensive realm of technological progress and its applications across a diverse range of fields.

Cyber-Physical Systems: Applications, Challenges, and Research Directions presents an exhaustive examination of the ever-evolving environment wherein the tangible and intangible realms intersect. Crafted by seasoned scholars in the field, this publication delves into a variety of practical implementations, illuminating the obstacles they entail and providing valuable insights into prospective avenues of investigation. Encompassing a wide array of sectors including healthcare, transportation, and infrastructure, it deftly navigates the intricate interplay between the digital and physical constituents. By addressing challenges and proposing innovative research avenues, the book serves as a valuable resource for researchers, practitioners, and enthusiasts seeking a deeper understanding of the evolving realm of CPS.

In the end, the editors want to extend their heartfelt thanks to the authors, reviewers, and Apple Academic Press for their invaluable contributions in bringing this book to fruition.

—Dr. Rajanikanth Aluvalu
Symbiosis Institute of Technology, Hyderabad, India

Dr. Samarjeet Borah
Sikkim Manipal University, Sikkim, India

Dr. Lalit Garg
University of Malta, Malta

INTRODUCTION

A cyber-physical system (CPS) refers to a system in which physical processes and operations are integrated with computing and communication elements to fulfill various technological and societal requirements. Essentially, it is the fusion of the physical world with the virtual or digital world. CPS combines computational algorithms with physical processes to enhance monitoring, control, and optimization of the overall system.

The key characteristics of CPSs include interconnected components that are basically the physical components; sensors for collecting data from the physical environment and actuators to affect physical processes; computational capabilities for processing data, making decisions, and controlling physical processes; real-time operation to respond quickly to changes in the physical environment; and finally the integration of cyber and physical components while creating a synergistic relationship between the virtual and physical aspects.

A CPS incorporates several connected aspects that merge the physical and virtual worlds. Physical processes ranging from industrial machinery to biological systems are at its core. These are monitored and controlled through the integration of sensors and actuators. Computational elements are used to process the data collected from the physical domain. Communication infrastructure facilitations are employed for a seamless data exchange between physical and other components. The sensors provide feedback to the control systems to regulate and optimize physical processes. In CPS, human interaction is crucial, necessitating user-friendly interfaces. It can be said that a CPS presents a multidimensional framework for advanced monitoring, control, and optimization of complex systems.

CPS plays a critical role in addressing various challenges of modern society. It offers innovative solutions that enhance efficiency, safety, and sustainability. In smart cities, CPS applications optimize traffic flow, energy consumption, and public services. It also mitigates urban congestion and reduces environmental impact. In the healthcare sector, CPS enables remote patient monitoring, tailored treatments, and timely interventions. In an industry context, it enhances automation thereby streamlining processes and minimizing resource wastage. Enabling precision farming in agriculture is an added advantage of CPS which optimizes resource usage and minimizes environmental impact. It provides the foundation for intelligent solutions in disaster management, climate monitoring, and resource allocation, and serves as a transformative technology. Nowadays CPS offers adaptive and interconnected solutions that address the complex challenges facing modern societies.

As can be seen, the research landscape in CPSs is quite dynamic and continually evolving. Investigators are actively exploring new avenues to improve the design, security, as well as performance of CPS across various domains. AI and machine learning integration with CPS is enabling systems to learn and adapt to dynamic environments. Still, the security concerns remain significant; however, efforts are being made to develop robust mechanisms to protect against cyber threats. As CPS applications continue to thrive, the research community is dedicated to concentrating on the emerging challenges and unlocking the full potential of these systems for the benefit of society.

The future research directions of CPSs are hovering to dive into several critical areas to advance the capabilities and applications of these integrated systems. Enhancing security concerns against evolving cyber threats will always remain a priority. Researchers may likely investigate some novel cryptographical techniques, intrusion detection systems (IDS), and secure-by-design practices. Future research may also concentrate on developing more energy-efficient CPS solutions by optimizing resource utilization and minimizing environmental impact. Various emerging technologies like 5G, 6G, and the exploration of decentralized approaches, such as edge and fog computing may also be integrated to shape the landscape of CPS research further.

CHAPTER 1

AN OVERVIEW OF COLLISION-FREE PATH PLANNING TECHNIQUES FOR INDUSTRIAL AUTONOMOUS ROBOTS

C. B. KOLANUR, JYOTI BALI, SHILPA TANVASHI, and ARUNKUMAR GIRIYAPUR

KLE Technological University, Hubballi, India

ABSTRACT

One of the most important research questions in industrial and humanoid robotics in dynamic situations is collision and occlusion-free path planning (COFPP). The effectiveness and efficiency of the path planning algorithms are assessed under a variety of restrictions, such as dynamic and static environments with various levels of complexity. This chapter carries out a comparative analysis of collision-free path planning in industrial robots. Discussion is held regarding the effectiveness of various collision-free path planning algorithms, including Dijkstra, A*, Probabilistic Road Map, RRT, and RRT*, as well as their benefits and drawbacks.

1.1 INTRODUCTION

The use of occlusion and collision-free path planning (OCFPP) techniques in various industrial activities, including industrial autonomous mobile

Cyber-Physical Systems: Applications, Challenges, and Research Directions. Rajanikanth Aluvalu, Samarjeet Borah, & Lalit Garg (Eds.)

DOI: 10.1201/9781779643421-1

robots, has great potential in line with Industry 4.0 criteria. One of the leading research questions in industrial robots is OCFPP. Humanoid robots are now widely used in fields such as education, business, agriculture, and search and rescue. In dynamic settings, OCFPP are essential for humanoid robots. Humanoid robots replace humans or outdated machinery in various industrial applications, significantly reducing human driver mistakes and neglect. Industrial applications use self-directed, collision-free robots in terms of architecture and operation [1–3].

The fundamental issue with object tracking and recognition is object collision. A potential research topic in the field of humanoid robotics is collision-free path planning (CFPP). Because they work quickly and can reach their destination in less time, industries are becoming increasingly reliant on robots. Without robotics, even the imaging industries appear to be off-limits. The usefulness of autonomous robots has grown significantly [4]. Robots are currently employed in a wide range of tasks, including warehouse work, industrial manufacturing, exploration, surveillance, medical applications, and demining in combat zones. Industrial robots typically operate with little to no prior awareness of their surroundings and any hazards, which makes the problem of CFPP difficult. Path-planning algorithms are used by robots, self-driving automobiles, and aircraft to discover the least-cost, safest, and most effective routes between a starting point and a destination. The collision-free approach is required for efficiency and more autonomy [5]. Many methods have been described for planning a humanoid robot's course without colliding with anything. Therefore, researchers working on industrial robots have primarily focused on CFPP.

In the first stages of development during the past ten years, improving path planning algorithm performance has proven to be more difficult. The current focus is on algorithms for graph searching, including Dijkstra, A*, D*, Voronoi diagram, visibility graph, and C-Space approach. Traditional techniques have been shown to be efficient and remedial. However, they have significant shortcomings, such as the fact that they are inappropriate for large-dimension applications and extremely expensive in more sophisticated contexts. The CFPP problem's objective is to assist mobile agents in locating a secure route from the desired or goal point (G) to the starting point (S). For mobile agents, it minimizes resource consumption along the shortest distance [6]. Object tracking or recognition may be calculated by automatic camera calibration for occlusion detection and depth information [3]. Robotics, video games, network routing, and DNA sequencing depend on finding the shortest path. Robots must therefore be aware of their location concerning the map [7].

In relation to a predetermined standard coordinate system, this study seeks to present an overview of safe CFPP approaches in a dynamic space. The robot's trajectory and speed can be reviewed using the collision time and position information. Planning a path without collisions enables movement along it at a set speed. For the coming obstacle, a path with the shortest possible distance can be found. Collision path planning is getting a robot to plan its route between the start and desired configurations while avoiding impediments in a working environment. The effectiveness and efficiency of the path planning algorithms are assessed under various restrictions, such as dynamically changing settings and static environments with varied levels of complexity [8–10].

1.2 PATH PLANNING

Path planning is the process of creating a route from a starting point (S) to an endpoint (G) using a full, partial, or dynamic map. For autonomous mobile robots, determining the best route between two points is necessary or crucial [11]. It reduces the amount of braking, turning, or other applications necessary along a given path. It is a significant robotics research issue. The task of path planning for an autonomous mobile robot involves many different activities [12]. By suggesting path planning in numerous domains, many problems are resolved.

Path planning can generally be used in a partially known and unknown organized environment by giving an appropriate algorithm. Without colliding, path planning allows the robot to change its vantage point. It cuts down on measurement time with only a short robot motion and can measure the entire object. It lowers the robot's energy use and the chance of damage. It involves linking a start configuration ("S") and a goal configuration ("G") while determining a continuous path and avoiding collision with recognized barriers [13].

Path planning could be classified into the following two categories:

- Model-based approach: Using the workspace data that the robot has already learnt, users can specify the three-dimensional models of items and their descriptions.

- Model-free approach: The user must describe details about the robot's surroundings using sensors, vision, or range in order to complete a task.

Model-free approach path planning is covered in this article. The environment is dynamic and unstructured, or the barriers are unknown in advance, making it completely unknown to the robot [14]. The robot must obtain or collect all environmental data in the form of sensor data, and then use these data algorithms to determine the shortest route [15]. The collision-free path-smoothing method is depicted in Figure 1.1.

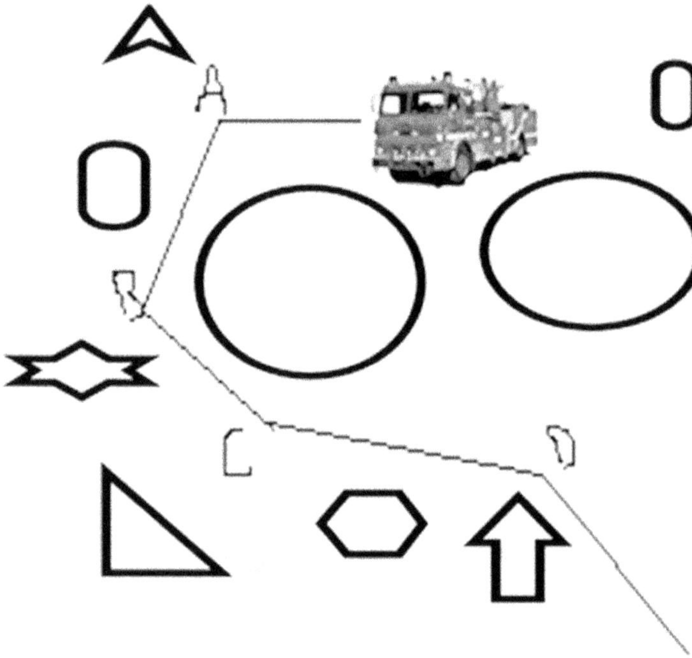

FIGURE 1.1 Collision-free path smoothing technique.

It is possible to determine the robot's shortest route around the impending barrier. There are essentially two methods: soft computing methods and graph-searching methods. Popular graph-searching techniques include the Visibility graph, VORONOI diagram, and Artificial Potential Fields. Neural Networks, Fuzzy Logic, Generic Algorithms, Ant Colony Optimization, Swarm Algorithms, and Hybrid Methods are standard Soft Computing methodologies [16]. Due to their success in intelligence and local minimum problems in collision-free route planning, the use of soft computing approaches in autonomous industrial robot path planning has grown.

1.3 COLLISION-FREE PATH PLANNING FOR INDUSTRIAL ROBOT

Industrial robots are capable of moving without colliding from one place to another in an environment with obstacles. Pseudo-obstructions are produced by natural obstacles, edges, or faces. It might also show up in intricate 3D settings where cutting-edge objects are being moved around [17]. Robot route planning involves using effective algorithms to determine a collision-free path from one place to another in order to solve issues with industrial robotics, computer animation, drug design, and automated surveillance. Planning a collision-free path enables moving along it at a set speed. There are numerous algorithms for path planning without collisions [18].

1.3.1 DIJKSTRA'S ALGORITHM

A routing protocol using Dijkstra's algorithm. It alters the forwarding table in routers. It only determines the shortest path between each source router node and all other routers in the network when the edge weights are non-negative, which ensures that the closest vertex is always marked [19]. This algorithm produces a collection of visited and unvisited nodes. All other node distances are infinity values, with the initial node at a distance of zero. The complexity of the calculation is correlated with the number of nodes and edges [20, 21].

The distance between the marking nodes and the end nodes is zero. It moves the current node of the visiting node, updates all nearby values, and then repeats the procedure for the incoming neighboring node with the closer distance value. Compared to other routes, the shortest way is indicated by this path line being designated as darkened, similar to ADCE in Figure 1.2. Any node cannot provide a faster route from A to E. The importance of the track is 6, which is the shortest path as compared to another way, and it demonstrates that the edge weights are, in fact, non-negative edges.

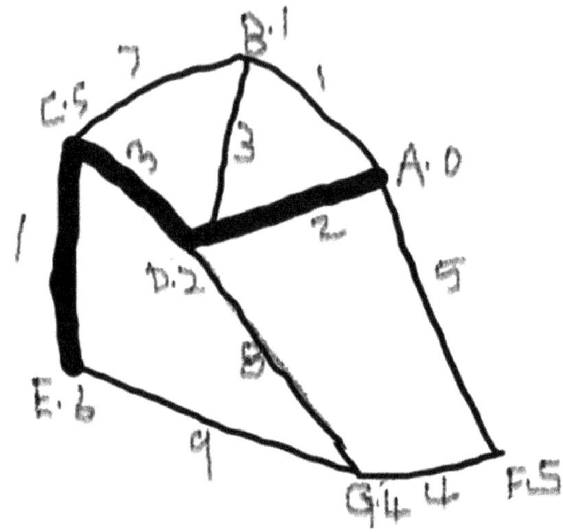

FIGURE 1.2 Dijkstra's algorithm shortest path-finding method.

1.3.2 A* ALGORITHM

Another kind of best-first-search path-finding algorithm is A*. Heuristics are added as an extension to the algorithm to prevent the search of unneeded nodes. To investigate options, it approximates a heuristic cost function [22]. The A* method starts by keeping track of all the reviewed nodes and their Euclidean distances from the target endpoint. According to Figure 1.3, this algorithm might result in paths that reduce the number of turning, braking, or other particular application needs.

1.3.3 PROBABILISTIC ROAD MAP ALGORITHM

The edges between the free and occupied space are constructed by the probabilistic road map, together with random samples. To create random nodes and edges, it iterates n times [23]. After repeatedly creating random nodes and edges, the graph develops to the desired beginning and ending points. The route is then decided from one end to the other [24]. In more complicated situations, as depicted in Figure 1.4, it is employed to prevent collisions between the robot and obstacles and give vision and a minimum occlusion path within the probabilistic roadmap.

FIGURE 1.3 A* shortest path-finding algorithm path.

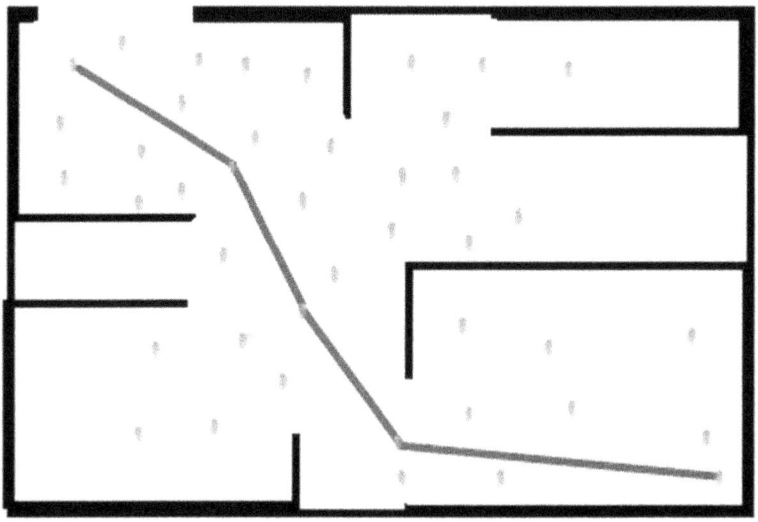

FIGURE 1.4 Path planning in environments of different complexity.

1.3.4 RAPIDLY EXPLORING RANDOM TREES (RRT) ALGORITHM

It is a potent tool for basic searching and sampling-based algorithms for convergence and algorithm speeds. This approach aims to find the shortest path by distance or matrix. The rapidly exploring random trees (RRT) algorithms randomly create the path points and use a cubic network to connect them to the closest free node [25]. A node within the desired region is generated using the RRT algorithm, or a limit is reached. This algorithm operates quickly. Figure 1.5 illustrates the RGBD sensor scanning in the MATLAB V-rep simulation engine to identify the target item and obstacles.

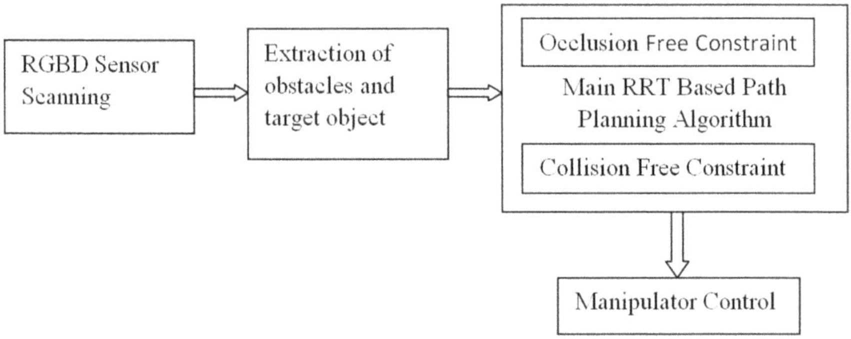

FIGURE 1.5 RRT procedure.

The segmentation and clustering approach for removing barriers and the target object is introduced in the second block. Following extraction, 3D global coordinates and dimensions are used to identify the target object and impediments. Additionally, the simulation is run using V-rep simulation, which takes in their sequences and activates the subsequent robot manipulator [26–30].

1.3.5 RRT EXTENSION PROCEDURE

The RRT method creates random samples and joins every node to create a distinct tree-type graph. Every node in the RRT technique is connected to a single parent node, and the tree root is rooted in a starting point [28, 29]. In two-dimensional configuration spaces, a tree is constructed in Figure 1.6. The system generates a random point at the configuration's base and once more determines whether any open space is there [30, 31]. After that, it seeks the closed point in the current tree once more before creating a new node by going along the tree's current vertex and a new random vertex in a straight line.

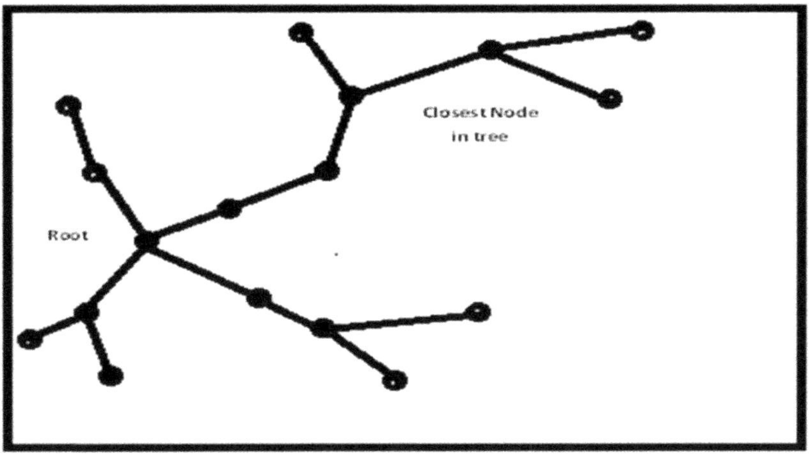

FIGURE 1.6 Rapidly random tree (nodes).

It will attempt to connect these two sites if the random configuration of the existing node and the distance from it in the tree are smaller [32]. This kind of method uses the effective displacement between two configuration space points to determine a trajectory free of collisions. Additionally, random samples of growing trees are produced. As a result, it traverses the open space, giving rise to the term Rapidly Randomly Tree [33]. Two trees can be formed to create a starting and an ending configuration: one is rooted in the stark configuration and the other is the ending or goal configuration [34].

The method's creation and connection of nodes from one node to another are depicted in Figure 1.7(a) [35, 36]. After that, the method tries to develop both the stark and end trees until they collide. Obstacles are present in the two-dimensional configuration space [37]. This approach works well in configuration areas with high dimensions. The Rapidly Random Tree approach gets beyond the limitations of the probabilistic Road Map algorithm, which only occasionally provides the best answer. The outcomes of the implementation V-rep simulation are displayed in Figure 1.7(b) [38].

1.4 OCCLUSION DETECTION FOR INTELLIGENT PATH PLANNING FOR INDUSTRIAL AUTONOMOUS ROBOTS

CFPP algorithms have recently offered industrial robots advanced navigational tactics. Occlusion detection offers an industrial robot an additional capability for better and more precise decision-making. The practice of locating hidden images using one or more sophisticated sensor setups is known as occlusion detection. For instance, a car may move in front of one person as two people cross paths while walking or running down the street. When a photo is taken from a fixed angle by another person, the car appears and then repeatedly vanishes, and because of the mismatched speeds, part of the car is obscured. Occlusion identification of such hidden image information is required in this case for decision-making. Robotic surgery, automated robot assembly and inspection, and many other machine vision-based procedures all depend on accurate occlusion detection. Such issues can be solved by incorporating a multivision system or a dynamic camera positioning system. A system that automatically detects hidden pictures using camera calibration is explored in more detail in the following section.

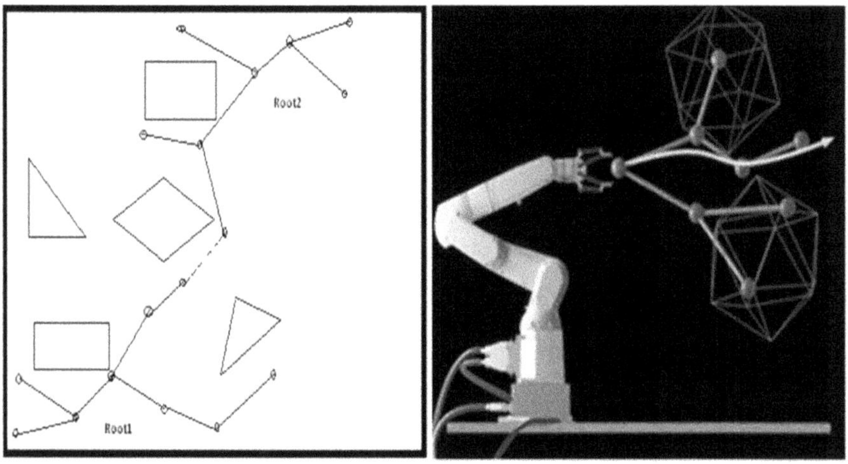

FIGURE 1.7 (a) Creating and connecting nodes in RRT, (b) simulated robot following the path planning.

1.4.1 AUTOMATIC CAMERA CALIBRATION-BASED OCCLUSION DETECTION SYSTEM

An occlusion detection algorithm is the Automatic Camera Calibration-Based Occlusion Detection system. It is used to identify video surveillance systems, locate occlusion, and enhance object tracking performance [39]. In this technique, moving objects are used to calibrate an automated red, green, and blue camera. Additionally, it is used to identify object depth data without the use of additional sensors. In addition, backdrop structure, occluded regions identification, and object depth estimates are included [40]. Therefore, this technique results in color distortion near the edge of any moving object. Artificial or automatic calibration patterns evaluate the vanishing points and extract lines from the image. Figure 1.8 displays the lines employed to calculate the camera parameters for object depth estimation [41].

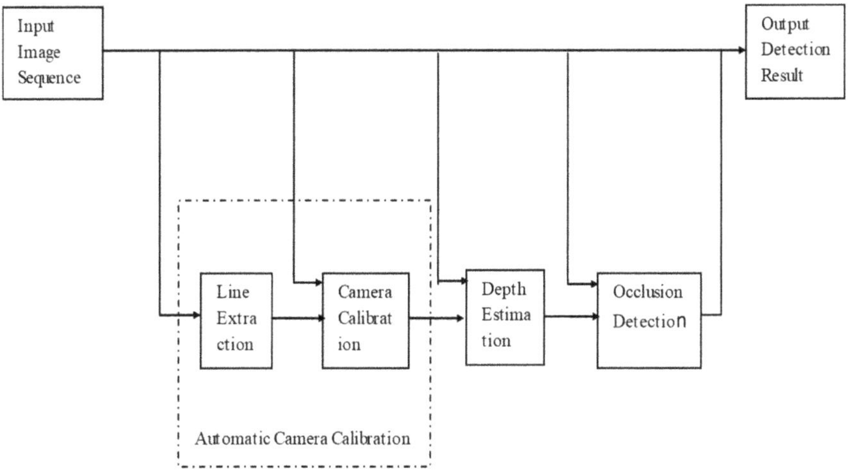

FIGURE 1.8 Object occlusion detection algorithm block diagram.

The block diagram shows the stages in processing the input image sequence beginning with camera calibration, depth estimation, and occlusion detection. Additionally, the effectiveness of the CFPP algorithms Dijkstra's, A*, RRT*, Genetic Algorithm (Dynamic), and Automatic Camera Calibration-Based occlusion detection algorithms are compared in terms of their dynamic environment, complexity, and computational cost. It is asserted that a superior algorithm for this is RRT* (RRT).

Table 1.1 provides a summary of the uses, benefits, and drawbacks related to various path-planning strategies. It is mentioned that, according to published literature, the RRT approach performs the best. The characteristics of OCFPP planning algorithms are shown in Table 1.2 in relation to the applicability, complexity, and computational cost in dynamic environments.

TABLE 1.1 A Comparison of Path Planning Techniques.

Sl No.	Type of Methods	Application	Advantages	Disadvantages
1.	Probabilistic road map-PRM	Occlusion and collision-free path planning	1. Convergence and retaining 2. Not so efficient in developing the shortest path as it is repetitive 3. Easy to implement 4. Applicable for many different types of motion planning problems	Provides the optimal solution occasionally
2.	Dijkstra's algorithm	Occlusion and collision-free path-finding	Minimal occlusion through the roadmap	Delay in finding the shortest distance path to the minimal occlusion path incurred
3.	A*	Shortest path-finding collision-free algorithm	Maps well to graph representation, flexibility, and versatility	Significant memory overhead, redundant points, and long computing time because it detects the least cost neighbor in its neighbors
4.	Automatic camera calibration-based occlusion detection	Occlusion and collision detection on	1. Multiple-camera video tracking 2. Moving object information 3. Tracking in a multiple-view surveillance system	Estimates depth information inaccurately in slanting ground
5.	Rapidly Exploring Random trees-RRT*	Occlusion and collision-free path planning 1. Fast and achieves the shortest path 2. Used in holonomic, non-holonomic, and kinodynamic problems with high DOF	1. Acheives convergence and algorithm with fast searching speed 2. Probabilistically optimal variant of RRT 3. Works efficiently in multiple query tasks 4. Built using the classical AI method making two random trees 5. RRT overcomes the drawbacks of PRM and RRT	

TABLE 1.2 Features of Collision and Occlusion-Free Path Planning Techniques.

Name of the Algorithm	Standards		
	Dynamic Environment Applicability	Complexity	Computation Cost
Dijkstra's	False	Low	High
Probabilistic road map (PRM) algorithm	True	Low	High
A*	False	Low	Medium
Rapidly Exploring Random Trees*(RRT*)	True	Low	High
Rapidly Exploring Random Trees (RRT)	True	Low	High
Automatic camera calibration-based occlusion detection algorithm	True	High	Medium

1.5 RESULTS AND DISCUSSION

The Dijkstra method, which delivers the shortest distance path to the minimal occlusion path, is the main form of the OCFPP algorithm, according to the study. A* algorithm is an improved form of the Dijkstra algorithm in finding the shortest path but is limited in that it detects the least-cost neighbor among the available neighbors while finding the shortest path. The probabilistic road map algorithm gives the optimal solutions occasionally as an OCFPP algorithm. These shortcomings of the PRM method are overcome by the RRT (RRT algorithm. The jagged and suboptimal path, which can be avoided by using the RRT* algorithm, is not improved by RRT, though. Rapidly Exploring Random Trees, or RRT*, is an algorithm with a quick search time. The method finds the shortest way; it is highly effective, simple-to-use, and suitable for many different kinds of path design. According to the most recent report, an automatic camera calibration-based occlusion detection algorithm provides multiple-view tracking surveillance systems, enabling sound decision-making.

1.6 CONCLUSION

The improvement in OCFPP algorithms has been increasingly adopted in various automated manufacturing systems and industrial and humanoid robots. The chapter discussed different collision and occlusion-free path-planning algorithms starting from the Dijkstra algorithm to the RRT* used for OCFPP. The advantages of the latest technology involving automatic camera calibration-based vision systems are discussed to aid accurate and efficient detection and decision-making for OCFPP. The proposed study provides insight into choosing OCFPP algorithms for autonomous mobility.

KEYWORDS

- **collision-free path planning**
- **industrial autonomous robots**
- **occlusion detection**
- **automatic camera calibration**
- **shortest path algorithm**

REFERENCES

1. Bali, J., Arpitha, H., Anushree, N., & Giriyapur, A. (2021). Power-efficient strategies for sensing in autonomous mobile robots, a critical requirement of I4. 0 standard. In *IOP Conference Series: Materials Science and Engineering* (Vol. 1187, No. 1, p. 012007). IOP Publishing. https://doi.org/10.1088/1757-899X/1187/1/012007.
2. Johnson, V. C., Bali, J. S., Kolanur, C. B., & Tanwashi, S. (2022). Industry 4.0: intelligent quality control and surface defect detection. 3C empresa. *Investigación y Pensamiento Crítico, 11*(2), 186–192. https://doi.org/10.17993/3cemp.2022.110250. 186-192.
3. Léonard, S., Croft, E. A., & Little, J. J. (2009). Planning collision-free and occlusion-free paths for industrial manipulators with eye-to-hand configuration. In *2009 IEEE/RSJ International Conference on Intelligent Robots and Systems* (pp. 5083–5088). IEEE.
4. Kujala, T., Kircher, K., & Ahlström, C. (2023). A review of occlusion as a tool to assess attentional demand in driving. *Human Factors, 65*(5), 792–808. https://doi.org/10.1177/00187208211010953, 2021.
5. Kamil, F., Tang, S. H., Khaksar, W., Zulkifli, N., & Ahmad, S. A. (2015). A review on motion planning and obstacle avoidance approaches in dynamic environments. *Advances in Robotics & Automation, 4*(2), 134–142. https://doi.org/10.4172/2168-9695.1000134, 2015.

6. Khatib, O. (1985). Real-time obstacle avoidance for manipulators and mobile robots. In *Proceedings. 1985 IEEE International Conference on Robotics and Automation* (Vol. 2, pp. 500–505). IEEE. https://doi.org/10.1109/ROBOT.1985.1087247.
7. Narksri, P., Takeuchi, E., Ninomiya, Y., & Takeda, K. (2021). Deadlock-free planner for occluded intersections using estimated visibility of hidden vehicles. *Electronics, 10*(4), 411. https://doi.org/10.3390/electronics10040411, 2021.
8. Do Nascimento, T. P., Costa, P., Costa, P. G., Moreira, A. P., & Conceição, A. G. S. (2013). A set of novel modifications to improve algorithms from the A* family applied in mobile robotics. *Journal of the Brazilian Computer Society, 19*, 167–179. https://doi.org/10.1007/s13173-012-0091-5, 2013.
9. Buniyamin, N., Ngah, W. W., Sariff, N., & Mohamad, Z. (2011). A simple local path planning algorithm for autonomous mobile robots. *International Journal of Systems Applications, Engineering & Development, 5*(2), 151–159.
10. Agarwal, V., Tapaswi, S., & Chanak, P. (2021). A survey on path planning techniques for mobile sink in IoT-enabled wireless sensor networks. *Wireless Personal Communications, 119*(1), 211–238. https://doi.org/10.1007/s11277-021-08204-w, 2021.
11. Keshmiri, S., & Payandeh, S. (2009). An overview of mobile robotic agents motion planning in dynamic environments. In *Proceedings of the 14th IASTED International Conference, Robotics and Applications (RA20), MA, Boston* (pp. 152–159).
12. Bijli, M., & Kumar, N. (2021). Autonomous navigation of mobile robot with obstacle avoidance: a review. In *Futuristic Trends in Network and Communication Technologies: Third International Conference, FTNCT 2020, Taganrog, Russia, October 14–16, 2020, Revised Selected Papers, Part II 3* (pp. 305–316). Springer. https://doi.org/10.1007/978-981-16-1483-5_28, 2021.
13. Nilavar, A. C., Bali, J., Huggi, A., Giriyapur, A., & Ashwini, G. K. (2021). Implementation of a navigational path planning algorithm for an autonomous mobile robot. In *2021 IEEE India Council International Subsections Conference (INDISCON)* (pp. 1–5). IEEE. https://doi.org/10.1109/INDISCON53343.2021.9582210.
14. Wee, S. G., Kim, Y. G., An, J., Lee, D. H., & Lee, S. G. (2015). Composite local path planning for multi-robot obstacle avoidance and formation navigation. *Advances in Computer Science: An International Journal, 4*(5), 61–68.
15. Mujumdar, A., & Padhi, R. (2011). Evolving philosophies on autonomous obstacle/collision avoidance of unmanned aerial vehicles. *Journal of Aerospace Computing, Information, and Communication, 8*(2), 17–41.
16. Garrido, S., Moreno, L., & Blanco, D. (2009). Exploration of 2D and 3D environments using Voronoi transform and fast marching method. *Journal of Intelligent and Robotic Systems, 55*, 55–80.
17. Baumann, M. A., Dupuis, D. C., Léonard, S., Croft, E. A., & Little, J. J. (2008). Occlusion-free path planning with a probabilistic roadmap. In *2008 IEEE/RSJ International Conference on Intelligent Robots and Systems* (pp. 2151–2156). IEEE.
18. Castro, D., Nunes, U., & Ruano, A. (2004). Feature extraction for moving objects tracking system in indoor environments. *IFAC Proceedings Volumes, 37*(8), 96–101.
19. Bhavesh, V. A. (2015). Comparison of various obstacle avoidance algorithms, *International Journal of Engineering Research & Technology (IJERT), 4*(12), http://dx.doi.org/10.17577/IJERTV4IS120636.

20. Mei, X., Ling, H., Wu, Y., Blasch, E., & Bai, L. (2011). Minimum error bounded efficient ℓ 1 tracker with occlusion detection. In *CVPR 2011* (pp. 1257–1264). IEEE. https://doi.org/10.1109/CVPR.2011.5995421, 2011.
21. Shin, H., & Chae, J. (2020). A performance review of collision-free path planning algorithms. *Electronics, 9*(2), 316.
22. Sun, J., Li, Y., Kang, S. B., & Shum, H. Y. (2005). Symmetric stereo matching for occlusion handling. In *2005 IEEE Computer Society Conference on Computer Vision and Pattern Recognition (CVPR'05)* (Vol. 2, pp. 399–406). IEEE.
23. Elbanhawi, M., & Simic, M. (2014). Sampling-based robot motion planning: A review. *IEEE Access, 2*, 56–77. https://doi.org/10.1109/ACCESS.2014.2302442.
24. Otte, M. W., Knights, D., Pfeiffer III, J. J., Mulligan, J., & Grudic, G. (2008). New techniques for path planning in image space.
25. Kim, Y. J., Wang, J. H., Park, S. Y., Lee, J. Y., Kim, J. J., & Lee, J. J. (2014). A RRT-based collision-free and occlusion-free path planning method for a 7DOF manipulator. In *2014 IEEE International Conference on Mechatronics and Automation* (pp. 1017–1021). IEEE.
26. Park, J. S., & Manocha, D. (2020). HMPO: Human motion prediction in occluded environments for safe motion planning. https://doi.org/10.48550/arXiv.2006.00424.
27. Lindemann, S. R., & LaValle, S. M. (2005). Current issues in sampling-based motion planning. In *Robotics Research. The Eleventh International Symposium: With 303 Figures* (pp. 36–54). Springer Berlin Heidelberg.
28. Huggi, A. S., Nilavar, A. C., Bali, J., Giriyapur, A., & Ashwini, G. K. (2021). Implementation of sensor fusion for a mobile robot application. In *2021 International Conference on Recent Trends on Electronics, Information, Communication & Technology (RTEICT)* (pp. 15–20). IEEE. https://doi.org/10.1109/RTEICT52294.2021.9573730, 2021.
29. Kunchev, V., Jain, L., Ivancevic, V., & Finn, A. (2006). Path planning and obstacle avoidance for autonomous mobile robots: a review. In *Knowledge-Based Intelligent Information and Engineering Systems: 10th International Conference, KES 2006, Bournemouth, UK, October 9-11, 2006. Proceedings, Part II 10* (pp. 537–544). Springer. https://doi.org/10.1007/11893004_70, 2006.
30. Dolgov, D., Thrun, S., Montemerlo, M., & Diebel, J. (2010). Path planning for autonomous vehicles in unknown semi-structured environments. *The International Journal of Robotics Research, 29*(5), 485–501.
31. Fulgenzi, C. (2009). *Autonomous navigation in dynamic uncertain environment using probabilistic models of perception and collision risk prediction* (Doctoral dissertation, Institut National Polytechnique de Grenoble-INPG).
32. Masehian, E., & Naseri, A. (2010). Mobile robot online motion planning using generalized Voronoi graphs.
33. González-Banos, H. H., Hsu, D., & Latombe, J. C. (2018). Motion planning: recent developments. In *Autonomous Mobile Robots* (pp. 373–416).
34. Singh, R., Lal, S., & Vanualailai, J. (2013). Obstacle avoidance via social mediation in cooperative payload transportation. In *2013 IEEE Intelligent Vehicles Symposium (IV)* (pp. 352–357). IEEE.
35. Chen, J. H., & Song, K. T. (2018). Collision-free motion planning for human-robot collaborative safety under cartesian constraint. In *2018 IEEE International Conference on Robotics and Automation (ICRA)* (pp. 4348–4354). IEEE.

36. Gupta, A., & Padhi, R. (2010). Nonlinear geometric and differential geometric guidance of UAVs with vision sensing for reactive obstacle avoidance. *DTIC Document.*
37. Hentout, A., Maoudj, A., Yahiaoui, D., & Aouache, M. (2019). RRT-A*-BT approach for optimal collision-free path planning for mobile robots. *Algerian Journal of Signals and Systems, 4*(2), 39–50.
38. Bennewitz, M., & Burgard, W. (2000). A probabilistic method for planning collision-free trajectories of multiple mobile robots. In *14th European Conference on Artificial Intelligence.*
39. DeSouza, G. N., & Kak, A. C. (2002). Vision for mobile robot navigation: A survey. *IEEE Transactions on Pattern Analysis and Machine Intelligence, 24*(2), 237–267.
40. Im, J., Jung, J., & Paik, J. (2015). Single camera-based depth estimation and improved continuously adaptive mean shift algorithm for tracking occluded objects. In *Advances in Multimedia Information Processing—PCM 2015: 16th Pacific-Rim Conference on Multimedia, Gwangju, South Korea, September 16-18, 2015, Proceedings, Part II 16* (pp. 246–252). Springer International Publishing.
41. Jung, J., Yoon, I., & Paik, J. (2016). Object occlusion detection using automatic camera calibration for a wide-area video surveillance system. *Sensors, 16*(7), 982.

CHAPTER 2

ROBOTIC MACHINE DESIGN AND IMPLEMENTATION OF NOISE CANCELLATION TECHNIQUES IN MICROPHONES

T. TIRUPAL, T. DHEERAJ, B. DASHARATH, B. HEMANTH, and K. JAGADEEP

Department of ECE, G. Pullaiah College of Engineering and Technology, Kurnool, Andhra Pradesh, India

ABSTRACT

The active noise cancellation (ANC) system is composed of microphones that are capable of detecting both internal and external noise. Additionally, the device incorporates an ANC system, which effectively mitigates noise. Furthermore, an internal speaker is integrated into the device to counteract external noises by employing sound wave cancellation techniques. This process can be likened to the addition of a positive two externally and a negative two internally, resulting in a net sum of zero. The adaptive robotic machine utilizes microphones and speakers to automatically adjust its noise cancellation capabilities in response to the surrounding environment. This chapter investigates the existing and recent studies on adaptive filter techniques in ANC systems. Many applications demand adaptive algorithms that converge swiftly since the change in incoming signals might be quite fast. This study examines the application of robotic cyber machine adaptive least mean square (LMS) algorithms for

Cyber-Physical Systems: Applications, Challenges, and Research Directions. Rajanikanth Aluvalu, Samarjeet Borah, & Lalit Garg (Eds.)

DOI: 10.1201/9781779643421-2

noise reduction in voice signals. This leads to improved efficiency and decreased computational complexity. The normalized LMS (NLMS) approach provides a noise suppression of approximately 13 dB, resulting in significant improvement in speech quality. In contrast, the LMS technique only offers a noise suppression of up to 10 dB. Signal-to-noise ratio (SNR) was measured using several methods under various blocking-kind matrices, step-sizes, and noise sources. The adaptive noise-canceling technology utilizes a configuration consisting of two microphones. Adaptive ANC is a technology that utilizes microphones and speakers to automatically adjust to the surrounding environment. This advanced form of ANC utilizes digital technology to adjust the level of noise cancellation based on the surrounding environment. ANC utilizes noise cancellation technology to eliminate unwanted background noise.

2.1 INTRODUCTION

This chapter outlines the utilization of neural oscillators for the purpose of mitigating background vibration noise in meshing plastic gears, thereby enabling the detection of gear-failure indicators. Several factors, including the fundamental frequency of mesh creation, particularly mesh harmonics and modulating sidebands, affect the acceleration response of the gear shaft's impact on the housing. Noise from rolling elements, vibration from motors, harmonic distortion, and other factors all contribute to the overall picture. Removing unwanted frequency components from the complicated response can help improve the detection of gear faults. The repetition intervals of both of these redundant components are significantly influenced by variations in driving torque. A new filtering system has been put in place as a direct result of the synchronization. The oscillators in your brain can automatically adjust their frequency to match the frequency shift of the redundant part. A proposed feed-forward system using neural oscillators is shown in Figure 2.1.

Adaptive filters are commonly employed for various purposes, including the identification of unfamiliar systems. For instance, they can be utilized to determine the frequency response of an auditorium or to analyze the characteristics of an unknown communications channel. Two additional applications include echo cancellation [7] and channel identification. Real-time noise suppression technology currently employs artificial intelligence (AI) to automatically remove background noise.

AI-based noise suppression utilizes speech recognition and analyzes the audio feed to effectively reduce noise.

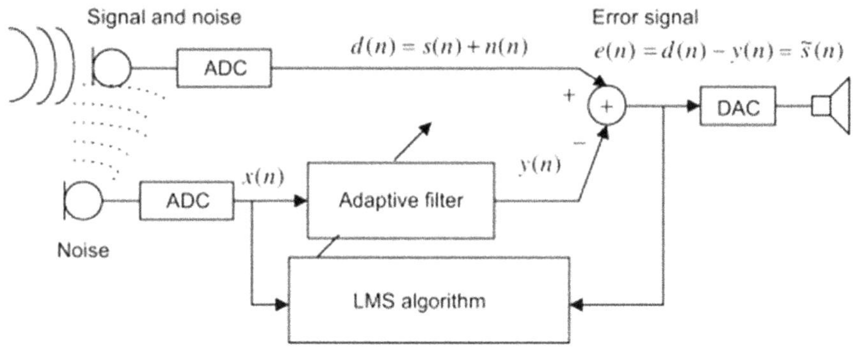

FIGURE 2.1 Schematic representation of noise suppression.

Digital filters, also known as adaptive filters, modify their coefficients to approach their desired state [5]. The cost function and optimization criterion used in comparing the input to the adaptive filter with the intended signal is the mean squared error (MSE) signal (Figure 2.1). The MSE decreases as the filter optimizes its coefficients, ultimately converging to zero. Now that we know how to tweak the filter, we can calculate its coefficients. According to Refs. [11,14], the filtered output $y(k)$ is quite comparable to the required input $d(k)$. To deal with changes in the filtering environment or the properties of the incoming data, the filter adjusts its coefficients.

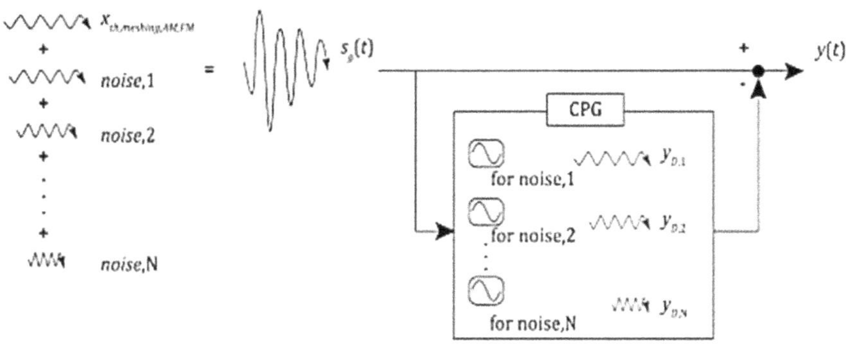

FIGURE 2.2 Types of noises process.

The system under consideration comprises several neural oscillators and incorporates a signal that is measured. The quantity of oscillators is equal to the quantity of superfluous components necessitating elimination. Every oscillator in your brain automatically tunes itself to the precise frequency needed by each part. The output of oscillators with a 180° phase shift relative to the input is blended with the measured response to dampen the impact of unwanted factors. The system is depicted in this manner.

$$\begin{cases} X_{D,k} = F_{D,K}\left(X_{D,k}\right) + E_{D,k} S_g\left(t\right) \\ y\left(t\right) = S_g\left(t\right) - \sum_{k=1}^{n} \beta_{D.K} y_{D,k}\left(t - L_K\right) \end{cases} \tag{2.1}$$

In this context, L_K represents the delay time and $\beta_{D,K}$ represents the arbitrary output gain of an oscillator.

2.2 PROPOSED METHOD

Adaptive filters distinguish themselves from conventional filter design methods by their independence from prior information or fixed filter coefficients. The parameters of an adaptive filter can be changed as needed. The coefficients of adaptive filters such as FIR and IIR filters, as well as lattice and transform domain filters, are adjusted to reduce the impact of error signals. The LMS method and the NLMS method are commonly employed adaptive filters in transversal filter applications [4]. Sound is produced through fluctuations in air pressure or vibrations. Objects cause the air surrounding them to vibrate due to their own vibrations. The relationship between pitch and vibration frequency is direct, meaning that as the pitch of a sound increases, so does its vibration frequency. Similarly, the relationship between wave amplitude and loudness is also direct, indicating that as the wave amplitude of a sound increases, its loudness also increases. A sine wave is the most basic way to demonstrate this concept.

2.2.1 FOURIER THEOREM

Fourier's theorem states that any sound, no matter how complex, can be described using a sequence of sine waves.

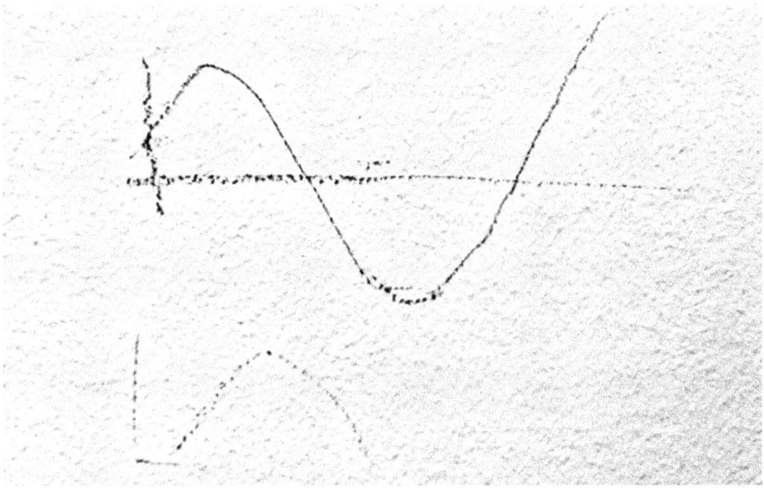

FIGURE 2.3 Air pressure vibration.

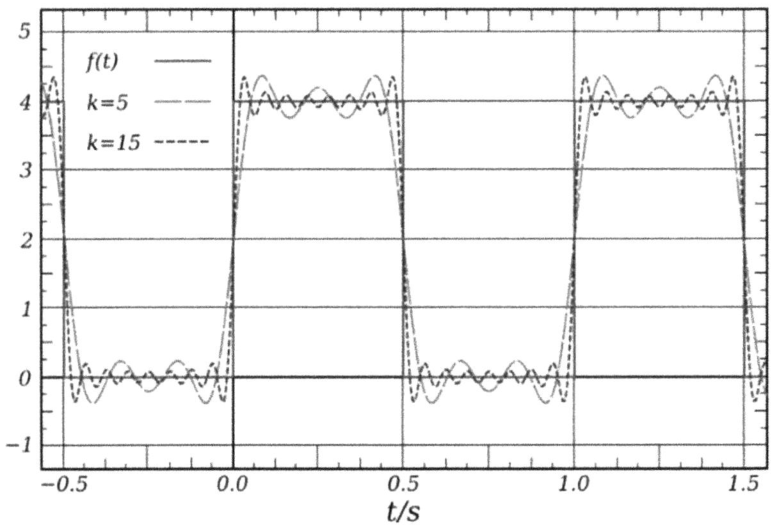

FIGURE 2.4 Various forms of sound waves can be represented by sine waves.

2.2.2 *CANCELLATION OF ACTIVE NOISE*

An externally attached microphone is used in the active noise cancelling feed-forward system. This microphone is mounted to the earphone. The use of an in-ear microphone is a key component in feedback active noise cancellation (ANC) technology [13]. By fusing together feed-forward ANC and feedback ANC, we get the hybrid active noise control system.

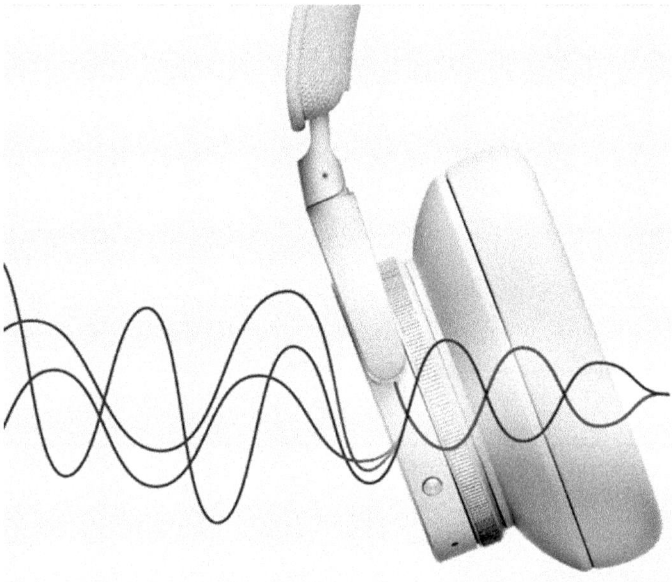

FIGURE 2.5 Noise-out in earbud.

ANC headphones typically include a miniature microphone that captures ambient sounds. The Fourier transform equation is employed to decompose complex sounds into their constituent sine waves result in a 180° anticlockwise shift of the antinoise wave in contrast to the ambient din. Phase cancellation is a frequent name for this method.

PNC necessitates well-designed ear cups to effectively mitigate distracting noises [6]. Figure 2.2 demonstrates the application of this technology to both in-ear and over-ear headphones, with the earpiece effectively attenuating ambient noise.

2.2.3 *NOISE CANCELLATION TECHNIQUES*

The ANC technology employs microphones and speakers to minimize background and ambient noises. Over-ear headphones commonly incorporate this type [8]. Advancements in technology have facilitated the development of authentic wireless earbuds. The adaptive active noise canceller is capable of automatically adjusting to the user's surroundings by utilizing microphones and speakers [9, 15]. This advanced iteration of ANC implements dynamic adjustment of noise cancellation intensity based on the surrounding environment. Adjustable ANC enables users to manually adjust the levels of noise cancellation in order to minimize the presence of ambient noise. This option is ideal for individuals seeking full autonomy and authority. Transparency mode enables seamless awareness of the surrounding environment while listening to music or wearing earbuds, eliminating the need to pause music playback or remove the earbuds. The adjustable transparency mode feature enables users to regulate the degree to which they can perceive their surroundings while simultaneously listening to music, without the need to disable the audio playback. The Adjustable Own Voice feature enables users to personalize the level of their own voice they hear when using ANC, such as during conference calls.

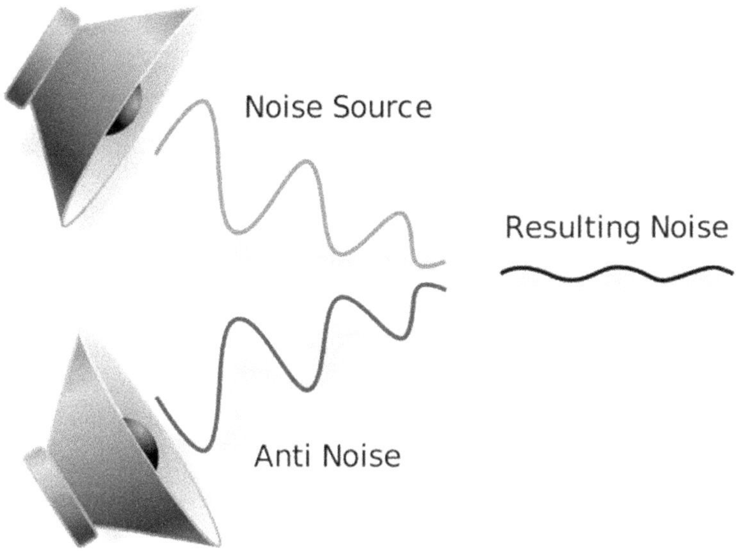

FIGURE 2.6 Phase cancellation.

An adaptive filter is a filter whose correction is determined using, you guessed it, adaptive methods. There are many ways to categorize adaptive algorithms, depending on factors such as convergence speed, stability, and robustness. Methods such as LMS, normalized LMS, leaky LMS, and RLS are the foundational methods that are used frequently.

2.3 SUGGESTED TECHNIQUE

2.3.1 ALGORITHM FOR LMS

The LMS approach is well recognized by experts in the field as an adaptive filter that can streamline a variety of processes that are associated with signal processing. It's a novel application of stochastic gradient descent in the form of a filter for machine learning [11, 12].

An adaptive filter like LMS algorithms can replicate a desirable filter by determining the filter coefficients that minimize the MSE between the intended and obtained signals [21]. Stochastic gradient descent recomputes the filter depending on the most recent error.

Currently, the most popular algorithm is the LMS algorithm. It is most often used in classification techniques, echo cancellation, channel quantization, voice coding, and noise cancellation [1]. The LMS algorithm approximates the steepest descent approach using the theory of Wiener filters. Convergence is preferred because of its dependability, low cost, and ease of use, despite its slow pace. The MSE is minimized as the primary goal of a steepest descent adaptive filter.

$$\xi(n) = E\left\{\left|X(n)\right|^2\right\} \tag{2.2}$$

The vector estimate w_n is determined to minimize the value of $\xi(n)$. The steepest descent occurs in the direction opposite to the gradient vector, as the gradient vector indicates the direction of the steepest ascent.

For the updated Steepest Descent formulae, please refer to the following.

$$W_{n+1} = W_n - \mu \nabla\xi(n) \tag{2.3}$$

Here μ denotes step size

$$\nabla \xi(n) = -E\left\{e(n)X^*(n)\right\} \tag{2.4}$$

and as a result, the most challenging decent algorithms have

$$W_{n+1} = W_n + \mu E\left\{e(n)X^*(n)\right\} \tag{2.5}$$

In its current iteration, the LMS is essentially a streamlined version of steepest descent techniques. This is the predicted value when using a single sampling distribution.

$$E\{e(n)X^*(n)\} = e(n)X^*(n) \tag{2.6}$$

the updated equation of it is

$$W_{n+1} = W_n + \mu e(n)X^*(n) \tag{2.7}$$

The equation is referred to as the LMS algorithm. The LMS method necessitates coefficients for computing the error. Filter order of the LMS algorithm determines the step-size limitation. In this context,

$$0 < \mu < \frac{2}{(P+1)E\{|X(n)|^2\}} \tag{2.8}$$

It is important that the LMS meet convergence to the mean requirements. The *P*-th order LMS approach can be thought of as.

Parameters:

P = filter-order
μ = step-size
Initialization:
$\hat{h}(0)$ = [zeros(p)]
Computation: For $n = 0, 1, 2, \ldots$
$x(n) = [x(n), x(n-1), \ldots, x(n-p+1)]^T$
$e(n) = d(n) - \hat{h}^H(n) \times (n)$
$\hat{h}(n+1) = \hat{h}(n) + \mu e^*(n) \times (n)$

In order to iteratively calculate results, the LMS algorithm makes use of the "method of steepest descent" to tweak filter weights. Through the convergence notion, the LMS method generates useful curves for practise [16]. These concepts have been developed through extensive efforts to improve machine learning models, input–output relationships, training and testing procedures, and the pursuit of convergence. Convergence refers to the recursive learning process reaching a cohesive end outcome without deviation.

2.3.2 ADAPTIVE FILTER-BASED NOISE CANCELLATION

Single-channel adaptive feedback ANC systems filter away ambient noise by filtering the source [18]. An input signal's noise can be greatly reduced by using a noise cancelling system. Figure 2.1 provides a diagrammatic illustration of a basic single-channel feedback ANC system, which is taken from Ref. [10] and illustrated there.

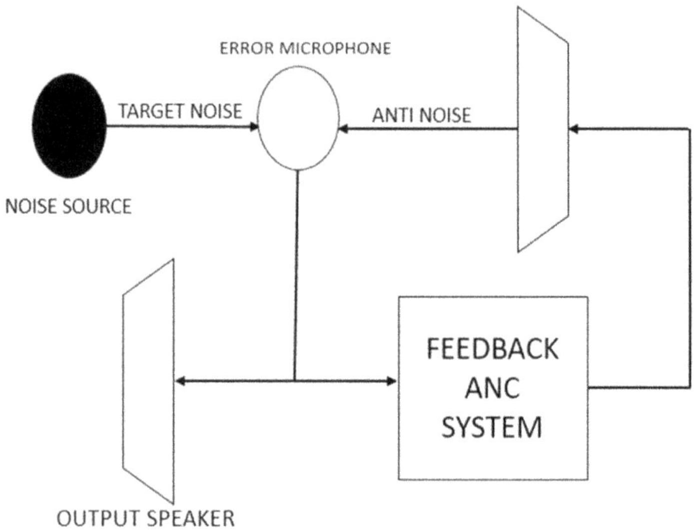

FIGURE 2.7 Active noise suppression with a single feedback channel.

2.3.3 NLMS ALGORITHM

In the least mean squares (LMS) algorithm, picking the step size is a challenging problem. Using a time-varying step size, which can be generated in the following way, is a feasible method.

$$\mu = \frac{\beta}{X(n)^2} \tag{2.9}$$

Standard step size β can take on values between 0 and 2, inclusive. By modifying the scaling factor, a revised equation using the LMS factors is obtained.

$$W_{n+1} = W_n + \frac{\beta}{X(n)^2} e(n) X^*(n) \tag{2.10}$$

Complication of system identification is seen in acoustic echo cancellation. In this context, normalized LMS (NLMS) is a popular adaptive filter. It solves the widespread problem of how to achieve both rapid convergence/tracking and low levels of misadjustment. The step size of this algorithm needs to be controlled [17] so that it can achieve these competing goals.

The antinoise waveform is comparable to the desired noise waveform. As a result, the resulting wave will be much weaker. The graphic depicts how the microphone picks up the weaker waveform. The input ANC system generates counter noise in anticipation of the arrival of the noise target [3].

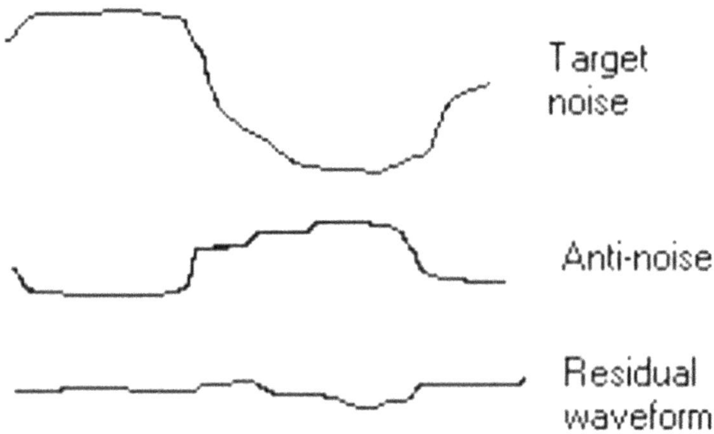

FIGURE 2.8 Explain how active noise reduction works.

The LMS step size normalization via the NLMS method solves the gradient noise enhancement problem.

$$W_{n+1} = W_n + \frac{\beta}{\varepsilon + \left\| X(n)^2 \right\|} e(n) X^*(n) \tag{2.11}$$

2.3.4 *ALGORITHM FOR RLS*

The Kalman filtering method serves as the basis for the recursive least squares method. Throughout the course of an iterative process, the weights of any and all signals that lie along the route of a gradient squared amplitude error signal are modified. The construction of the filter is quite similar to that of the LMS approach, with one key difference being the absence of an adaptive procedure. Here is the updated equation with the new coefficient values.

$$W_n = W_{n-1} + K^H(n)e(n) \tag{2.12}$$

This is the error function, e(n), where K^H is the gain vector.

2.3.5 *IMPROVEMENT IN QUALITY*

Quality depends on SNR [20]. System performance is determined by input–output SNR fluctuations. Calculating the SNR using the input signal and noise is typical. An adaptive system prevents using the output SNR. Formula for output SNR.

$$\text{Output SNR} = 10\log_{10}\frac{\text{var}(E_s)}{\text{var}(E_n)} \tag{2.13}$$

2.4 EXPERIMENTAL RESULTS

Using the NLMS method can lead to improved approximation of the desired signal, as shown in Figure 2.9. Both the LMS and RLS methods provide useful comparisons to this one.

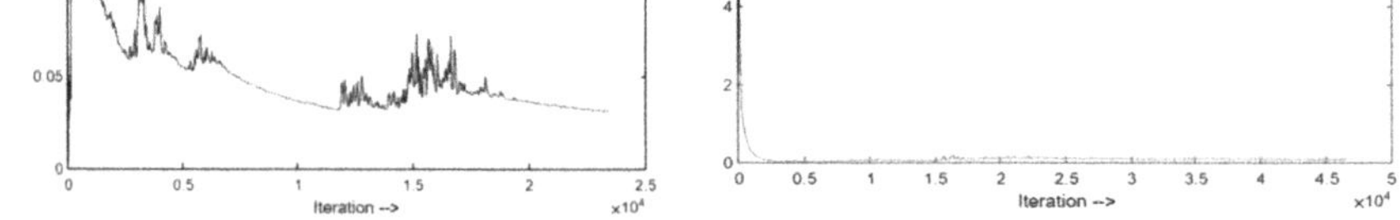

FIGURE 2.9 The LMS and NLMS algorithms attenuate filter impulse responses by 15 dB with a learning curve for filter orders of 50.

In order to compare the sound attenuation across microphones with 15 and 30 dB attenuation, we plotted the voice signals for both microphones on a single graph. The sound that is picked up by the microphone has a very slight lag. After transferring two microphones to separate locations, the output SNR and the SNR gain can be determined [19].

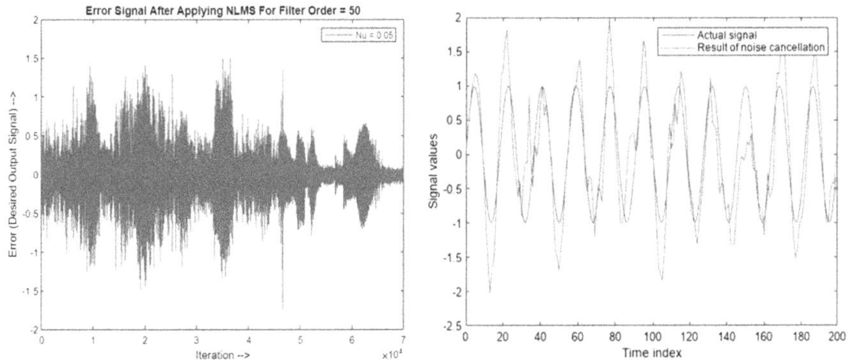

FIGURE 2.10 The NLMS method was applied to the signal waveform, and the learning curve showed an erroneous signal at a filter order of 50. The EAC recorded mic signal. A temporal graph of ANC deactivation and revival.

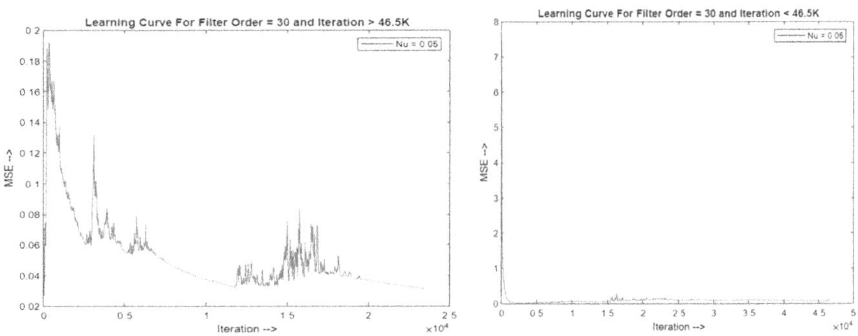

FIGURE 2.11 LMS and NLMS algorithms dampen filter impulse responses by 15 dB with a learning curve for filter orders of 30.

In the final test, the algorithm is subjected to impulsive noise, a particularly challenging type of noise. In this experiment, we used two different noise generators. There is a striking resemblance between the waveforms of these two artificial sources and those of real-world disturbances brought on by electromagnetic compatibility difficulties in electronics [2].

The optimal method for adaptive noise cancellation when a reference signal is not available. The stable MSE error efficiency of the proposed filter is higher than that of the LMS and NLMS filters, while maintaining the same convergence rate. Additionally, the SNR ratio after filtration is improved compared to the LMS and NLMS filters, resulting in enhanced adaptive noise cancellation performance. The use of an adaptive filter is recommended when the filter needs to be adjusted to account for progressively changing signal circumstances or system characteristics. An adaptive linear combiner, often called a variable-length infinite-interval (FIR) filter, is a simple and powerful filtering method. The process factors that impact the qualities of conventional FIR and IIR filters are assumed to be known. The potential for change exists, but it is expected that the nature of this change will be recognized.

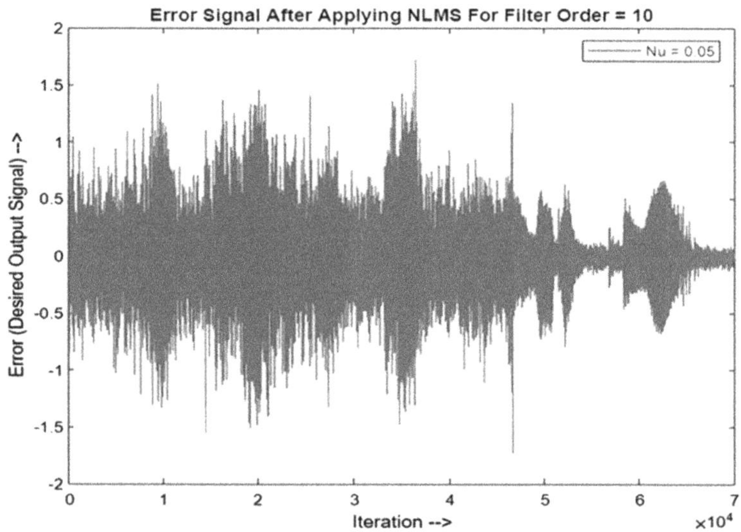

FIGURE 2.12 Applying the normalized least mean squares (NLMS) algorithm to the raw signal waveform yields the error signal depicted in the learning curve for a filter order of 10.

2.5 CONCLUSION AND FUTURE SCOPE

The microphone is safeguarded from ambient noise through the use of passive absorbers and ANC, and it is also designed to prevent mechanical damage. Furthermore, even in its simplest form, the implementation of a spectrum subtraction technique significantly reduces the level of residual background noise. The hybrid noise reduction system incorporates a spectrum-subtracting algorithm and noise reduction from the environment. In conclusion, we developed an offline model of the procedure, enabling its future real-time execution. The system demonstrates extensibility in various noisy environments through the use of RLS and LLMS.

The article exhibits a substandard improvement quality of less than 18 dB, which indicates the requirement for additional enhancement in future study. Noise cancellation is currently achieved by the use of this method, which may be found in mobile communication devices as well as headphones; nevertheless, the approach is connected with high costs. Through the comparison of multiple techniques, it is possible to determine the most budget-friendly option for systems including mobile phones, headsets, phones, video processors, and noise cancellers.

KEYWORDS

- **ANC**
- **LMS**
- **microphones**
- **NLMS**
- **SNR**

REFERENCES

1. Jain, D., & Beniwal, P. (2022). A novel LMS algorithm applied to adaptive noise cancellation with varying parameters. *ISSN, 10*, 2321–9653.
2. Kapoor, J., Pathak, A., Rai, M., & Mishra, G. R. (2022). Speech quality enhancement through noise cancellation using an adaptive algorithm. *IAENG International Journal of Computer Science, 49*(3), 653–665.
3. Bokde, P. R., & Choudhari, N. K. (2015). Implementation of adaptive filtering algorithms for removal of noise from ECG signal. *International Journal of Computer Technology & Applications, 6*(1), 51–56.
4. Pérez-Martínez, I., Torres-San Miguel, C. R., Rodríguez-Martínez, R., & Murillo-Martínez, C. (2020, September). Microcracks UHMWPE assessment using the wavelet transform. In *Journal of Physics: Conference Series* (Vol. 1633, No. 1, p. 012167). IOP Publishing.
5. Stergiopoulos, S. (Ed.). (2017). *Advanced signal processing: theory and implementation for sonar, radar, and non-invasive medical diagnostic systems*. CRC Press.
6. Dixit, S., & Nagaria, D. (2017). LMS adaptive filters for noise cancellation: A review. *International Journal of Electrical and Computer Engineering (IJECE), 7*(5), 2520–2529.

7. Tirupal, T., Dheeraj, T., Dasharath, B., Hemanth, B., & Jagadeep, K. (2023, May). Implementation and design of noise cancellation using nlms algorithm in microphones. In *International Conference on Data Analytics and Insights* (pp. 51–64). Springer Nature Singapore.
8. Tajammul, A., Gupta, D. K., & Gupta, V. K. (2015). Analysis of noise signal cancellation using LMS, NLMS and RLS algorithms of adaptive filter. *Advanced Research in Electrical and Electronic Engineering, 2*(14), 50–53.
9. Dewasthale, M. M., & Kharadkar, R. D. (2014, January). Acoustic noise cancellation using adaptive filters: A survey. In *2014 International Conference on Electronic Systems, Signal Processing and Computing Technologies* (pp. 12–16). IEEE.
10. Ilyas, M. Z., Noor, A. O. A., Ishak, K. A., Hussain, A., & Samad, S. A. (2008, December). Normalized least mean square adaptive noise cancellation filtering for speaker verification in noisy environments. In *2008 International Conference on Electronic Design* (pp. 1–4). IEEE.
11. Huang, Y., Benesty, J., & Chen, J. (2006). *Acoustic MIMO signal processing*. Springer Science & Business Media.
12. Ifeachor, E. C., & Jervis, B. W. (2002). *Digital signal processing: a practical approach*. Pearson Education.
13. Lee, D. J., Jeon, W. H., & Chung, K. H. (2002, January). Development and application of fan noise prediction method to axial and centrifugal fan. In *Fluids Engineering Division Summer Meeting* (Vol. 36169, pp. 987–992).
14. Douglas, S. C., & Rupp, M. (2017). Convergence issues in the LMS adaptive filter. In *Digital signal processing fundamentals* (pp. 485–506). CRC Press.
15. Hayes, M. H. (1996). *Statistical digital signal processing and modeling*. John Wiley & Sons.
16. Hayes, M. H. (1996). *Statistical digital signal processing and modeling*. John Wiley & Sons.
17. Laakso, T. I. (1996). Splitting the unit delay. *IEEE Signal Processing Magazine, 13*(1), 30–60.
18. Adams, R., & Kwan, T. (1993). Theory and VLSI architectures for asynchronous sample-rate converters. *Journal of the Audio Engineering Society, 41*(7/8), 539–555.
19. Pyfer, M., & Ansari, R. (1987, April). The design and application of optimal FIR fractional-slope phase filters. In *ICASSP '87. IEEE International Conference on Acoustics, Speech, and Signal Processing* (Vol. 12, pp. 896–899). IEEE.
20. Tirupal, T., Dheeraj, T., Dasharath, B., Hemanth, B., & Jagadeep, K. (2023, May). Implementation and design of noise cancellation using NLMS algorithm in microphones. In *International Conference on Data Analytics and Insights* (pp. 51–64). Springer Nature Singapore.
21. Kim, M. S., Lee, K. W., Lim, J. W., Kim, D. H., & Hong, S. G. (2023). Ranking roughly tourist destinations using BERT-based semantic search. In *Sentiment Analysis and Deep Learning: Proceedings of ICSADL 2022* (pp. 1–11). Springer Nature Singapore.

CHAPTER 3

THD REDUCTION OF PI-CONTROLLED NINE-LEVEL INVERTERS AND DATA VISUALIZATION

N. KAVITHA and N. P. G. BHAVANI

Department of Electronics and Communication Engineering, Saveetha Institute of Medical and Technical Sciences, Chennai, Tamil Nadu, India

ABSTRACT

Multilevel inverters have been an interesting area of research in recent years. Multilevel inverter researchers started with proposing various topologies to reduce the space when implemented for various applications. Further, research extended to increasing the levels of the inverters. Higher the level, better was the output. Total harmonics are reduced when the level of the inverters is increased. In this context, this chapter proposes a suitable closed-loop control for such a nine-level inverter, which ensures constant output voltage and current. Apart from providing constant output voltage and current, controlled output contains fewer harmonics. A nine-level inverter is simulated and total harmonic distortion (THD) is calculated. Further, a closed-loop proportional-integral controller-based nine-level inverter is simulated and THD is recorded. A combination of proportional and integral controllers increases efficiency. THD reduction is observed in a closed-loop proportional integral controlled nine-level inverter when compared to a nine-level inverter. A statistical package for social sciences is used to visualize the data obtained. THD reduction is clearly shown in a graphical depiction.

Cyber-Physical Systems: Applications, Challenges, and Research Directions. Rajanikanth Aluvalu, Samarjeet Borah, & Lalit Garg (Eds.)

DOI: 10.1201/9781779643421-3

3.1 INTRODUCTION

In the current scenario, in an economically emerging country like India, renewable energy sources are utilized, and this utilization is gradually increasing. To meet the demand or to supply the load appropriately, multilevel inverters are designed in recent years. With the advent of electric vehicles and the deterioration of resources, the need for power generation has increased. Again, with the increase in demand for power converters, the need for designing multilevel inverters is greater. The design of inverters not only involves an increase in levels to meet the requirements but also focuses on reducing the number of switches. Overall, it provides a topology that supports the system. A multilevel inverter (MLI) reduces the non-sine area of the generated output wave shape, which reduces interference of magnetizing circuits in the vicinity [14]. Multilevel inverters originated from the design of arrays that use multiple sources [11]. In this chapter, cascaded H bridge nine-level inverter topology (CHBMLI) [12] is simulated using MATLAB SIMULINK. Further, a proportional-integral (PI) controller is designed to maintain output voltage and to reduce the harmonics. The PI controller is simple and finds its application in control industries and, hence, it is widely used to obtain output free of distortions and ripples [6]. The controller will compare a fixed reference value with the voltage and current obtained from the actual inverter which in turn controls the switching signals to the switches in the multilevel inverter [4, 9]. A major advantage of using the controller along with a multilevel inverter is that it produces high power, high voltage output. Switching losses and harmonics are eliminated when a closed-loop control is employed [26]. This chapter focuses on achieving closed-loop control of a nine-level inverter using PI controller. For the fractional-order systems, PI controllers show better efficiency [16, 17]. When a closed loop is employed, better output voltage and current control are achieved. Furthermore, a reduction in harmonics of the output voltage obtained is observed.

3.2 RELATED WORKS

Multilevel inverters are used in industrial applications such as steel production, food industry, robotics, motor drives, variable frequency drives [14], and so on [2]. MLI basically employs DC sources to produce AC output of multiple-step voltages [11]. Multilevel inverters produce harmonic-free output [2]. Voltage stresses in the switches are reduced to a great extent when using multilevel inverters. Also, *dv/dt* is reduced using MLI [11, 19]. To achieve a closed-loop control, the multilevel inverter chosen here is a cascaded H bridge nine-level inverter involving 16 switches [12]. Whenever we combine controllers, we achieve better efficiency [5]. The PI controller reduces differences in voltages and maintains smooth operation [3]. The reference and actual signals are compared and, further, based on this, gating pulses are generated for the switches [15, 20]. We can see that this type of controlled inverter is widely used for industrial applications [6]. When we use a PI controller, total harmonic distortion (THD) is reduced drastically [13]. It is one of the primitive and simple form of controller [4]. The proposed system is simulated and output voltage and current waveforms are recorded [8]. THD is the measure of harmonics present in the output waveforms recorded [10]. Once we employ the PI controller, we achieve closed-loop control of the nine-level inverter and we can observe a reduction in harmonics [13]. THD reduction is perfectly achieved by generating gating pulses of cascaded H bridge inverter using a PI controller [27].

3.3 CASCADED H BRIDGE NINE-LEVEL INVERTER

Based on the existing topologies such as neutral point diode clamped (NPC), flying capacitor, and cascaded H bridge [22], a cascaded H-bridge multilevel inverter is chosen mainly to maintain high-quality output even with less number of switches [21, 25]. When inverters are connected in series to produce high-quality output, it is termed cascaded inverters. To form a nine-level H bridge inverter, four single-level inverters cascaded. Each inverter has four switches that convert DC input voltage to AC output voltage [12]. Each single bridge of the full-bridge inverter produces +Vdc, 0, and −Vdc [4]. The system taken for closed-loop control consists of four identical DC sources [12]. Each bridge consists of four insulated gate bipolar transistors [23]. Sine and triangular waves are compared to generate a sinusoidal pulse-width modulated (SPWM) signal. AC drives preferably use PWM in order to achieve the best efficiency as per the required output [18].

Since four inverters are connected in series, the actual voltage is equal to the sum of the voltages of all four H-bridge modules. In a nine-level inverter, there are 16 switches, of which two switches are turned on or off at a voltage level of Vdc/4 [5]. One voltage level at +ve, −ve half cycles, and 0 voltage levels are generated in every unit [13]. The switching sequence of nine-level cascaded inverter is shown in Table 3.1. Figure 3.1 shows the MATLAB/SIMULINK model of a nine-level inverter.

TABLE 3.1 Switching Sequence of Nine-Level Inverter.

Switching Sequence	Output
SW1, SW2, SW5, SW7, SW9, SW11, SW13, SW15	Vdc/4
SW1, SW2, SW5, SW6, SW9, SW11, SW13, SW15	2Vdc/4
SW1, SW2, SW5, SW6, SW9, SW10, SW13, SW15	3Vdc/4
SW1, SW2, SW5, SW6, SW9, SW10, SW13, SW14	Vdc
SW1, SW3, SW5, SW7, SW9, SW11, SW13, SW14	0
SW3, SW5, SW5, SW7, SW9, SW11, SW13, SW14	−Vdc/4
SW3, SW5, SW7, SW8, SW9, SW11, SW13, SW14	−2Vdc/4
SW3, SW5, SW7, SW8, SW11, SW12, SW13, SW14	−3Vdc/4
SW3, SW5, SW5, SW7, SW11, SW12, SW15, SW16	−Vdc

3.4 CASCADED H BRIDGE NINE-LEVEL INVERTER

PI is a combination of proportional and integral controllers. When the power supply and feedback control loop reduces the error, a combination of controllers provides higher efficiency [1]. PI controllers are used in many industries to regulate temperature, pressure, and so forth. Figure 3.2 depicts the basic topology of PI controller-based CHBMLI. The actual voltage of the multilevel inverter is fed to the PI controller, where it is compared against the fixed voltage [3]. This difference controls the switching signals given to the switches [26].

Traditional methods of fault detection such as wavelet transform and fast Fourier transform give only the time and frequency components whenever a fault occurs in a multilevel inverter [24]. A PI controller is one of the widely used controllers and is easy to implement. This controller tunes the inverter output to achieve steady-state operation. A combination of proportional and integral controllers enables us to design an efficient system. For a majority of industrial applications, PI controllers are more appropriate [4].

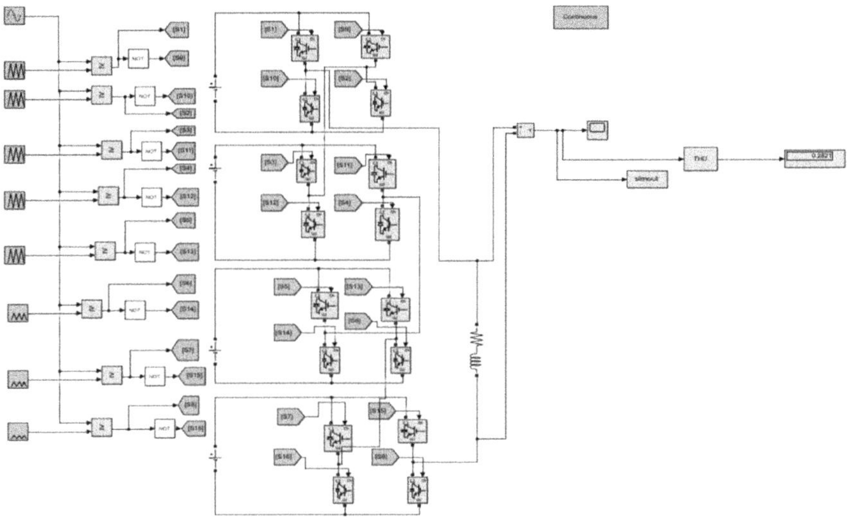

FIGURE 3.1 MATLAB/SIMULINK model of cascaded H bridge nine-level inverter.

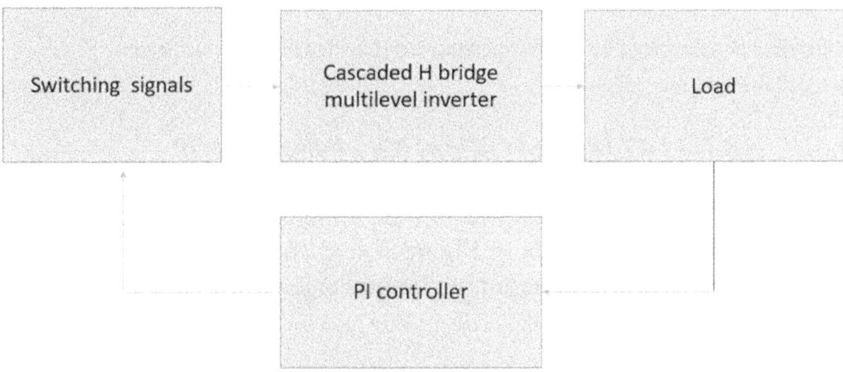

FIGURE 3.2 Basic topology of PI controller-based nine-level inverter.

3.5 PROPOSED PI CONTROLLER-BASED NINE-LEVEL INVERTER

CHBMLI nine-level inverter controlled by PI controller is shown in Figure 3.3 [5]. Voltage is compared with the reference signal and the difference is fed to the switching pulses in order to achieve constant output voltage [8].

Using MATLAB/SIMULINK voltage, current control of a nine-level inverter is performed using PI controller.

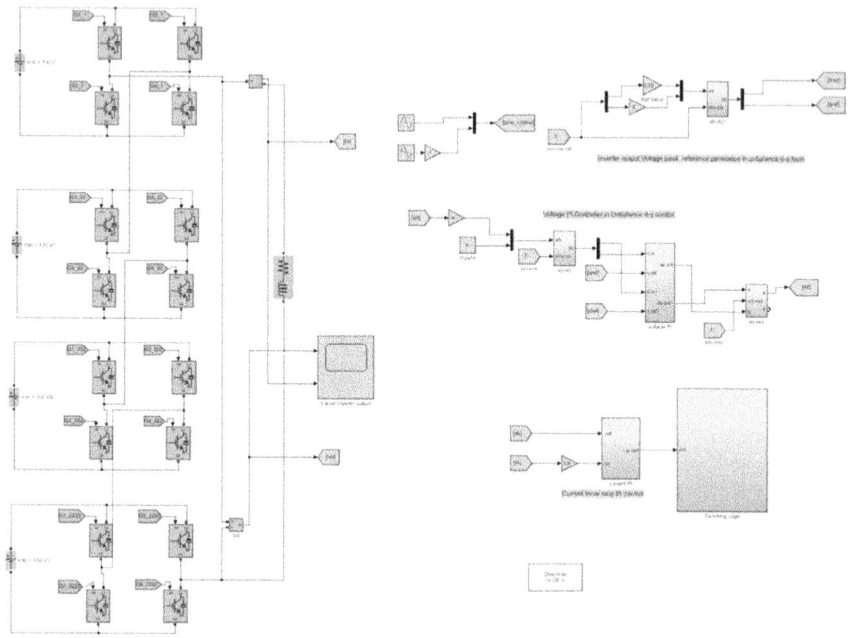

FIGURE 3.3 Proposed PI-controlled cascaded H bridge nine-level inverter.

3.5.1 AB TO DQ TRANSFORMATION INNER LOOP

A reference signal is transformed into dq components using the ab to dq transformation block shown in Figure 3.4; d and q represent direct and quadrature components. These transformed signals are fed to the voltage control loop as dref and qref.

3.5.2 VOLTAGE PI CONTROLLER IN UNBALANCED DQ CONTROL

The voltage of the multilevel inverter is transformed to dq using a transformation block and is fed to the voltage PI controller. Previously, dref and qref are also fed to this controller. The PI controller incorporates proportional and integral blocks. So, the reference dq signal is compared with voltage fed from a multilevel inverter. Figure 3.5 shows the voltage PI controller in unbalanced d-q control. The compared output which is in dq form is again transformed to ab form and hence fed to the current control loop.

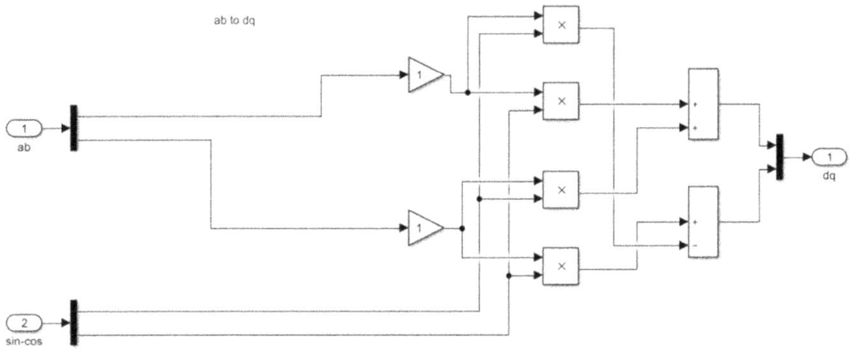

FIGURE 3.4 ab to dq transformation inner loop.

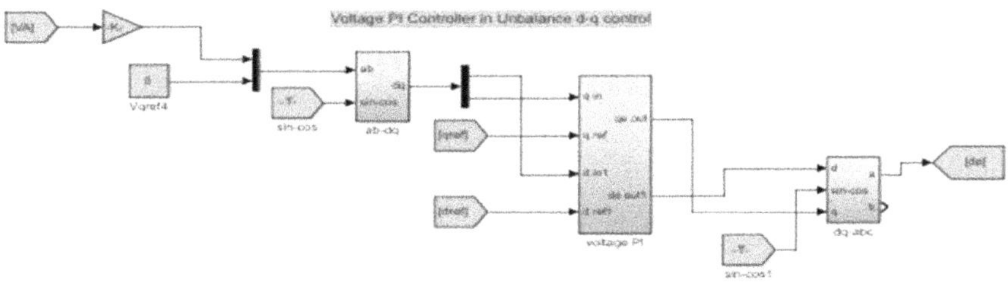

FIGURE 3.5 Voltage PI controller unbalanced dq control.

3.5.3 *VOLTAGE PI CONTROLLER INNER LOOP*

The voltage controller inner loop contains proportional and integral blocks. The reference voltage and the voltage fed from multilevel inverters are compared here and fed to PI. Output qe add de are obtained. Figure 3.6 shows voltage-based PI controller inner loop.

3.5.4 *CURRENT CONTROLLER INNER LOOP*

The reference current signal and the original current signal output from the nine-level inverter are compared here and fed to the PI. Compared output is given to the switches further. Figure 3.7 shows the current inner loop.

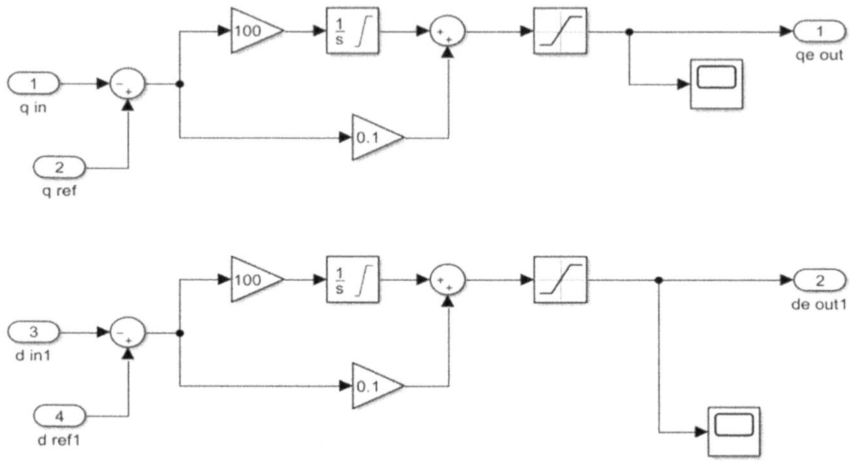

FIGURE 3.6 Voltage PI controller inner loop.

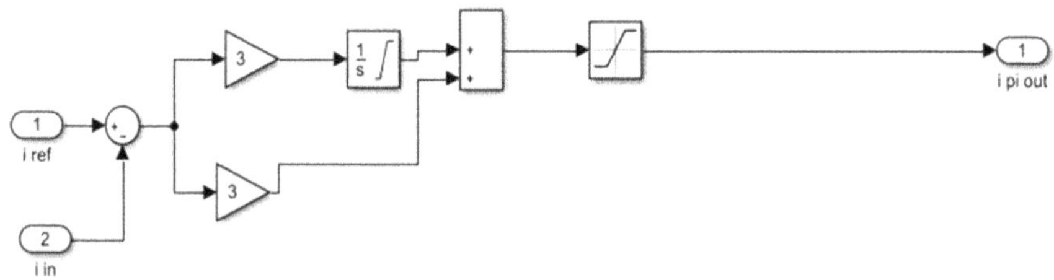

FIGURE 3.7 Current controller inner loop.

3.6 RESULTS AND DISCUSSION

3.6.1 NINE-LEVEL OUTPUT WAVEFORM OF THE INVERTER

Figure 3.8 shows the output voltage waveform of a nine-level inverter output.

3.6.2 CONTROLLED VOLTAGE AND CURRENT OF PI CONTROLLER

Figure 3.9 shows the controlled voltage and current of a PI controller-based nine-level inverter.

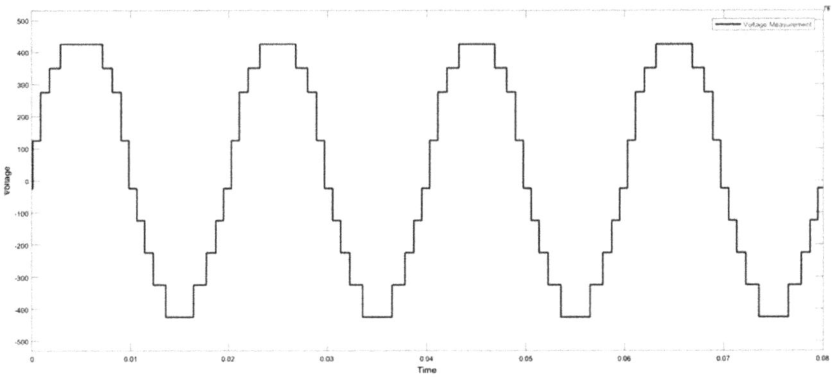

FIGURE 3.8 Nine-level output waveform of the inverter.

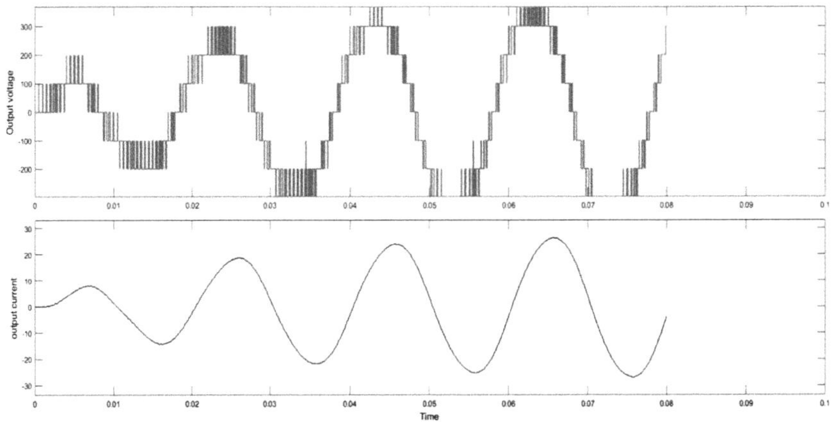

FIGURE 3.9 Controlled voltage and current of PI controller.

TABLE 3.2 PI Controlled Nine-Level Inverter Parameter.

Parameters	Values
Rise time	0.0215
Settling time	0.0799
Settling min	−399.9618
Settling max	399.9618
Overshoot	33.3065
Undershoot	133.3067
Peak	399.9618
Peak time	0.0710

Table 3.2 lists the various parameters of the recorded output waveform. The amount of time the system takes to reach 90% of the steady state value is rise time. For the PI-controlled output of nine-level inverter rise time is recorded as 0.0215. A peak value of voltage is recorded as 399.9618 and the peak time taken to reach the value is recorded as 0.0710.

3.6.3 STATISTICAL TOOL-BASED COMPARISON CHARTS

Table 3.3 shows the comparison of THDs. THD is a measure of harmonics recorded in the output. The ratio of total harmonics with respect to fundamental harmonics is THD [10]. The table gives the details of the THD value of the CHBMLI nine-level inverter as recorded in the above simulation of Figure 3.2. It is compared against the PI-controlled CHBMLI's THD of Figure 3.10.

TABLE 3.3 Comparative Table for Total Harmonic Distortion of Nine-Level Inverter Parameter.

Inverter	Total Harmonic Distortion
CHBMLI nine-level inverter	0.0215
CHBMLI nine-level inverter with PI control	0.0799

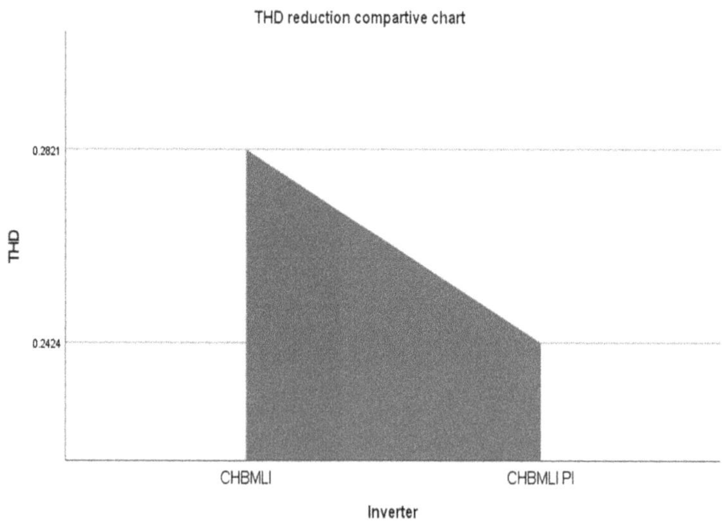

FIGURE 3.10 Comparative chart for THD reduction using SPSS.

The above data is visualized using a statistical package for social sciences. Figure 3.10 shows a comparative chart depicting THD reduction using the SPSS tool. THD drastically reduces when PI controller-based inverter is incorporated into the circuit.

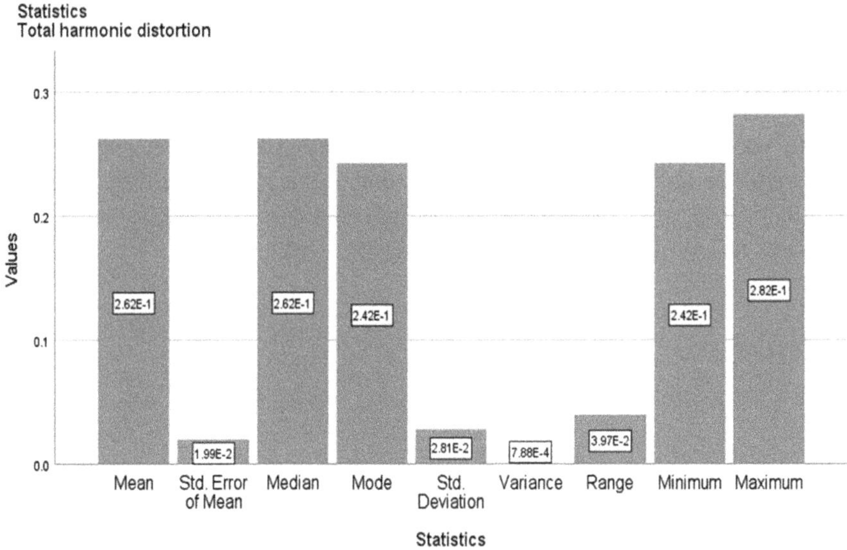

FIGURE 3.11 Statistical chart for comparing THD values using SPSS.

TABLE 3.4 Statistical Parameters of THD Comparison Using SPSS.

Parameters	Values
Mean	0.262
Standard error of mean	0.01985
Median	0.2622
Mode	0.2424
Variance	0.001
Range	0.0397
Minimum value	0.02424
Maximum value	0.2821

Figure 3.11 gives a clear picture of the statistical values calculated for the compared THDs. Table 3.4 clearly depicts the parameter values calculated using the statistical tool.

3.7 CONCLUSION

Thus, the PI controller-based nine-level inverter is designed to produce controlled output voltage and current. THD is calculated for a nine-level inverter and PI-controlled nine-level inverter. It is found that there is a reduction in the THD when we control the inverter parameters like voltage and current using the PI controller.

Parameters such as rise time, peak time, and settling time are calculated using MATLAB. THD comparison is visualized using SPSS. Further, mean, variance, and standard deviation values are deducted using SPSS and are all visualized using SPSS and suitable bar charts. Further, research will be extended to compare PI-controlled inverters with fuzzy and neural network-based nine-level inverters in future works.

KEYWORDS

- **cascaded H bridge multilevel inverter**
- **closed-loop control**
- **proportional integral controller**

REFERENCES

1. Malathi, M., Maalmarugan, J., & Hemavathy, S. (2022). Closed loop control of grid connected PV generation system fed with cascaded H bridge inverter. *International Transactions on Electrical Energy Systems*, 8588215.
2. Babaie, M., Sharifzadeh, M., & Al-Haddad, K. (2019). Adaptive ANN based single PI controller for nine-level PUC inverter. In *IEEE Electrical Power and Energy Conference (EPEC)*, Montreal, QC, Canada.
3. Tejaswi, G., Ramesh Kumar, S., & Ravi Kumar, Ch. (2019). PV-FC systems fed multilevel inverter by employing PI controller. *International Journal of Innovative Technology and Exploring Engineering (IJITEE), 9*(1).
4. Karthikeyan, D., Khiani, S. R., Hepzibah, A. A., Jothilingam, U., Ramkumar, R., Prathaban, A. V., & Vijiyakumar, K. (2023). PI controller for photovoltaic-fed novel multilevel inverter topologies. *Indonesian Journal of Electrical Engineering and Computer Science, 29*(3).
5. Palanisamy, R. Dev, S. S., Rana, A., Karthikeyan, D., & Selvabharathi, D. (2017). Simulation of novel 9-level inverter topology with reduced number of switches.

International Journal of Electrical Engineering and Technology (IJEET), 11(3), 58–65.
6. Kannan, C., Mohanty, N. K., & Selvarasu, R. (2017). A new topology for cascaded H-bridge multi-level inverter with PI and fuzzy control. *Science Direct Energy Procedia, 117*, 917–926.
7. Malathi, M., Maalmarugan, J., & Hemavathy, S. (2018). Closed loop control of grid connected PV generation system fed with cascaded H bridge multilevel inverter. *GRD Journal for Engineering, 3*(7), 124–131. *National Conference on Emerging Trends in Electrical, Electronics and Computer Engineering.*
8. Sathish Kumar, K., Belwin Edward, J., Chinmonyo, K., & Rushambwa, M. C. (2018). Implementation of single phase cascaded H-bridge nine-level inverter and simulation study with PI controller. In *International Conference on Smart Grid and Clean Energy Technologies (ICSGCE)* (pp. 228–233), Kajang, Malaysia.
9. Ali, M., Din, Z., Solomin, E., Cheema, K. M., Milyani, A. H., & Che, Z. (2021). Open switch fault diagnosis of cascade H-bridge multi-level inverter in distributed power generators by machine learning algorithms. *Energy Reports, 7*, 8929–8942.
10. Siva Priya, A., & Kalaiarasi, N. (2022). An extensive review on fault detection and fault tolerant control of multilevel inverter with applications. *International Journal of Renewable Energy Research, 12*(2).
11. Bhaumik, T., Bhaumik, S., Paul, S., & Mitra, S. (2022). A comparative study between different types of multilevel inverter. *International Research Journal of Engineering and Technology (IRJET), 9*(6).
12. Mathews, M., Ramesh, B., & Sreedhar, T. (2016). Minimization of THD in nine level cascaded H-bridge inverter using artificial neural networks. *International Journal of Innovative Works in Engineering and Technology (IJIWET), 4*(1), 99–110.
13. Selokar, M. S., & Sankhe, M. S. (2021). Implementation of ANN based closed loop multilevel inverter. *International Journal of Mechanical Engineering, 6*(2).
14. Suman, K., Agarwal, R., Agarwal, D., & Kumar, A. (2022). Modified reduced count cascaded nine-level inverter topology. *International Journal of Engineering Research and Applications, 12*(7), 179–185.
15. Sivagamasundari, M. S., & Melba Mary, P. (2016). Real time implementation of PI and PID controlled cascaded H bridge eleven level inverter using SPWM. *Journal of Microelectronics, Electronic Components and Materials, 46*(2), 65–73.
16. Sathishkumar, S., Ramkumar, R., Sivarajeswari, S., Ramya, D., Subburaj, T., Sankoh, M., & Marinkovic, D. Z. (2022). Performance enhancement of a three phase boost-cascaded fifteen level inverter using the PI controller. *Mathematical Problems in Engineering, Hindawi, 2022*, 1–17.
17. Venkateswara Rao, R., Ramesh Kumar, K., & Narasimha Raju, V. S. N. (2017). Design of PI controller for seven level symmetrical MLI with minimal quantity of switches plus snubber circuit. *International Journal of Engineering Trends and Technology (IJETT), 47*(8).
18. Dhineshkumar, K., & Subramani, C. (2018). Design and implementation of nine-level multilevel inverter. *Journal of Physics: Conference Series, National Conference on Mathematical Techniques And Applications, 1000.*
19. Sanju, B., Rejoice, T. P., Grace, M., & Mohanan, A. (2019). THD analysis in cascaded H-bridge multilevel inverter. *Global Research and Development Journal*

For Engineering [ERTEE' 19] (pp. 1–4), KTU(Kerala Technological University), Adishankara Institute of Engineering and Technology, Kalady, Kerala.

20. Anil Kumar, V., & Mouttou, A. (2020). Improved performance with fractional order control for asymmetrical cascaded H-bridge multilevel inverter. *Bulletin of Electrical Engineering and Informatics*, *9*(4), 1335–1344.
21. Al-Samawi, A. A., & Trabelsi, H. (2022). New nine-level cascade multilevel inverter with a minimum number of switches for PV systems. *Energies*, *15*(16), 5857.
22. Benioub, R., Adnane, M., & Itaka, K. (2017). Comparison study and simulation of the main multilevel inverter topologies for different output voltage levels. *International Journal on Electrical Engineering and Informatics*, *9*(3).
23. Sim, S. Y., Chia, C. K., Utomo, W. M., Goh, H. H., Buswig, Y. M. Y., Lim, A. J. M. S., Kek, S. L., Bohari, A. A., & Cham, C. L. (2018). Enhance cascaded H-bridge multilevel inverter with artificial intelligence contro. *Indonesian Journal of Electrical Engineering and Computer Science*, *11*(1), 105–112.
24. Veerendra, A. S. Shah, A. A., Mohamed, M. R., Sekhar, C. P., & Leung, P. (2021). Wavelet transform based fault identification and reconfiguration for a reduced switch multilevel inverter fed induction motor drive. *Electronics*, *10*(9), 1023.
25. Balasundaram, G., Arumugam, S., & Dinakaran, C. (2015). Design and simulation of single phase cascaded multilevel grid connected inverter using photovoltaic system. *Journal of Chemical and Pharmaceutical Science*, *8*(2), 441–449.
26. Sivagamasundari, M. S., & Melba Mary, P. (2016). Real time implementation of PI and PID controlled cascaded H-bridge eleven level inverter using SPWM. *Journal of Microelectronics, Electronic Components and Materials, 46*(2), 65–73.
27. Praveena, K., & Swarnasri, K. (2022). Closed loop PI control of PV tied grid system with multi carrier PWM based modular 5-level converter. *International Journal of Intelligent Systems and Applications in Engineering*, *10*(3), 403–410.

CHAPTER 4

LEGAL DIMENSIONS OF ACCIDENTAL LIABILITY OF AUTONOMOUS VEHICLES

PARIMITA DASH and AMARENDRA PATTNAIK

School of Law, Kalinga Institute of Industrial Technology (a Deemed to be University), Bhubaneswar, Odisha, India

ABSTRACT

Every vehicle needs a driver. We can't imagine a moving vehicle on the road without a driver. But days are not far away when we will see driverless vehicles (autonomous vehicles or AVs) on the road. Technology will enable such vehicles to think and act intelligently. These vehicles will move on the road based on predefined algorithms. As there will be no driver, there will be no human errors during driving. Even when the occupier of the car is drunk or distracted, there will be no safety concerns. People who don't know how to drive can use these vehicles as passengers. Drivers engaged in long-distance driving don't have to suffer the fatigue of driving. They can leave the vehicle in the auto-pilot mode and relax. These vehicles can also be deployed in military actions. Higher efficiency of the transport sector will help the economy grow. Therefore, we should adopt AVs as and when they are available in the market. But, before the adoption and use of AVs, the traditional legal framework has to be suitably modified to take care of issues that may arise because of this new technology. One of the important issues is accidental liability. This article examines the adequacy of existing laws in India to address accidents caused by AVs.

Cyber-Physical Systems: Applications, Challenges, and Research Directions. Rajanikanth Aluvalu, Samarjeet Borah, & Lalit Garg (Eds.)

DOI: 10.1201/9781779643421-4

4.1 INTRODUCTION

Transportation and mobility are essential parts of human life. We need to commute from one place to the other for work, business, and leisure. Goods need to be transported and drivers need to work without fatigue [1]. Various types of vehicles are available for such transport. But in all these vehicles, there is a need for a driver. He drives the vehicle from one point to the other and navigates the roads using his driving skill, knowledge of the road, and intelligence when faced with complex or unexpected events. Despite all the training and care during driving, accidents do happen, causing injuries to the driver and the passengers. It is not possible to eliminate human errors while driving. Sometimes drivers drive recklessly under the influence of intoxication and cause huge damage to the life and property of the passengers and the public. Autonomous vehicles (AVs) are an answer to these problems. These are vehicles that don't need a driver. It moves from one point to the other based on algorithms or logic in the pilot mode. This has been possible because of decades of research in the USA, Europe, and a few other developing countries. AV is now being developed and tested across the USA. Manufacturers and testing agencies are doing trial runs to check their capability to do what it is expected to do. It will take some time before we can see a fleet of AVs on the road, but this is a disruptive idea whose time has come. People have read these ideas in fiction and seen these happening in sci-fi movies. Now the reality is at their doorstep. Once they see the benefits of AVs, there will be no looking back. People will accept AVs as a workable and effective solution to their transport needs. However, there is always anxiety before adoption, particularly because there are many legal and safety issues of AVs that have not been addressed. The law always lags behind technology, but then we need to work on that front now.

Li [2] underscores that AVs are no more futuristic fiction. It is present and we need to be ready with the laws to take care of this new era of driverless vehicles. Mohan [3] thinks that the adoption and penetration of driverless vehicles (DLVs) in India will not exceed 5% of all vehicles by 2030 due to the incidence of high cost or affordability issues. However, the author agrees that DLVs are a good idea compared to the existing high cost of metro transport. He also underscores that DLVs can be very effective for goods transport, but we need to have separate corridors and infrastructure for such vehicles.

Two important considerations about AVs are reliability and safety. Reliability means: can AVs repeat good performance with similar efficiency and accuracy? Safety is how good AV is in protecting the life of the public and the life of the user. It is not easy to accept any claim that AV technology can understand all kinds of real-life situations like a human being and can act wisely. A machine has its limitations and that is what creates concern in the mind of the people. This is the reason why the debate around AVs is very polarized, with each one sticking to his point of view. However, we must understand that AVs have both opportunities and risks [4]. Opportunities that AVs offer need to be explored. Its development, trial, and deployment should be expedited. At the same time, the risks associated with AVs must be mitigated. That requires a lot of field trials and development of the technology to a level where there is no risk to the public or the users.

Globally, at least ten auto manufacturers have made big investments in the development, testing, and deployment of AVs. They are doing so in their captive R&D set-ups or have partnered with research organizations or universities for the purpose. Ford has tied up with Stanford and MIT for research in this area. Volvo has already completed testing of a fleet of vehicles with a pilot vehicle. Google is also working on AVs though it is not an automobile manufacturer. Some people view such investments as too futuristic, but there are places where such vehicles are in use with a driver as a standby. Wallich [5] narrates how AV models are becoming progressively more effective. He refers to a time when a safety pilot sits behind the wheels in case the automatic pilot malfunctions. But today there is no need for a driver. The benefits that will accrue to the agencies in the transportation sector are huge. The savings will not be limited to drivers' wages only. Because of efficient operation, there will be good savings in fuel costs. The number of road accidents will be reduced and that will save thousands of precious lives. The movement of vehicles will be more precise and will lead to lower chances of accidents and possible damage to life and property. As such, the insurance costs will be reduced. People can do a lot of other work when they travel by AVs. They can read, listen to music, watch a movie, or simply plug into Wi-Fi and attend to work or business. But for AVs to become a reality and achieve a critical mass of use, there is a need for the development of infrastructure. Dedicated roads for AVs will be a welcome initiative. Interestingly, manufacturers of vehicles for water transport are also not far behind in the race for AVs. They are better placed to deploy the vehicles as the waterways are relatively less congested with comparatively lower chances of accidents.

AVs use GPS, censors, Wi-Fi, digital networks, and computer systems for navigation and other essential acts during their movement. In the absence of a driver, machines are in charge of all actions and their decisions are based on data and algorithms. Technology makes life easy if there is no outside interference. But AVs are exposed to the risk of cyber-attacks [6]. People and agencies with *mala fide* intentions can get into the software system of the AV and can misguide its actions resulting in accidents and damage to life and property. And if it is a case of connected vehicles that are moving as a fleet, the damage would be more intense. Therefore, the technology used for AVs should be robust enough to insulate itself from all kinds of cyber-attacks. This is not a one-time affair. Constant research is needed to update the technology that can hedge any attempt to cyber-attack. The state should appropriate cyber security standards for compliance by all auto manufacturers.

The three broad issues about AVs that need attention and debate are:

1. Development, testing, and deployment of AVs.
2. Licensing of manufacturers/third-party autonomous technology makers for producing AVs or converting regular vehicles into AVs through the use of technology.
3. Legal issues arising out of accidents caused by AVs, particularly on the matter of establishing liability.

In this article, the authors have tried to delve into the legal issues arising out of accidents caused by AVs and the determination of liability, penalty, and punishment. They have looked at the existing laws that can be applied to AVs and then have evaluated if there is a need for some modification in laws to address newer kinds of issues that have arisen because of AVs.

4.2 OBJECTIVES

1. To delve into applicable Indian Laws in the context of AV accidents and their liability.
2. To find out if the existing laws are adequate to address all issues arising out of AV accidents.

4.3 METHODOLOGY

The systematic literature review method was followed in this chapter. Nearly 60 articles were searched with keywords and were considered at the initial stage of the research. But later on, only 21 most relevant articles were studied in-depth to understand how the laws are evolving across the globe. Legislation related to AV accidents in important jurisdictions was studied in depth. Indian laws related to automobile accidents were also studied. From the above, the authors tried to find out if Indian laws related to automobile accidents also need changes to take care of new situations that will arise with the introduction of AVs.

4.4 LITERATURE REVIEW

The systematic literature review was done on the following thrust areas—ethical choice by AVs, accidental liability of AVs as followed in different jurisdictions, and accidental liability and Indian laws.

4.4.1 ETHICAL CHOICE BY AVS

AVs are dependent on AI, robotics, automation, and algorithms. Machines don't understand morality. They act as per logic. Huang [7] in his article, delves into the moral questions that arise about AV. When a vehicle is driven by a human being, the person behind the wheel takes a decision that he thinks is correct and reasonable at that moment based on his morals and ethics. Whether that decision is right or wrong is a matter of debate. But in the case of AV without a driver, the technology decides any evaluation of moral context. The author mentions that without a moral quotient, many of the actions of an AV can be questioned in a court of law. AV acts as per logic and is devoid of morality. While there is a positive argument that AVs will reduce human errors and also accidents caused by human errors, it will not be well placed to decide on moral questions. One example of the moral question is if in one scenario AV can save either the passenger or the public on the road, then whom should it save? As per the technology, it will act as per a pre-defined rule and that rule is not flexible from one situation to the other. It will act similarly in all contexts, while a human being will act differently in different contexts. The author underscores the need for AVs to address this issue of morality in decision-making. Keeling [8] in his article, tries to find out the basis of moral judgments by the AVs. He says one way to take a moral issue is by following utilitarian principles. The other way such issues can be addressed is by following the legal method of the doctrine of necessity. The author mentions two situations for explaining the dilemma of an AV—one, if there is a certainty of injuring at least one person, how it should act. Should it affect less number of people in a minor way or a smaller number of people in a major way? The second question is if more than one person gets injured, then how will the impact of the AV be distributed among the people in the case? The author raises a question as to how these issues can be managed by an AV and he calls this a moral design issue. The author concludes that a restricted Pareto principle can be used to address the dilemma of justified harm. In real-life scenarios, human beings break the law to be more efficient and productive. This is also true in the case of the use of vehicles. In this regard, it is important to understand if AVs should also compromise on the laws to be efficient and productive. Authors believe that AVs have a role to play in eliminating the tendency of drivers to cut corners, putting others in jeopardy. AVs should always follow the rules. This means the algorithms that guide and control AVs should have no provision for violation of laws. Himmelreich [9] in his narrative says that ethics is not an independent issue and that AVs should not be programmed to take decisions in isolation. He says that there are several other stakeholders in a traffic situation and when an AV takes a decision, it must consider the collective interest of all the stakeholders. Nyholm and Smids [10] mention the importance of the right programming of the AVs so that they can make proper decisions in situations of risk and uncertainty. Etzioni and Etzioni [11] on the topic of ethics, mention that the ethics of the machine is equal to the ethics of the man behind the machine. The author underscores that ethical dilemmas can't be solved by looking at extreme choices. In reality, there will be some outlier considerations, but an algorithm should address the general contexts of the dilemma and the moral choice that the AV has to make. Pagallo [12] mentions the idea of AI "Personhood" with legal status and a clear list of rights and obligations. The author says that the use of robots and AI will keep increasing in the future and therefore clarity must prevail about the legal status of a robot that acts as an AI. As of today, it has no legal status, neither like a "natural person" nor as a "legal non-living entity". While the EU has been deliberating about giving robots operated through AI a legal status, manufacturers and third-party modifiers are not in favor of the idea. The point of contention is how a robot will be treated in a court of law and whether it will be liable for both civil and criminal liability. If it is criminal liability, then who will serve the sentence—the manufacturer, the owner, or the user? The author underscores that just because the damage was caused by a robot, it should not be dismissed. Clarity of accountability and liability of robots must be there so that the affected people can be duly compensated and the people behind the robot and AI are suitably penalized [36-40].

4.4.2 ACCIDENTAL LIABILITY OF AVS AS FOLLOWED IN DIFFERENT JURISDICTIONS

Laws have not been able to keep pace with technological change. Technology brings its complexity. An AV can be on the road without any driver in the vehicle. AV can also be used on the road with a driver inside the vehicle who needs to take charge in case of an emergency. He will be alerted by the technology installed in the car about the impending dangers. Therefore, in case an AV is plying on the road without a driver and it meets with an accident causing damage to life and property, there is a logic that the manufacturer is liable for the damage, presuming that the passengers inside the car have caused no fault to the vehicle. On the other hand, if there is a driver inside the AV and there has been an alert before the accident, then the driver or the owner is liable under the master–agent principle. While the above explanation sounds very logical, there is no unanimity on the matter. Courts have differed on liability. Car manufacturers are not willing to bear the liability once the vehicles are licensed for production and are sold after following due diligence during production. Lohmann [13] mentions that AVs need no human driver, the cause of accidents should be attributed to the manufacturer and not the holder (owner) of the car. On the other hand, Ljungholm [14] says that the driver's accidental responsibility and liability will be reduced drastically with technology taking over AV in pilot mode. AVs are in a nascent stage and manufacturers are investing a lot of money to come up with reliable AVs. If the law imposes penalties on them where they are not strictly liable, they will feel discouraged and will shy away from investment in AVs.

Therefore, the state must amend the laws taking into account the liability of the party in the scenario. Hanna [15] mentions that one of the major causes of automobile accidents is drunken driving. He says that there will be a sure reduction of accidental fatalities, injuries, and damage if more AVs ply on the road. He adds that there will be AVs where the driver will be on a standby arrangement, in that case, there is a possibility of drunken driving in some situations, leading to accidents. Therefore, the laws have to be suitably modified to take care of such behavior of drivers and consequent accidents. Nixon [16] mentions that laws related to automobile accidents have to change with time as more and more technologically superior vehicles come to the market and other environmental changes. Regarding the responsibility of "drive-it-yourself" [17] cars, the law of the USA mandates that the agencies who are in such business must insure the vehicle against the risk of accidents and damage caused to life and property. Brodsky [18] says that disparate traffic laws may make designing and programming cars difficult, and uncertain liability frameworks may create economic obstacles for manufacturers or consumers, depending on where liability will fall. Seidenberg [19] cites that, to spur progress, the Obama administration has asked the Department of Transportation to propose complete national testing and safety standards. But as far as the question of accountability and liability goes, we might already be homing in on an answer, one that points to a shift in how the root cause of damage is assessed: When a computerized driver replaces a human one, experts say the companies behind the software and hardware sit in the legal liability chain, not the car owner or the person's insurance company. Eventually, and inevitably, the carmakers will have to take the blame. On another interesting dimension of law, [20] talks about the criminal liability of robots in general and robots used in AVs in particular. The author says criminal law provides for punishment if the culpability of a human being is established as they are considered to be morally responsible agents. But robots, though they act like humans, are not humans. They are not well-placed to conduct themselves morally. Therefore, we can't consider them as an agency for punishment. It is the man behind that robot who has to be punished if he has done some wrong with clear intention. Braham and Rabin [21] mention that in the long future, driver-centric liability will be replaced by technology-centric liability, whether it is the manufacturer or the third party who converted the vehicle to AV. But till that time, there will be mixed liability and the legal system should be able to handle that transition with the existing laws.

The Vienna Convention on Road Traffic, 1968, has been the guiding international legal instrument when it comes to road traffic, safety, and liability regimes in case of damages or injuries sustained in road accidents. This is adopted by most countries in the world and hence has the force of international consensus which requires a human driver as a requirement for all types of vehicles on the roads who will be in a position to directly control such vehicles all the time he is sitting in the capacity of a driver of such vehicles. This requirement, which was adopted internationally as a part of the Vienna Convention, made it almost impossible for countries to carry out road testing of driver-less or AVs, even though many countries had already conceived this idea and had started researching the technologies making the AVs into a reality by pumping a lot of infrastructure and resources in this domain of research. But soon, countries realizing the potential of this domain of research, which could be instrumental in changing the face of the automobile sector and making it vastly futuristic, consented to bring an amendment to the Vienna Convention [22] which made driving a vehicle without a driver legally permissible as long as the vehicle can be overridden and switched off with human intervention and this allowed the countries adopting this convention to be able to not only research on the AI-based AVs but also to go for their road testing. However, as per this convention, such road testing could be done only for vehicles that are not fully autonomous but would still require human intervention to take charge of the vehicle when a need arises. Still, this amendment in the Vienna Convention is considered a landmark as it allowed the countries to explore the different facets of this evolving technology behind driverless vehicles, if not for fully AVs but at least for partially AVs.

Liability Regime for AVs: The Global Scenario

As the USA was never a party to the Vienna Convention, it was never under the obligation to adopt the requirements of the Vienna Convention and hence, the USA has been leading this race of not only injecting resources and investments in designing such autonomous technologies and manufacturing driver-less vehicles but also going for its road trials and deliberating on having federal laws in many of its states to address various legal issues in connection with the AVs [23]. California also has allowed the use of AVs on public roads but with a requirement for human drivers so that when a need arises, the car can be taken under the control of a human driver. As many as 29 states in the USA have successfully enacted federal laws about AVs to date, which clearly shows that the USA is leading the domain of AVs in the world. However, it is equally a matter of concern that even though the USA has funded heavily into the research and development of AI-based AVs, it has not given due consideration to developing legal regimes setting out the contours of accidental liability in the case of AVs and it remains an evolving area requiring much debates and deliberations in the USA leaving rooms for it to look into the accidental liability regimes developed by other countries in case of failure or malfunction of the systems of the AI-driven AVs [24].

Germany has allowed its federal states to the trails of AVs provided the vehicle will have a human driver who would be in a position to take charge of the vehicle when a need arises to maintain safety and prevent any untoward happening on the roads. Germany is one of those very few countries that has set a clear liability regime in the case of AVs within its legal framework [25] on the principles of Person v Property damages and nondiscriminatory damages by clearly opining that as per the German code of Ethics, Human life will always have an overriding effect of the property and animal life. Germany also clearly set out in its framework [26] that a black box must be installed on the AV with a wireless system that runs on a GPS algorithm tracking the vehicle's speed, operational features, space traveled, and so forth and the recording of this black box will be very crucial in fixing the liability in case of an accident caused by such AV. In this context, Germany's liability regime regarding AVs seems to be clear in terms of its enforcement aspect and hence has attracted attention from other countries like China. While formulating its comprehensive policy guidelines related to the operation of AVs on Chinese roads, the Chinese government has adopted a significant part of Germany's Code of Ethics enumerated under the German Road Traffic (Amendment) Act, 2017, specifically around liability fixation.

Britain has also been a front-runner in terms of not only boosting research and development in the AVs sector but also in legislating laws to regulate the legal issues that would arise out of the practical execution of AVs on roads in general and the issues relating to fixation of accidental liability, in particular. The British law on the given subject of AVs [27] is observed as a comprehensive legislation in having stringent provisions regarding fixing liability in the case of accidents and deaths caused by AVs. As per the provision of the Act, in case an accident is caused by an autonomous car, the liability for the accident will be borne by the owner of such an AV [28]. Furthermore, as per the provision of this legislation, if the AV is insured, and any accident is caused by such an insured AV, where the reason for the accident was found to be unauthorized use of the software used in such a car or failure to update the software, then the liability of the insurance company decreases and the liability in such a case for the insurance company is only fixed under the secondary liability less grave than the primary liability which is fixed on the owner based on the notion that it is the primary duty of the owner of an AV not to use any of the software of such vehicle beyond authorization or instructed guidance and also to update the software as and when it is required as instructed by the manufacturer. In case the owner fails to take due reasonable care and caution in undertaking these two mentioned acts, then as per British law, the primary/direct liability will fall on the owner in the case of any accidents caused by such owned AVs.

A comprehensive study of the international liability regime about AVs throws light on the fact that presently few countries have their sui generis legislative and regulatory frameworks to fix liability in case of accidental claims in connection with AVs and there is no international legal consensus on this area of law yet. Hence, to have better clarity in enforcing laws relating to AVs, there is a serious need for an international legal order/instrument that will guide the contracting countries to legislate their domestic laws in this regard under some established international mandate which will be a welcoming step towards achieving parity in terms of legal and regulatory enforcement in connection with liability claims caused by AVs.

4.4.3 ACCIDENTAL LIABILITY AND INDIAN LAWS

When it comes to self-driving or autonomous cars, India lacks any dedicated legislation. The existing major legislation relating to motor vehicles and accidental claims that India has, that is, the Motor Vehicles Act, 1939, doesn't have any specific provisions relating to AVs, and the general provisions of this Act, cannot be adequately implemented for AVs as the interpretation of those general provisions is ambiguous when it comes to the special needs and issues that these AVs pose. The latest major amendment in the Motor Vehicle Act, 1939 [29], also is mostly silent and doesn't provide anything on the issues of AVs. The major difficulty in setting out the provisions for liability fixation in the case of AVs has been observed to be the concept of personhood and Agency which have been two guiding principles under Indian Jurisprudence for any legislation that sets out Civil or Criminal liability against any claims.

The concept of "personhood' refers to the identification of the person whether natural or legal/artificial, which can be held liable for any liability. The concept of "personhood" is discussed under the "primary liability" regime under Indian Jurisprudence which states that a person (whether natural or artificial/legal entity) who is supposed to be in a capacity to directly control any particular act or event, doesn't do so or fails to take reasonable care in doing so proper and because of such nonperformance of the act or failure to perform the act with due care, any accident is caused then the person shall be directly or primarily liable for any liability arising out of such accident. The concept of "agency" refers to a relationship between two stakeholders in a given act or event where one party is in a position to exercise control over the other, where the latter is performing the act and the former is in a capacity to exercise control, guide, or take reasonable care and caution in the act, but fails to exercise due control or fails to take reasonable care and caution then such person cannot be absolved of any liability and the liability that would be fixed on such a person is that of a liability under a "master-servant", "employer-employee," or "agency" relationship popularly known as Secondary Liability under the law of Torts under Indian Jurisprudence.

To arrive at observations with clarity in the area of accidental liability in the case of AVs in India, provisions of the following Indian legislation need to be examined.

4.4.4 MOTOR VEHICLES ACT, 1939

The legal position of AVs under the Motor Vehicles Act, of 1939, can be examined by interpreting a few important provisions of the said Act. First, the provision [30] makes it a mandate for a driver to own a driving license to be able to drive any vehicle legally under this Act and secondly, it also mandates that the drivers of any vehicle must have attained 18 years of age to be entitled under the law to be able to drive on India roads. Now, when it comes to AVs, which are AI-driven and are called driverless vehicles, how this provision of the Act will be interpreted remains a great area of debate and deliberation. To date in India, under no law, AI has got proper legal status as a natural/legal person, and hence, in the absence of any such accepted and clear legal status for AI, whether AI can be interpreted to be a person/driver to be entitled to drive a vehicle legally on Indian roads remains a question and that requires serious attention from the law and policymakers in India, and accordingly amendment to this basic provision of the MV Act, 1939, can be brought to aptly govern the issues of vehicles driven by AI. Another provision [31] of the Act empowers the state governments to "make special regulations for driving motor vehicles" and hence if the need arises, governments can make special laws keeping given the special issues concerning AVs by making appropriate by-laws or bring suitable amendments to this existing law.

In the landmark case of Manushri Raha versus B. L. Gupta [32], the Supreme Court for the first time introduced the concept of "No Fault" Liability as a principle of Strict Liability in addition to the already existing "Fault Liability" principle under the MV Act, 1939 [33], to bring the insurance companies under the radar and make them pay for the damages as a matter of strict liability. A similar provision can be explored in the case of liability fixation on the manufacturers of AVs in case of any accidental claims as a matter of strict liability on the principle of "No Fault" liability. Motor Vehicle Amendment Bill (MV Bill), 2016, was introduced with a proposition of new provision [34] that will empower the state governments to exempt certain kinds of vehicles such as electric vehicles, AVs, and other such vehicles that run on special technological systems from the provision of the Motor Vehicle Act, 1939, and make appropriate measures for the use and operation of such special kinds of vehicles and not bring them under the general provisions of the Motor Vehicles Act, 1939. However, the bill is still pending and, once approved, this could be a welcome step toward setting out some special new legal order for AVs in India.

4.4.5 CONSUMER PROTECTION ACT, 1986

India has its wholesome law on consumer protection already in place which deals with a lot of issues about settlement of consumer claims. Hence, the courts in India often feel the need to go for a harmonious construction between the provisions of the Motor Vehicles Act, 1939, and the Consumer Protection Act, 1986 while deciding on the matter of fixation of accidental liability in the case of motor vehicle accidents in India. Various matters concerning the liability fixation in case of accidental claims both in the case of general vehicles and AVs upon the manufacturer of the vehicles will fall under the ambit of consumer protection laws such as negligence, manufacturing defects, design defects, software, and automated system defects, especially, in case of AVs, failure to warn, breach of warranty, misrepresentation, and so forth. Hence, the need to make consumer protection laws compatible with the enforcement of provisions about AVs is the need of the hour. Implementation of better awareness among the consumers and other involved stakeholders regarding the various legal complexities involved in the operation of AVs should be the prime focus of the authorities under the consumer protection laws in addition to introducing such relevant provisions under the existing Consumer Protection Act, 1986, by the legislators which shall be compatible with the special needs of these AVs.

4.4.6 INFORMATION TECHNOLOGY ACT, 2000

The Information Technology Act, of 2000, is the law governing data and information protection in cyberspace in India. Under this law, the "concept of property" has been attached to data, and data is considered the property of the owner or creator of such data. Data again has been given a very vast interpretation by various courts in India through a series of judicial pronouncements to include any algorithm, digital logic systems, software, encryption-enabled technology, or any encrypted codes in general under the provision of this law. The IT Act of 2000, provides for the protection of data and information in cyberspace against any misappropriation or unauthorized use under the provision related to Hacking [35] and fixes liability on the person hacking such private or secured data by unauthorized means in cyberspace. In light of AVs, the scope of the provision of Sec. 66 of the IT Act, 2000, needs to be revisited to include the liability of a person hacking the secured algorithm of AI-run technological systems embedded in the AVs to take remote charge of an AV and deliberately cause an accident by such AVs resulting in loss of life or property.

These are some of the very important areas under different existing legislation in India for some in-depth legal research which to date remains unexplored and needs to be taken up seriously by the law and policymakers before India gathers momentum to introduce AVs on Indian roads.

4.5 CONCLUSION

AVs based on cutting-edge AI technology will appear on the Indian market sooner or later. AVs will usher in a new era in transportation and logistics. Technology changes very fast and often laws fail to keep pace with it. The existing Motor Vehicle and other laws in India need to have provisions to handle various kinds of accidental damages that may be caused by AVs. In a given situation, laws must give clarity as to who is responsible for paying the damages—the owner, the user, third-party modifier, the automobile manufacturer, or a hacker. It is important to underscore here that without proper laws in place, allowing the use of AVs on the roads will be suicidal. Only when the state is ready with the right kind of legislation it should allow the use of AVs on the road. There must be more debate and discourse around the topic of AVs and AI. The development of roads or dedicated corridors for AVs will go a long way in getting the real benefit from the deployment of AVs in the Indian market. AV is a disruptive idea, and we must be legally ready to welcome this new technology to our roads.

KEYWORDS

- **accidental liability**
- **autonomous vehicles**
- **driverless vehicles**
- **motor-vehicle laws**
- **smart transportation**

REFERENCES

1. Turnbull, E. C. (2017). Hours of service of drivers: Harnessing autonomous technology for safer operations. *Jurimetrics*, *58*(1), 105–125. https://www.jstor.org/stable/26426452
2. Li, V. (2018). The dangers of digi-things: Writing the laws when driverless cars (or other computerized products) take a wrong turn. *ABA Journal*, *104*(3), 38–47. https://www.jstor.org/stable/26516273
3. Mohan, D. (2016). Driverless vehicles and their future in India. *Economic and Political Weekly*, *51*(32), 106–113. https://www.jstor.org/stable/44004704

4. Danks, D., & London, A. J. (2018). How driverless cars think. *Issues in Science and Technology*, *34*(4), 8–10. https://www.jstor.org/stable/26597970
5. Wallich, P. (1992). Smart wheels. *Scientific American*, *267*(4), 23–23. https://www.jstor.org/stable/24939247
6. Watney, C., & Draffin, C. (2017). Addressing new challenges in automotive cybersecurity. *R Street Institute, JSTOR*. http://www.jstor.org/stable/resrep19133 (accessed November 9, 2024)
7. Huang, B. I. (2019). Law's halo and the moral machine. *Columbia Law Review*, *119*(7), 1811–1828. https://www.jstor.org/stable/26810850
8. Keeling, G. (2018). Legal necessity, pareto efficiency and justified killing in autonomous vehicle collisions. *Ethical Theory and Moral Practice*, *21*(2), 413–427. https://www.jstor.org/stable/44955614
9. Himmelreich, J. (2018). Never mind the trolley: The ethics of autonomous vehicles in mundane situations. *Ethical Theory and Moral Practice*, *21*(3), 669–684. https://www.jstor.org/stable/44955658
10. Nyholm, S., & Smids, J. (2016). The ethics of accident-algorithms for self-driving cars: An applied trolley problem? *Ethical Theory and Moral Practice*, *19*(5), 1275–1289. https://www.jstor.org/stable/44955471
11. Etzioni, A., & Etzioni, O. (2017). Incorporating ethics into artificial intelligence. *The Journal of Ethics*, *21*(4), 403–418. https://www.jstor.org/stable/45204573
12. Pagallo, U. (2018). Apples, oranges, robots: Four misunderstandings in today's debate on the legal status of AI systems. *Philosophical Transactions: Mathematical, Physical and Engineering Sciences*, *376*(2133), 1–16. https://www.jstor.org/stable/26601845
13. Lohmann, M. F. (2016). Liability issues concerning self-driving vehicles. *European Journal of Risk Regulation*, *7*(2), 335–340. https://www.jstor.org/stable/24769813
14. Ljungholm, D. P. (2019). Autonomous car regulation in the smart transportation infrastructure: Ethical issues, legal liabilities, and privacy concerns. *Geopolitics, History, and International Relations*, *11*(2), 7–12. https://www.jstor.org/stable/26810362
15. Hanna, K. L. (2015). Old laws, new tricks: Drunk driving and autonomous vehicles. *Jurimetrics*, *55*(2), 275–289. https://www.jstor.org/stable/24395573
16. Nixon, R. M. (1936). Changing rules of liability in automobile accident litigation. *Law and Contemporary Problems*, *3*(4), 476–490. https://doi.org/10.2307/1189341
17. Constitutional Law. (1932). Validity of ordinance requiring liability insurance from owners of "Drive-it-Yourself" automobiles. *Virginia Law Review*, *18*(7), 796–797. https://doi.org/10.2307/1066039
18. Brodsky, J. S. (2016). Autonomous vehicle regulation: How an uncertain legal landscape may hit the brakes on self-driving cars. *Berkeley Technology Law Journal*, *31*(2), 851–878. https://www.jstor.org/stable/26377774
19. Seidenberg, S. (2017). Behind the wheel: Who's to blame when self-driving cars crash? *ABA Journal*, *103*(7), 18–19. https://www.jstor.org/stable/26515997
20. Gless, S., Silverman, E., & Weigend, T. (2016). If robots cause harm, who is to blame? self-driving cars and criminal liability. *New Criminal Law Review: An International and Interdisciplinary Journal*, *19*(3), 412–436. https://www.jstor.org/stable/26417695
21. Abraham, K. S., & Rabin, R. L. (2019). Automated vehicles and manufacturer responsibility for accidents: A new legal regime for a new era. *Virginia Law Review*, *105*(1), 127–171. https://www.jstor.org/stable/26842234

22. Amendment to Article 8 of the Vienna Convention, 1968, brought in the year 2014.
23. Nevada became the first state in the world to have allowed autonomous vehicles on public roads in the year 2011, which was much earlier than the amendment brought in the Vienna Convention on Traffic, 1968 in the year 2014 and its actual implementation, thereafter. Available at https://blog.ipleaders.in/legal-issues-related-autonomous-vehicles/ (accessed December 3, 2022)
24. Autonomous Vehicles/Self Driving Vehicles: Enacted legislation: Enacted Legislation, Available at: https://www.ncsl.org/research/transportation/autonomous-vehicles-self-driving-vehicles-enacted-legislation.aspx (accessed December 3, 2022)
25. German Road Traffic (Amendment) Act. (2017).
26. Section 63(A) German Road Traffic (Amendment) Act. (2017).
27. The Automated and Electric Vehicles Act. (2018).
28. As per Sec 2(2) of The Automated and Electric Vehicles Act. (2018)
29. Motor Vehicle (Amendment) Act. (2019)
30. Section 3 (I) of the Motor Vehicle Act.
31. Sec-133 of the Motor Vehicles Act. (1939).
32. 1977 AIR 1158
33. Sec. 144 of the Motor Vehicles Act. (1939).
34. The proposition in the form of Provision Sec. 2B to the Motor Vehicles Act. (1939).
35. Sec.66 of the Information Technology Act. (2002).
36. Polkowski, Z., Mishra, S. K., Mishra, B. K., Borah, S., & Mohanty, A. (2023). Impact of internet of things in social and agricultural domains in rural sector: a case study. *International Journal of Cloud Computing*, 90–105. Inderscience Publishers. https://doi.org/10.1504/IJCC.2023.129776
37. Prasad, S. S., Mishra, J. P., Mishra, S. K., & Borah, S. (2020). Challenges toward blockchain and renewable energy linked to IoT: A survey, advances in energy technology. *Advances in sustainability science and technology advances in energy technology* (pp. 49–55). Singapore: Springer Nature.
38. Nath, M. P., Priyadarshini, S. B. B., Mishra, D., & Borah, S. (2020). A comprehensive study of contemporary IoT technologies and varied machine learning (ML) schemes. *Soft computing techniques and applications* (pp. 623–634). Singapore: Springer Nature.
39. Mishra, J. P., Borah, S., & Mishra, S. K. (2021). Performance of scalable data linked to internet of things: A case study. In S. Borah, R. Pradhan, N. Dey, & P. Gupta (Eds.), *Soft computing techniques and applications: Advances in intelligent systems and computing* (Vol. 1248). Singapore: Springer. https://doi.org/10.1007/978-981-15-7394-1_54
40. Verma, U., Sohani, M., Borah, S., Nagwanshi, K. K., & Pathak, S. (2022). Role of fog-assisted internet of things-enabled system for managing the impact of COVID-19. In S. Borah, S. K. Mishra, B. K. Mishra, V. E. Balas & Z. Polkowski (Eds.), *Advances in data science and management: Lecture notes on data engineering and communications technologies* (Vol. 86). Singapore: Springer. https://doi.org/10.1007/978-981-16-5685-9_39

CHAPTER 5

SELECTIVE WEIGHTING HISTOGRAM SHIFTING FOR IMAGES IN CYBER-PHYSICAL SYSTEMS

RAVI UYYALA and S. CHINA RAMU

Department of CSE, Chaitanya Bharathi Institute of Technology (CBIT), Hyderabad, India

ABSTRACT

One of the more recent reversible data hiding (RDH) techniques for greater embedding capacity with little distortion is prediction error expansion (PEE). The initial step in a PEE-based technique is to use a pixel prediction strategy to forecast the original pixel. The literature has a number of pixel prediction strategies. To get a more precise accurately anticipate the existing pixel, a selective weighting-based prediction has been proposed in this study initially, assisted by gradient estimation. Later, Adaptive Novel Histogram (AH Bin Shifting) was used to construct embedding. Each pixel's local complexity for data embedding is calculated by its adjacent neighboring pixel's variance. Experimental findings indicate that the suggested approach has done better than a few mentioned efforts in the literature. This technique has been explored to ensure the integrity of the original images in cyber-physical systems.

5.1 INTRODUCTION

The objective of reversible data hiding (RDH) [1] technology is to achieve covert communication while maintaining the integrity of the load image

Cyber-Physical Systems: Applications, Challenges, and Research Directions. Rajanikanth Aluvalu, Samarjeet Borah, & Lalit Garg (Eds.)

DOI: 10.1201/9781779643421-5

for over two decades. Since their creation, RDH approaches in the plain-text domain have led to the greatest achievements in research. Nearly all the RDH methods may be divided into three groups [1]: (i) DE, (ii) HBS, and (iii) prediction error expansion (PEE). Despite the ease of use of the Internet and the widespread usage of personal computers, there is a growing risk of information leakage on the Internet. In order to implement RDH, RDH now needs to figure out how to use images. As everyone knows, a key factor in classical RDH is the load image's context. However, images are never exposed to any correlation or redundancy. RDH is often called lossless data hiding since the marked media can be used to rebuild the cover media. Through reversible pixel value change, data is appended to the primary data. In Ref. [2], the RDH approach was first presented. This topic has been progressed by other RDH approaches.

In 2003, Tian introduced the DE approach, which expands on the pixel difference to include data. Tian's approach, in contrast to the compression-dependent RDH schemes, may give a substantially greater embedding payload with minimal distortion [3].

A single piece of extra information is added to the primary data by examining the differences between the pixels that are right next to each other. An alternative method is the HBS-based RDH, which was explained in Ref. [4]. Using the highest and lowest bins of the histogram, the single data was entered into the HBS technique. An RDH method based on several histograms is suggested in Ref. [5]. In order to compute the multiple histograms, various image complexity values are investigated. RDH utilizing the PEE techniques is another type of these techniques that has been determined to be better and superior to others [6].

PEE-dependent methods for RDH examine a collection of contextual pixel within a certain neighborhood to forecast the pixel. The PE of the appropriate pixel includes hidden data bits. As a result of PEE-dependent method's supremacy, researchers' focus has grown. That motivates me to look into PEE-based RDH techniques deeper. Consequently, the study's objective as it is given in this work is to look into suitable pixel prediction systems in order to create new RDH methods. New contextual pixel techniques have been presented within the scope of RDH. With the help of accurate pixel prediction algorithms, the compromise between the ability to deform and embed has evolved. RDH's effectiveness is context-dependent.

With respect to the PVO, RDH's ability differs. There has been a noticeable difference in pixel values using the PVO-based method. Following the segmentation of the image into nonoverlapping blocks, multiple PVO methods are used to predict one or more pixels inside each block [7]. In PVO approaches, pixel ordering is more effective.

Reference pixel (RP)-based approaches are another kind of PEE-based approaches. Several RP and non-RP techniques are listed in Refs. [8, 9]. To make a prediction, the GAP took advantage of the gradient's strengths in different directions [6]. The RDH technique also makes use of the GAP predictor to forecast a pixel's value. As for other gradient-based predictors, gradient-based selective weighting (GBSW) and extended GBSW are two more options [10]. Gradient-based methods have been evolved from the image coding in the literature. For a better understanding of the image information in coding gradients is used [11–13]. Multilayers are utilized to anticipate the current pixel using the perceptron-based predictor [14]. Many RDH techniques are presented in Refs. [15, 16].

To improve the contextual pixel prediction, we provide in this study a forecast using gradients employing a 4 × 4 localized situation. Later, a novel AHBS technique was applied in order to more effectively replace the main content with the data.

This chapter is structured as follows: In Section 5.2, prediction using gradients is examined within the framework of the 4 × 4 neighborhood. The embedding and extraction process with AHBS is covered in Section 5.3. Section 5.5 provides an illustration of the conclusion after Section 5.4 outlines the experimental results.

5.2 GRADIENT-BASED SELECTIVE WEIGHTING (GBSW)-BASED PREDICTION

This article suggests a novel method for data reversibility hiding that uses gradient-dependent pixel prediction. In the present work, gradients are computed in diagonal, antidiagonal, vertical, and horizontal directions with a 4 × 4 neighborhood and a 15-pixel context.

In most cases, one can use gradients to locate the prominent edges in the vicinity of the central pixel. The firm edge allows for more accurate forecasting. In this study, we employ a pair of improved gradients to make our value estimates. The current pixel (x) in Table 5.1 is analyzed alongside its immediate 15-pixel surroundings. Specifically, in the new scenario when predictions are being made, the four cardinal directions are as follows: diagonal, antidiagonal, vertical, and horizontal.

TABLE 5.1 Fifteen-Pixel-Wide Neighborhood-Based 4 × 4 Context.

P	Q	R	S
T	X (m, n)	U	V
W	Y	Z	E
F	G	H	I

$$DH = 2 \times |T\text{-}U| + 2 \times |U\text{-}V| + |P\text{-}Q| + |Q\text{-}R| + |R\text{-}S| + |W\text{-}Y| + |Y\text{-}Z| + |Z\text{-}E| + 1 \quad (5.1)$$

$$DV = 2 \times |Q\text{-}Y| + 2 \times |Y\text{-}G| + |P\text{-}T| + |T\text{-}W| + |W\text{-}F| + |R\text{-}U| + |U\text{-}Z| + |Z\text{-}H| + 1 \quad (5.2)$$

$$DD = 2 \times |P\text{-}Z| + 2 \times |Z\text{-}I| + |T\text{-}Y| + |Y\text{-}H| + |Q\text{-}U| + |U\text{-}E| + |W\text{-}G| + |R\text{-}V| + 1 \quad (5.3)$$

$$DA = 2 \times |R\text{-}W| + |T\text{-}Q| + |Y\text{-}U| + |G\text{-}Z| + |Z\text{-}V| + |H\text{-}E| + 1 \quad (5.4)$$

It is possible to calculate the horizontal (gh), vertical (gv), diagonal (gd), and antidiagonal (ga) gradients by using the following formulas:

$$gh = \lfloor DH/\ 10 + 0.5 \rfloor,\ gv = \lfloor DV\ /10 + 0.5 \rfloor.$$

$$gd = \lfloor DD/\ 10 + 0.5 \rfloor,\ ga = \lfloor DA\ /7 + 0.5 \rfloor. \quad (5.5)$$

5.2.1 *LINEAR PREDICTORS*

Linear predicted values in the diagonal xd', antidiagonal xa', horizontal xh', and vertical xv' axes are determined using the linear equations

$$xh' = \lfloor (T + U) / 2 + 0.5 \rfloor,\ xv' = \lfloor (Q + Y) / 2 + 0.5 \rfloor \quad (5.6)$$

$$xd' = \lfloor (2(P + Z) + Q + Y + T + U) / 8 + 0.5 \rfloor \quad (5.7)$$

$$xa' = \lfloor (2(R + W) + Q + Y + T + U) / 8 + 0.5 \rfloor \quad (5.8)$$

where the numbers on the right-hand sides of the above equations correspond to those in Table 5.1. After that, the gradients are ranked from highest to lowest, The resulting rank order can be written as $r0 \leq r1 \leq r2 \leq r3$. Take r0 and r1 as the origin, and l0 and l1 as the linearly anticipated values in those directions. With gradients and linear pixels in hand, the anticipated value is then calculated as

$$x' = \lfloor((l_0 \times r_1 + l_1 \times r_0) / (r_0 + r_1)) + 0.5\rfloor \quad (5.9)$$

The variance of the contextual pixel is computed using the right and the bottom of the present pixel (W, Y, Z, and U) is used as a method for determining a pixel's local complexity (L). During both the embedding and the extraction process, these pixel values are unaltered from their original state.

5.3 EMBEDDING AND EXTRACTION

The pixel's expected error, or prediction error (pe), is calculated by subtracting the pixel's main value from its estimated or anticipated value. Inspired by the research in Ref. [17], we use a novel modifying approach. The possibility of enlarging only one or two bits of data into the main body has been explored in this study. The image's complexity is used to help decipher its primary content. One or two bits of data can be inserted using L in the suggested work. The primary goal of computing the L is to ascertain the texture of the image enclosing the current pixels. The ranges $(-1, 0)$ and $(1, 0)$ are used to store the confidential information. The primary reason for using only two categories is to minimize the distortion of the primary content. The core content has only been slightly distorted during the embedding process if PE is less. It is hoped that the modest PE will result in little changes from the embedded pixel to the original pixel. Figure 5.1 presents a sample of the PE histogram for reference.

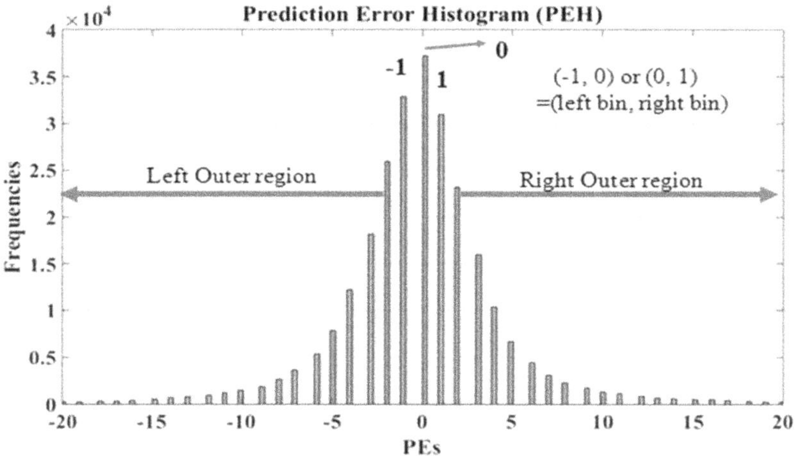

FIGURE 5.1 Adaptive HBS.

The following are some of the most important aspects of this embedding approach that are discussed here:

1. Depending on the pixel's L, the pixel's pe will have either 1 or 2 bits of information. The L of a pixel determines how harsh its environment is. Here, the variance of pixel values in (W, Y, Z, and U) is used to make an estimate.
2. A pixel's size is increased by two bits if its L value is less than or equal to TL and its pe value falls within the range [TP, +TP].
3. Pixels are extended by one bit if their pe values fall inside the range [TP, +TP] and L is greater than TL.
4. Pixels are shifted to the left or right by 4 TP and 4 TP + 3 if L is less than or equal to TL and pe is outside the range [−TP, +TP].
5. Similarly, pixels are shifted to the left or right by 2 TP and 2 TP + 1 if L is greater than TL and pe is outside the range [−TP, +TP].
6. Both the L and PE threshold values are denoted by their respective symbols (TL and TP).
7. Based on HBS, the suggested approach solely employs the (−1, 0) and (1, 0) bins for either embedding two bits of data or only one bit, depending on the case.
8. Only pixels from two consecutive PE-histogram bins are used to increase the size of the hidden information. In most cases, the best predictor will place the greatest value in the bin where the predictive effectiveness is 0.
9. Therefore, we choose (−1 and 0) or (0 and 1) as the pes for two adjacent bins.
10. For the opposite side to extract the primary data, the following information is deemed supplemental. (i) Final hidden information insertion pixel position, (ii) what the total number of bits are in the finished value, (iii) TL, (iv) which of the two conditions—OF or UF, (v) shortened length of OF/UF map indicating where things are, and (vi) a shortened version of the OF/UF diagram [8].

The following are some of the most important aspects of extraction:

1. To begin with, we investigate the LSBs of the pixels in the first few tagged image rows to glean the supplementary data.
2. The location of the final extended pixel can be determined using the new data. It is possible to recover the original cover image's pixel values by starting at this pixel and working backward through sequence of raster scans to retrieve informational byte string.

3. Exploration of that pixel yields two bits of information if L is less than or equal to TL and pe is in the range [4TP + 4TP + 3].
4. When extracting one bit of information from a pixel, if L > TL and pe belongs to [2TP, +2TP + 1], then that pixel is investigated.
5. Pixels are shifted to the right by 4TP or the left by 4TP+3 if L is less than or equal to TL and pe does not fall within [−4TP, +4TP + 3].
6. If L is greater than TL and PE is beyond the range [−TP, +TP], then the pixel is shifted to the right by 2TP or the left by 2TP + 1.
7. Finally, an LSB-substitution technique moves the actual LSBs to the top rows of the image.
8. The projected or estimated value should be the same during both the secret data's insertion and extraction.

5.4 EXPERIMENTAL RESULTS

Standard 512-by-512-pixel test images like the ones in Figure 5.2 are used for the tests. The values (i) either one or zero and (ii) either zero or one are used in the tests. Both the scenarios described above have embedding capabilities that are comparable to one another and do not notably contrast one another. PSNR for both the main image and its annotated counterpart are also comparable between the two cases (at the maximum allowable image embedding size). All the images presented in this work are standard test images for RDH.

FIGURE 5.2 Test images.

Estimates of the PSNR comparisons of varying payload sizes are made between the main image and labeled for each of the previously described experimental scenarios. Such PSNRs are presented in Figure 5.3, which can be found here. The threshold value of L (TL) is varied throughout the course of the tests, taking on values such as zero, one, two, four, eight, and sixteen. When TL has a value of zero, this indicates that the embedding is a single bit. When TL is greater, there is room for more data to be embed. With the help of single-bit embedding, also known as TL = 0, a comparison is made between the work that has been proposed and other approaches that are already in use. In the tests that were presented, the value of the L threshold TL was varied as follows: Zero, One, Two, Four, Eight, and Sixteen. This was done in order to maintain control over the adaptive embedding process. When TL is zero, it indicates that the prediction error for a pixel carries a payload of one bit of information. When the TL value is increased, it indicates that additional pixels are able to store two bits of payload. As the value of TL increases, so does its embedding potential. Increases in TL value are accompanied by a decline in PSNR. Table 5.2 reveals how much weight various payloads can be embedded with PSNR values for selected images.

TABLE 5.2 TL = 0 and TL = 16 Payload.

Image	TL = 0		TL = 16	
Lena	0.17	48.63	0.23	41.60
Lake	0.13	48.67	0.17	42.89
Elaine	0.11	48.60	0.13	43.70
Boat	0.10	49.27	0.13	43.13

This chapter compares the rhombus average (RA) [18], DI [19], and SB [20] with the proposed RDH technique making use of a 4 × 4 grid to calculate gradients. Figure 5.4 presents the PSNRs between the main images and their counterparts.. These values are presented for a variety of payload sizes. Figure 5.5 presents the labeled ones of the present method with a range of sizes. The RDH approach here outperforms the standard methods of comparison. Because the PSNR values at (−1, 0) and (1, 0) are so similar, we compared the proposed methodology with (−1, 0) to the other methods.

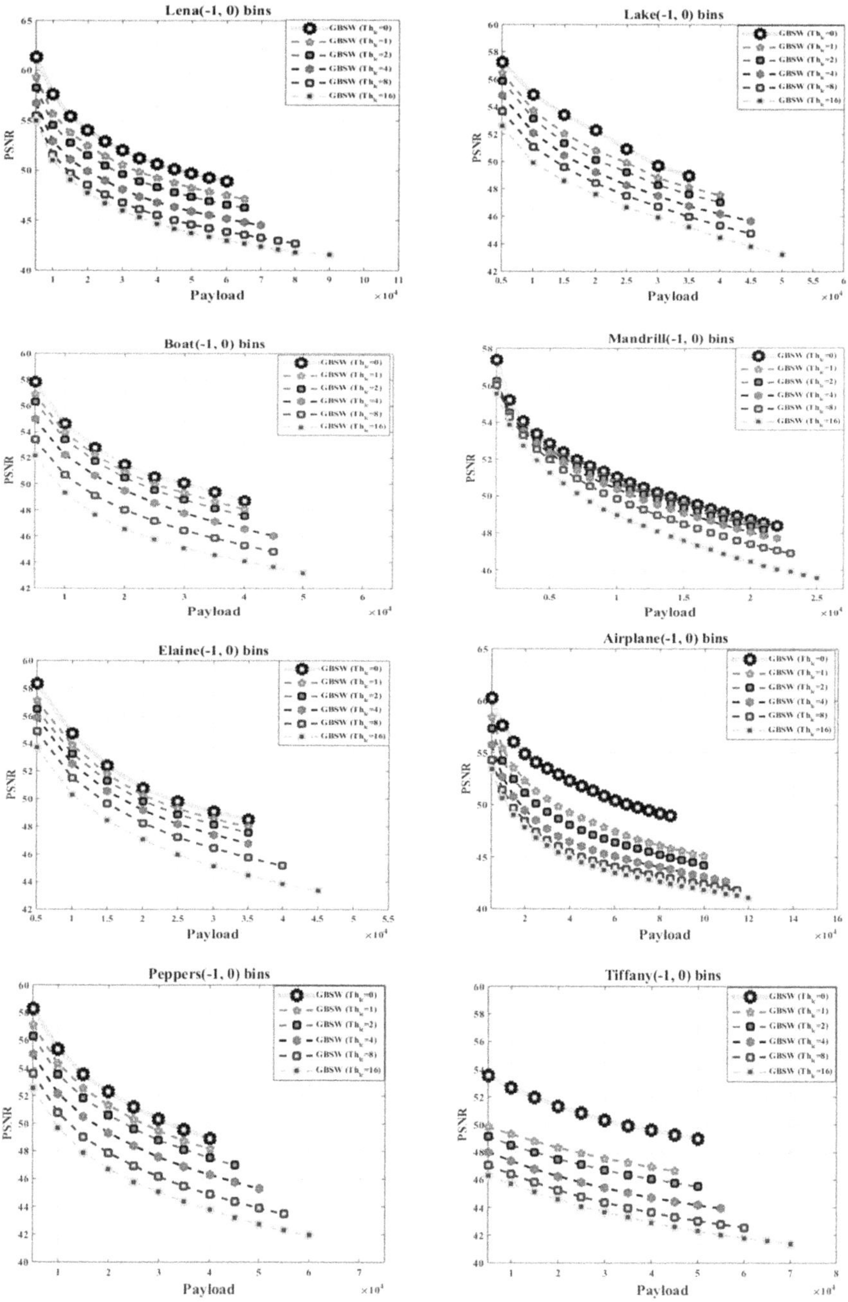

FIGURE 5.3 Variations in TL and their effects on performance.

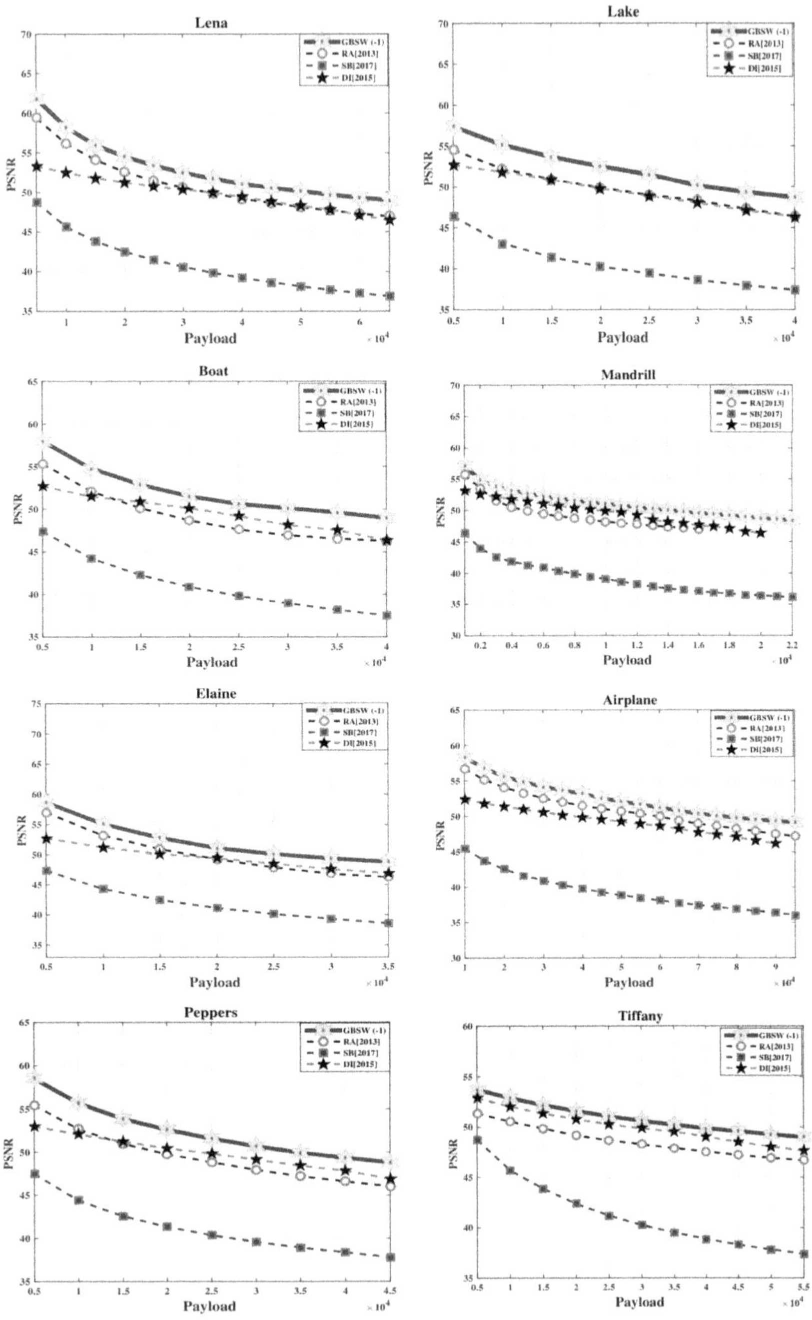

FIGURE 5.4 Analyzing other methods' performance.

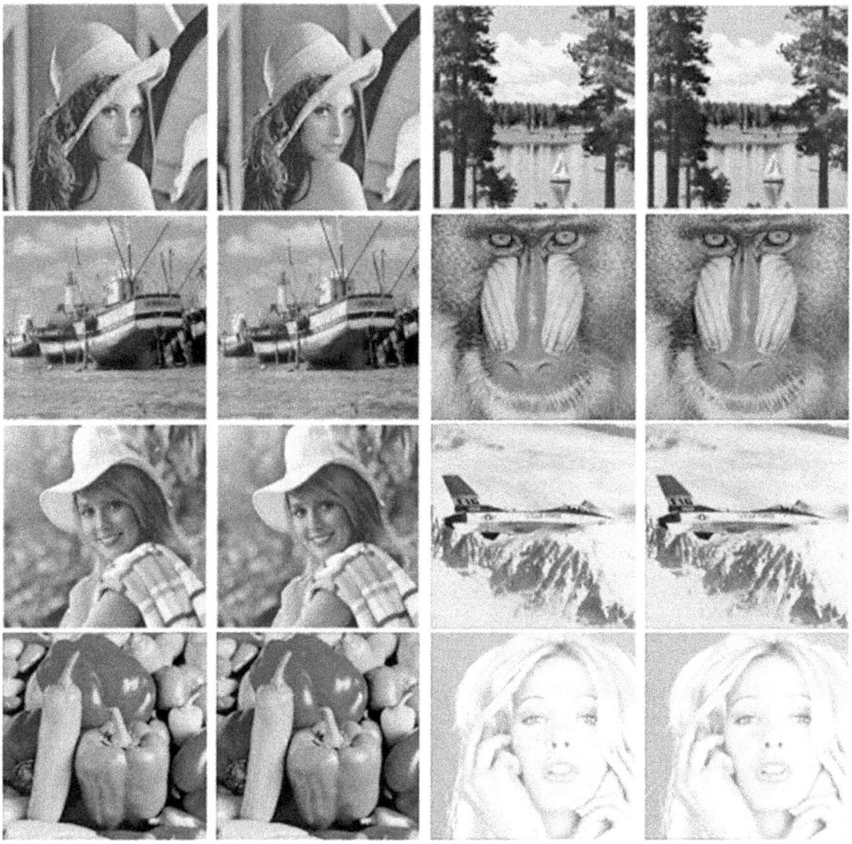

FIGURE 5.5 The end result of applying the RDH method to several sample images.
Source: Original and marked images for a payload of 5000 bits.

5.5 CONCLUSION

In this study, we present a new selective weighting method for RDH called gradient-based selective weighting. The forecast was worked out using a 4×4 neighborhood setting with gradients in each direction. Using the adaptive HBS technique, the information was progressively inserted into the histogram's divisions. The proposed method outperforms a number of existing solutions, including HBS and gradient-based approaches.

The RDH technique that has been suggested is fragile, just like the RDH techniques that already exist. As a result, the resistance of the suggested technique to multiple assaults upon the indicated image is not typically explored for these strategies. Such RDH approaches are distinct from robust data hiding systems in that the unrevealed data is rendered inaccessible whenever there is even a slight variation in the value of the pixel that has been marked. Because of this, it is possible to determine whether or not the original cover image has been altered. Consequently, the fundamental objective of the RDH approach is to preserve the authenticity of the main content. Therefore, the proposed method can be used on cyber-physical systems, even in situations where making even a tiny modification to the original images is not encouraged.

KEYWORDS

- **adaptive HBS**
- **adaptive novel histogram**
- **cyber-physical systems**
- **RDH**
- **gradient-based prediction**

REFERENCES

1. Shi, Y. Q., Li, X., Zhang, X., Wu, H. T., & Ma, B. (2016). Reversible data hiding: Advances in the past two decades. *IEEE Access, 4*, 3210–3237.
2. Honsinger, C. W., Jones, P. W., Rabbani, M., & Stoffel, J. C. (2001). *U.S. Patent No. 6,278,791*. U.S. Patent and Trademark Office: Washington, DC.
3. Tian, J. (2003). Reversible data embedding using a difference expansion. *IEEE Transactions on Circuits and Systems for Video Technology, 13*(8), 890–896.
4. Ni, Z., Shi, Y. Q., Ansari, N., & Su, W. (2006). Reversible data hiding. *IEEE Transactions on Circuits and Systems for Video Technology*, *16*(3), 354–362.
5. Li, X., Zhang, W., Gui, X., & Yang, B. (2015). Efficient reversible data hiding based on multiple histograms modification. *IEEE Transactions on Information Forensics and Security*, *10*(9), 2016–2027.
6. Thodi, D. M., & Rodríguez, J. J. (2007). Expansion embedding techniques for reversible watermarking. *IEEE Transactions on Image Processing*, *16*(3), 721–730.
7. Li, X., Li, J., Li, B., & Yang, B. (2013). High-fidelity reversible data hiding scheme based on pixel-value-ordering and prediction-error expansion. *Signal Processing*, *93*(1), 198–205.

8. Uyyala, R., Pal, R., & VNK Prasad, M. (2019). Reversible data hiding using B-tree triangular decomposition based prediction. *IET Image Processing, 13*(11), 1986–1997.
9. Uyyala, R., & Pal, R. (2018, July). Reversible data hiding based on the random distribution of reference pixels. In *2018 IEEE Region Ten Symposium (Tensymp)* (pp. 225–230). IEEE.
10. Uyyala, R., Pal, R., & Prasad, M. V. (2016, February). Gradient dependent reversible watermarking with low embedding impact. In *2016 3rd International Conference on Signal Processing and Integrated Networks (SPIN)* (pp. 184–189). IEEE.
11. Uyyala, R., & Pal, R. (2020, February). Reversible data hiding using improved gradient based prediction and adaptive histogram bin shifting. In *2020 7th International Conference on Signal Processing and Integrated Networks (SPIN)* (pp. 720–726). IEEE.
12. Uyyala, R., Prasad, M. V., & Pal, R. (2016, December). Selected context dependent prediction for reversible watermarking with optimal embedding. In *Proceedings of International Conference on Computer Vision and Image Processing: CVIP 2016* (Vol. 2, pp. 35–46). Springer Singapore.
13. Uyyala, R., & Pal, R. (2020, July). Reversible data hiding with selected directional context based prediction using 8-neighborhood. In *2020 IEEE International Conference on Electronics, Computing and Communication Technologies (CONECCT)* (pp. 1–6). IEEE.
14. Bhandari, A., Sharma, S., Uyyala, R., Pal, R., & Verma, M. (2020, July). Reversible data hiding using multi-layer perceptron based pixel prediction. In *Proceedings of the 11th International Conference on Advances in Information Technology* (pp. 1–8).
15. Uyyala, R., & Subramaniam, M. (2023, March). Gradient adaptive planar prediction for high-fidelity images. In *2023 International Conference on Wireless Communications Signal Processing and Networking (WiSPNET)* (pp. 1–6). IEEE.
16. Li, L., Yao, Y., & Yu, N. (2023). High-fidelity video reversible data hiding using joint spatial and temporal prediction. *Signal Processing, 208*, 108970.
17. Hong, W., Chen, T. S., & Shiu, C. W. (2009). Reversible data hiding for high quality images using modification of prediction errors. *Journal of Systems and Software, 82*(11), 1833–1842.
18. Coltuc, D., & Dragoi, I. C. (2013, June). Context embedding for raster-scan rhombus based reversible watermarking. In *Proceedings of the First ACM Workshop on Information Hiding and Multimedia Security* (pp. 215–220).
19. Hong, W., Chen, T. S., & Chen, J. (2015). Reversible data hiding using Delaunay triangulation and selective embedment. *Information Sciences, 308*, 140–154.
20. Wang, W., Ye, J., Wang, T., & Wang, W. (2017). Reversible data hiding scheme based on significant-bit-difference expansion. *IET Image Processing, 11*(11), 1002–1014.

CHAPTER 6

ASSESSING THE WHOLE-BODY VIBRATION AND THE LEVEL OF VIBRATION TRANSMITTED THROUGH THE SEATS OF GRADER OPERATORS

SANDEEP KUMAR, G. VRISHANK REDDY, and R.V.S. SRIKASYAP SAGAR

Meluha International School, Hyderabad, India

ABSTRACT

This research looks into the effects of whole body vibration (WBV) and how it is transferred through seat effective amplitude transmissibility (SEAT) when motor graders are used in surface coal mines. Three models of graders from three different companies (M1, M2, and M3) were analyzed, with six graders used from each model, totaling 18 graders. Data was collected using SV 38V and SV 151 tri-axial accelerometers by Svantek. Results showed a sizable difference between the models in terms of WBV exposure, with the M1-grader having the highest in the lateral direction, followed by M2-graders in the longitudinal direction and M3-graders in the vertical axis direction. The vertical WBV exposures exceeded the ISO action limit of 0.5 m/s^2. The transmissibility of vibration to the operator's seat was 94% for the M1-grader, 89% for M2-graders, and 83% for M3-graders, with M3-graders showing better vibration attenuation.

Cyber-Physical Systems: Applications, Challenges, and Research Directions. Rajanikanth Aluvalu, Samarjeet Borah, & Lalit Garg (Eds.)

DOI: 10.1201/9781779643421-6

6.1 INTRODUCTION

Professional drivers are more likely to develop MSDs such as low back pain which can have direct and indirect costs. Whole body vibration (WBV) is suggested to be the major cause for low back pain [2]. The way WBV is experienced depends on its direction, duration, magnitude, and frequency. The ISO 2631-1 (1997) Standard defines barycentric coordinate systems to evaluate WBV in field-based investigations. The frequency of the weighted root mean square acceleration (a_w) is used to characterize WBV exposure over a given period of time [9]. Furthermore, the transmissibility factor is used to measure the amount of vibration attenuation provided by a seat.

The design of seats in vehicles is essential to lowering the exposure to WBV. Paddan and Griffin's 2002 study found that a seat with a suspension system was more effective than a conventional seat in cutting down transmissibility. Motor graders come equipped with spring-loaded seats as the default seating option, reducing the operator's exposure to WBV. Nevertheless, it is not clear if air suspension seats have a similar performance across all grader models, as there could be some differences in how the seat and grader suspension systems interact. This research aims to measure the amount of WBV experienced by grader operators as well as the transmissibility of the vibrations from the footing to the seat of three graders made by three separate companies (hereafter called M1, M2, and M3) [3].

6.2 METHODS AND MATERIAL

6.2.1 STUDY DESIGN

Subjects: WBV exposure data of 18 operators were collected, by considering 6 machineries from each model type. The Human Factor details of the subjects are given in Table 6.1.

Study Site: The information was obtained from three different types of graders made by three diverse companies (labeled M1, M2, and M3) [4]. These graders operated in two different coal mines, known as Mine-I and Mine-II. Mine-I had a dumper shovel combination, while Mine-II used In-pit crusher conveying (ICC) technology [5].

TABLE 6.1 Human Factors Details of Subjects.

Personal/human factors		M1-Graders		M2-Graders		M3-Graders		
		Total no = 6		Total no = 6		Total no = 6		
		Number (%)	Mean (Standard Deviation)	Number (%)	Mean (Standard Deviation)	Number (%)	Mean (Standard Deviation)	Significance of p value
Age in years			44.5 (8.7)		45.3 (14.3)		45.6 (13.6)	0.83
Height in cm	≤ 168	2(33%)	169.5 (9.8)	1(16%)	167.8 (9.2)	2(33%)	168.5 (9.6)	0.03
	168–190	3(50%)		2(33%)		2(33%)		
	≥190	1(16%)		3(50%)		2(33%)		
Weight in kg	≤ 72	2(33%)	98.3 (36.2)	2(33%)	99.2 (14.7)	1(16%)	98.9 (32.8)	0.85
	72–96	2(33%)		3(50%)		2(33%)		
	≥ 96	2(33%)		1(16%)		3(50%)		
Body mass index (BMI) (kg/m^2)	18.3– 26.9	2(33%)	34.5 (9.6)	2(33%)	31.6 (3.8)	1(16%)	32.8 (9.1)	0.49
	27.0–29.9	3(50%)		2(33%)		2(33%)		
	≥ 30	1(16%)		2(33%)		3(50%)		

6.2.2 EXPOSURE TO WBV

The average frequency of the weighted root-mean-square (RMS) acceleration (a_w) indicated exposure to WBVs as per ISO 2631-1:1997 Standards [6]. In three directions—the x-axis (longitudinal), the y-axis (lateral), and the z-axis (vertical)—the vibration levels relating to exposure to WBV were measured [2]. These exposures were adjusted so that they represented an 8-h daily value (A(8)) and were then compared to the 0.5 m/s^2 limit, which was defined by ISO 2631-1:1997 [2].

6.2.3 SEAT EFFECTIVE AMPLITUDE TRANSMISSIBILITY

Calculating SEAT is a good way to assess the effectiveness of a seat. This is found out by seeing if the seat increases or decreases the vibrations from the floor. This is measured using Seat Effective Amplitude Transmissibility (SEAT), which is found by dividing the vibration of the seat by the vibration of the floor. If the SEAT value is 100%, it implies that the vibrations of the seat and the floor are the same, meaning that the seat does not provide any cushioning [2]. A lower SEAT rating shows that the seat has more vibration dampening.

6.2.4 DATA COLLECTION AND PROCESSING

In order to measure WBV, a tri-axial accelerometer named SV 38V and a data logger named SV106, which is compliant with the ISO 2631-1:1997 Vibration standards, were used [11]. This data logger is more energy-efficient than IEPE accelerometers, operating on four AA batteries for up to 12 h and functioning in a temperature range of −10 to 50°C. The data logger had a setting of 0.1 Hz–2 kHz (depending on the transducer) and a sampling rate of 6 kHz with 90 dB of dynamic range [10]. Additionally, a tri-axial accelerometer, called SV 151, was used to measure floor emission vibration in three directions. The mounting of the equipment was done in accordance with the ISO 2631 (1997) WBV guidelines. The accelerometers were calibrated according to the manufacturer's instructions prior to the measurements and each test session for the grader machinery was 20 min long. During the session, operators were prompted to follow their normal routines [3].

6.2.5 STATISTICAL ANALYSIS

The SPSS version 21 (IBM) software was used to analyze the data [7]. The mean and standard error of each grader was computed. ANOVA was used to evaluate the variation in WBV exposures among the three buses for variables such as SEAT values and A(8) exposures. A p value that exceeded 0.05 was indicative of a noteworthy divergence.

6.3 RESULT

6.3.1 EXPOSURE TO WBV

The values in Table 6.2 demonstrate the A(8) exposure levels based on the ISO 2631-1 criteria, with the x and y axes multiplied by 1.4 and the z axis by 1. The z-axis was the primary direction of A(8) exposure for all three graders, and the differences between them were significant (Table 6.2). The M2-grader had the highest $A(8)_x$, the M1-grader possessed the biggest $A(8)_y$, and the M3-grader had the most $A(8)_z$. All of the graders had A(8) exposures that exceeded the action limit of 0.5 m/s^2 stated in the ISO 2631-1.

6.3.2 SEAT EFFECTIVE AMPLITUDE TRANSMISSIBILITY

The performance of the airmatic seat is seen to be dependent on the grader type when looking at the SEAT-A(8) values in Table 6.2. The Volvo grader was found to have the most significant decrease in floor transmitted vibration when compared to the two other graders.

6.4 DISCUSSION

The dominant axis of vibration with respect to WBV was found to be different in all three types of graders (i.e., M1, M2, and M3). M2 graders reported highest A(8) in the x-direction. This may be due to the sudden penetration of the blade into the ground while cutting and grading operations in the longitudinal direction. This is also attributed to the fact that M2-grader experienced over bumps and the operator might have undergone

TABLE 6.2 Standard Estimation of WBV Measurements Along with SEAT Values for Different Graders.

		M1-Graders (Total No = 6) Mean (Standard Error)	M2-Graders (Total No = 6) Mean (Standard Error)	M3-Graders (Total No = 6) Mean (Standard Error)	Significance of p Value
Estimated vibration exposure values for 8 h duration [A(8)] m/s²	1.4 x-seat	0.45 (0.02)	0.46 (0.01)	0.45 (0.01)	0.004
	1.4 y-seat	0.49 (0.02)	0.45 (0.01)	0.43 (0.02)	< 0.0001
	z-seat	0.68 (0.01)	0.88 (0.01)	0.89 (0.02)	0.01
	z-floor	0.72 (0.01)	0.98 (0.01)	0.97 (0.02)	< 0.0001
SEAT- A(8)		94 (4) %	89 (7) %	83 (5) %	< 0.0001

higher angular longitudinal translations. The highest A(8) in the y-direction was reported by the M1-grader machinery. This is due to sudden jerking of the blade with high intensity, which makes the operator's seat swing in the lateral direction, which transmits vibration to the operator. The A(8) in the z-direction in case of M3-grader was the highest, and this could be due to mismatching between the seat and M3-grader suspension system.

It is significant to note here that the WBV exposures in different directions may yield a variety of spine injuries. This is further explained that WBV exposure in the z-direction leads to spine compression of operators and in the longitudinal direction leads to imparting shear force on the spine.

SEAT: The estimated seat values show that seats under use in grader machinery did not offer absolute isolation but substantially attenuated the WBV exposures. M2 and M1 graders transmitted 89% and 94% of the vibration felt at the footing to the operator's seat, respectively, while M3 had a minimum of 83% [8]. This performance could be due to an incongruity between the suspension of the grader seat suspension seat and the seat itself. Since the M1 grader is the older machinery among the three types of graders, the use of these older seats could have aggravated for lower vibration attenuation. Hence, M1 grader demonstrated poor performance. Similarly, M3 grader depicted better vibration attenuation compared to other two types of graders, which may be due to its better seat design.

6.5 CONCLUSIONS

Since the graders are in use even after its economic life, it is learnt that there lacks a regular monitoring/checkup of seats unlike any other parts of grader machinery. Hence, older seats are more prone to develop vibration than the newer ones. Further, the maintenance of graders is carried out whenever there is breakdown of its parts or as and when it is scheduled for the periodical maintenance. However, least importance is given to the maintenance or conditioning of its seats. This study demonstrated that the periodical maintenance of seats is very much necessary to mitigate the vibration transmission from the floor to seat. Future studies offer a scope for assessing WBV exposures and performance estimation of various types of seat designs in grader machinery.

KEYWORDS

- **grader machinery**
- **occupational ergonomics**
- **seat effective amplitude transmissibility**
- **transmissibility**
- **whole-body vibration**

REFERENCES

1. Salmoni, A. W., Cann, A. P., Gillin, E. K., & Eger, T. R. (2008). Case studies in whole-body vibration assessment in the transportation industry–Challenges in the field. *International Journal of Industrial Ergonomics, 38*(9-10), 783–791.
2. Thamsuwan, O., Blood, R. P., Lewis, C., Rynell, P. W., & Johnson, P.W. (2012, September). Whole body vibration exposure and seat effective amplitude transmissibility of air suspension seat in different bus designs. In *Proceedings of the Human Factors and Ergonomics Society Annual Meeting* (Vol. *56*(1), pp. 1218–1222). Sage CA: Los Angeles; CA: Sage Publications.
3. Cann, A. P., Gillin, E. K., Salmoni, A. W., Vi, P., Eger, T. R., De, F., Globales, V., Corps, D., & Partir, À. ( ). Transmissiblity of whole_body vibration from floor to seat experienced by scraper operators in the construction industry.
4. Okoro, T., Whitaker, R., Gardner, A., Maddison, P., Andrew, J. G., & Lemmey, A. (2016). Does an early home-based progressive resistance training program improve function following total hip replacement? Results of a randomized controlled study. *BMC Musculoskeletal Disorders, 17*(1), 1–9.
5. Satapathy, S. C., Raju, K. S., Molugaram, K., Krishnaiah, A., & Tsihrintzis, G. A. (Eds.). (2019). In *International Conference on Emerging Trends in Engineering (ICETE): Emerging Trends in Smart Modelling Systems and Design.*
6. Alfadhli, A., Darling, J., & Hillis, A. (2018). *Active seat suspensions for automotive applications* (Doctoral dissertation, University of Bath).
7. Thirunavukkarasu, A., Alotaibi, N. H., Al-Hazmi, A. H., Alenzi, M. J., Alshaalan, Z. M., Alruwaili, M. G., Alruwaili, T. A. M., Alanazi, H., & Alosaimi, T. H. (2021, December). Patients' perceptions and satisfaction with the outpatient telemedicine clinics during COVID-19 era in Saudi Arabia: A cross-sectional study. In *Healthcare* (Vol. *9*(2), p. 1739). MDPI.
8. Thamsuwan, O., Blood, R. P., Ching, R. P., Boyle, L., & Johnson, P. W. (2013). Whole body vibration exposures in bus drivers: A comparison between a high-floor coach and a low-floor city bus. *International Journal of Industrial Ergonomics, 43*(1), 9–17.

9. Jeripotula, S. K., Mangalpady, A., & Mandela, G. R. (2020). Musculoskeletal disorders among dozer operators exposed to whole-body vibration in Indian Surface Coal Mines. *Mining, Metallurgy & Exploration, 37*(2), 803–811.
10. Jeripotula, S. K., Mangalpady, A., & Mandela, G. R. (2019). Evaluation of whole-body vibration (WBV) of dozer operators based on job cycle. *Journal of The Institution of Engineers (India): Series D, 100*(2), 187–193.
11. Jeripotula, S. K. K., Mangalpady, A., & Mandela, G. R. R. (2020). Assessment of exposure to whole-body vibration of dozer operators based on postural variability. *Mining, Metallurgy & Exploration, 37*(2), 813–820.

CHAPTER 7

MIMO ANTENNA DESIGN: AN INCLUSIVE STUDY OF ANTENNA AND ISOLATION ENHANCEMENT TECHNIQUES FOR 5G COMMUNICATION

M. JYOTHIRMAI[1,2] and T. TIRUPAL[3]

[1]*Department of ECE, Jawaharlal Nehru Technological University Anantapur, Ananthapuramu, India*

[2]*Ravindra College of Engineering for Women, Kurnool, India*

[3]*Department of ECE, G. Pullaiah College of Engineering and Technology, Kurnool, India*

ABSTRACT

The Fifth Generation (5G) wireless technology lying in millimeter-wave band ranges from 30 to 300 GHz proves a better successor from its previous wireless communication in terms of data rates, geographical coverage, spectrum available, latency, multiapplication facility, and energy consumption. 5G has been given a specific range of band in the spectrum to be deployed globally. However, we all know the relation between the frequency and wavelength is inverse in nature. This is the major impact on the 5G communication, as the wavelength is shorter compared to 3G, 4G which were deployed at a lower microwave region. Fading and absorption are the main causes for the degradation in performance of 5G. To alleviate

Cyber-Physical Systems: Applications, Challenges, and Research Directions. Rajanikanth Aluvalu, Samarjeet Borah, & Lalit Garg (Eds.)

DOI: 10.1201/9781779643421-7

these drawbacks, multi-input multi-output (MIMO) seeks attraction by the researcher. As in MIMO, number of antennas will be deployed in both the transmitter and receiver sides which reduces the multipath fading and increases the channel capacity. But implementing the MIMO itself imposes diverse challenges in terms of design. This survey chapter provides an extensive study of various design approaches along with distinguished mutual coupling enhancement techniques which are present to improvise the performance parameters of MIMO antenna design such as envelope correlation coefficient (ECC), channel capacity loss (CCL), and so on for 5G communication.

7.1 INTRODUCTION

The wireless communication technology has seen a tremendous evolution in terms of performance and hardware aspect in the last four decades. Each successor wireless technology came as a possible solution to the hurdles which were not able to overpass by its predecessor technology. As an answer to the low data rates, coverage issue, mobile user accommodation, latency, and so on, 5G (Fifth Generation) evolved. The 5G technology is appealing technology having potential to surpass numerous concerns which can even trouble in near future. Not only fast connectivity, higher capacity, and high bit rate will be achieved but also the door to the virtual world will open for mankind which eventually led to the advancement of mankind. With the deployment of 5G, there is no uncertainty that there will be massive transformation in the various sectors such as healthcare, education, and industry. The third-generation partnership project (3GPP) is governing body for finalizing the standards for 5G. It has mentioned key scenarios for 5G, namely, eMBB, mMTC, and uRLLC depicted. For these scenarios, 5G can be operated at three different frequency bands to cater broad coverage, fast speed, interoperability, multiapplication facility, and much more. For any wireless technology, the antenna plays a decisive part for procuring the desired performance with specified dimension. As per the 5G performance parameters such as high gain, good bandwidth, better radiation pattern, and efficiency, procuring these levels is a tedious task for designing antenna.

However, the concept of traditional one-port antenna failed at most of the time when it comes for speedy network and better channel bandwidth.

Because of the extensive usage of numerous mobile devices and other electronic gadgets, channel capacity is reduced. Not only this, atmospheric absorption in the millimeter-wave frequency leads to the fading of transmission signal. As a solution to these stated issues, multi-port antenna system came into picture which is popularly known as multiple-input multiple-output (MIMO) system. This MIMO system is an emerging form of SISO which can be a proving contender for high-speed connection in the coming wireless generations [1]. In this case, the entire radiator that is deployed on either the transmitter or the receptor side will have separate input feed. Each element will radiate individually and hence gain, bandwidth, and other parameters will not degrade. But the main drawback while implementing a MIMO antenna is the mutual coupling among the radiators which results in diminishing the overall performance, and hence the true behavior of the MIMO system will not be procured. In the literature, several methodologies are present, which the author has incorporated in their designs as an attempt in reduction of isolation technique. One of the premier methods to attain better isolation among elements is the deploying metamaterial-surface-based MIMO designs [2]. Another decoupling structure is the split ring resonator (SRR) consisting of reactive loads (inductive and capacity) to obtain better isolation [3]. Instead of a physical decoupling structure, changing the orientation of elements can also lead to the improvement in mutual coupling which is prominently known as orthogonal arrangement [4]. For the bandwidth enhancement, partial ground plane or defected ground structure (DGS) is commonly used. However, applying DGS not only improves operation frequency bandwidth but also supports in reducing mutual coupling [5]. Similar techniques are available in previous works by which the complication in coupling can be overcome. In this literature survey, we have combined the MIMO antenna design methodology, MIMO parameters, and varied mutual coupling reduction techniques that can be incorporated to achieve the improved isolation. This survey mainly focuses on 5G communication, as multipath fading and absorption is more in this frequency spectrum for which MIMO antenna design is the possible solution. The review manuscript has been organized in different sections which broadly cover the approach for MIMO design, specific techniques relevant to the reduction of mutual coupling, and comparison table for convenience of the readers to implement their research to achieve optimum performance.

7.2 MIMO DESIGN PARAMETERS

In the modernized communication world, using the conventional method of single transmitter and single receiver will not cater to the present demand of high data rates, low latency, and enormous growth of mobile user in the allocated spectrum. Hence, the concept of multi-antenna at both Tx and Rx can lead to overcoming the drawback of multipath fading, reduced channel capacity, and low data rates. However, to get the desired results from the MIMO antenna, each element in the system must work independently on either side. To achieve this, the mutual coupling among the elements must be as low as possible by which the antenna parameters will not degrade. To evaluate the MIMO performance, some key diversity parameters needed to be obtained whose values are destined as per the practical application using the stated equations for respective parameters [6].

7.2.1 ENVELOPE CORRELATION COEFFICIENT (ECC)

The resemblance between the antenna elements can be determined by the envelope correlation coefficient (ECC) parameter. There are a couple of methods by which ECC can be evaluated. The first method is using the S-Matrix or S-Parameter whereas other methods use the radiation pattern of the antenna element. In the practical scenario, the value of ECC must be less than 0.5.

$$ECC = \frac{\left|S_{11}^{*}S_{12} + S_{21}^{*}S_{22}\right|^{2}}{(1-|S_{11}|^{2}-|S_{21}|^{2})(1-|S_{22}|^{2}-|S_{12}|^{2})} \tag{7.1}$$

7.2.2 DIVERSITY GAIN (DG)

The performance and reliability of any MIMO design can be justified by diversity gain (DG). Therefore, it is said that DG of MIMO antenna system is equivalent to 10 dB. The DG can be related to ECC by the following equation:

$$DG = 10 \times \sqrt{1-|ECC|^{2}} \tag{7.2}$$

7.2.3 CHANNEL CAPACITY LOSS (CCL)

Channel capacity loss (CCL) is the maximum range in the MIMO system from which the data can be sent with nearly zero losses in the transmission channel. CCL of 0.4 bits/s/Hz is defined for the MIMO system.

$$CCL = -log_2 \det (\vartheta^{\mu}) \tag{7.3}$$

7.2.4 MEAN EFFECTIVE GAIN (MEG)

The mean effective gain (MEG) of the MIMO system is defined as the ratio of average received power by MIMO antenna to the received power of isotropic antenna. As far as the practical environment is considered, the value of MEG should be less than 3 dB.

$$MEG_i = 0.5\left(1-\left|S_{ii}\right|^2-\left|S_{ij}\right|^2\right) \tag{7.4}$$

$$MEG_j = 0.5\left(1-\left|S_{ij}\right|^2-\left|S_{jj}\right|^2\right) \tag{7.5}$$

7.2.5 TOTAL ACTIVE REFLECTION COEFFICIENT (TARC)

Total active reflection coefficient (TARC) is defined as the ratio of the total reflected power from the radiator to the incident power on patch. Ideal value for the TARC is considered to be zero. The equation for the TARC is given below:

$$TARC = \frac{\sqrt{\sum_{i=1}^{N}\left|b_i\right|^2}}{\sqrt{\sum_{i=1}^{N}\left|a_i\right|^2}} \tag{7.6}$$

where $b_i = [s].a_1$; [s], [a], and [b] are known as scattering matrix, excitation vector, and scattering vector.

These parameters, especially the mutual coupling, can be enhanced by implementing the decoupling techniques reported in previous literatures. Isolation enhancement techniques covered in this chapter are electromagnetic band gap structure (EBG), neutralization lines (NL), metamaterial/SRR, and so on. The MIMO designs implemented using such a method are covered in detail in the following sections.

7.3 COMPARISON OF ISOLATION ENHANCEMENT TECHNIQUES IN MIMO ANTENNA DESIGN

In this review article on 5G MIMO antenna several aspects have been studied and analyzed thoroughly which increases the range of techniques and its ability. However, all the concepts have been focused related to MIMO antenna which is in trend and need of the hour in millimeters wave technology especially 5G and coming wireless generation. Furthermore, each technique has its own disadvantages as well as advantage with this review chapter it will help the readers to find the gap between techniques and can apply it according to the research requirement.

A brief comparison among distinct literatures based on MIMO antenna design for 5G on the basis of area occupied by the MIMO antenna design, number of antenna elements, operating frequency, isolation, return loss, gain, and ECC has been summarized in Table 7.1. Similarly, Table 7.2 showcases the positive and negative sides of isolation enrichment techniques in aspect of simplicity in design, placement over the antenna, and enhancement in the performance of MIMO antenna.

7.4 CONCLUSIONS

The MIMO antenna system is the solution to the underlying problems in the wireless communication. The need of the hour is not just the high data rates and low latency but answer to the multipath fading, channel capacity, wide geographical coverage, and multiple application implementations in the desired spectrum with matching the growth of mobile users. Furthermore, power consumption is another area of concern to which the wireless technology needed to overcome in the future. The concept of deploying a number of antennas at both the transmitter and the receiver can possibly overcome the issues stated above. However, in the designing of MIMO antenna one needs to focus on the effect which is popularly known as mutual coupling. In this literature survey, the authors have tried to perform an extensive study which emphasizes on the mutual coupling reduction techniques in the MIMO antennas.

As mankind is observing the complete transformation of communication world, the positive impact of MIMO can closely be seen in the 5G wireless technology. The deterioration in the MIMO parameters is because of antenna element's close proximity. The effect of diverse isolation

TABLE 7.1 Comparison of Referred Literatures in MIMO Antennas for 5G.

Ref.	MIMO Elements	MIMO Antenna Size	Resonating Frequency	Technique Used	Return Loss	Gain	Isolation	ECC	DG
[7]	2	$55 \times 110 \times 0.508$ mm^3	28 and 38 GHz	DGS	35 and 45 dB	7.88 and 9.49 dBi	29.4 and 27.3 dB	0.000013, 0.000038	-
[8]	4	$65 \times 65 \times 1.6$ mm^3	3.1–10.6 GHz	DGS	< 10 dB	-	>15 dB	0.025	≤10 dB
[9]	2	$20 \times 20 \times 0.787$ mm^3	28, 38, and 60 GHz	DGS	-	Between 3.5 and 8.5 dBi	-	0.0015 (for all bands)	≤10 dB
[10]	4	$30 \times 28 \times 0.508$ mm^3	27 GHz	DGS	20 dB	6.22 dBi	30 dB	< 0.05	≤10 dB
[11]	8	$150 \times 80 \times 0.7$ mm^3	3.3–5.82 GHz	DGS	-	-	20 dB	< 0.04	-
[12]	6	$134 \times 75 \times 0.8$ mm^3	3–5 GHz	EBG	9 dB	-	20 dB	0.0045	10 dB
[13]	2	$1.09\,\lambda_0 \times 0.56\,\lambda_0$	3.45 GHz	DRA-EBG	-	4.83 dBi	26 dB	0.03	9.8 dB
[14]	8	$124 \times 74 \times 6$ mm^3	3.3–3.6 GHz	NL, Ground slot	6 dB	-	15 dB	0.15	-
[15]	8	$150 \times 75 \times 7$ mm^3	3.5 and 5 GHz	NL	6 dB	-	11.5 dB	0.05	-
[16]	4	$30 \times 43 \times 0.787$ mm^3	26 GHz	Metasurface	-	10.27 dBi	45 dB	$< 0.1e^{-6}$	9.98 dB
[17]	2	$36 \times 66 \times 1.6$ mm^3	3.5 GHz	MTM resonators	45 dB	3.28 dBi	16 dB	0.025	> 9.95 dB
[18]	4	$14 \times 14 \times 0.8$ mm^3	28 and 38 GHz	SRR-MTM	30 dB	8.4 and 6.02 dBi	> 25 dB	0.005	< 10 dB

enhancement techniques such as EBG, NL, DGS, and metamaterial has been carried out meticulously to analyze the contribution toward the enrichment of isolation in MIMO antennas designed at the 5G spectrum. These specified isolation enhancement methodologies can produce a significant number of changes in the evaluated results depending upon type of elements chosen for designing the decoupling structure. Hence, one can incorporate any method presented in this literature survey for possible neutralization of the mutual coupling for the MIMO antennas and to procure the true behavior of the MIMO antenna system.

TABLE 7.2 Comparisons of Isolation Enhancement Techniques for MIMO Antenna System.

Technique name	Pros	Cons
Defected ground structure (DGS)	1. Simple in design 2. Easily manipulates the surface waves	1. Efficiency may reduce
Electromagnetic band gap (EBG)	1. Reduces size 2. Can suppress the waves within the band gap	1. Placement dependent
Neutralization lines (NL)	1. Simple in design	1. Unpredictable in placement 2. Less effective toward enhancing isolation
Metamaterial/ Split ring resonator (MTM-SRR)	1. Quite effective in improvement of isolation	1. Complex in designing 2. Increases overall area 3. High possibility of antenna pattern deterioration

KEYWORDS

- **channel capacity loss**
- **defected ground structure**
- **diversity gain**
- **envelope correlation coefficient**
- **mean effective gain**
- **multiple-input multiple-output antennas**
- **total active reflection coefficient**

REFERENCES

1. Sakli, H., Abdelhamid, C., Essid, C., & Sakli, N. (2021). Metamaterial-based antenna performance enhancement for MIMO system applications. *IEEE Access*, *9*, 38546–38556. DOI: 10.1109/ACCESS.2021.3063630
2. Xue, C. D., Zhang, X. Y., Cao, Y. F., Hou, Z., & Ding, C. F. (2017). MIMO antenna using hybrid electric and magnetic coupling for isolation enhancement. *IEEE Transactions on Antennas and Propagation*, *65*, 5162–5170. DOI: 10.1109/TAP.2017.2738033
3. Bhattacharya, A., & Roy, B. (2020). Investigations on an extremely compact MIMO antenna with enhanced isolation and bandwidth. *Microwave and Optical Technology Letters*, *62*, 845–851. DOI: https://doi.org/10.1002/mop.32084
4. Tiwari, R. N., Singh, P., Kumar, P., & Kanaujia, B. K. (2022). High isolation 4-port UWB MIMO antenna with novel decoupling structure for high speed and 5G communication. In *International Conference on Electromagnetics in Advanced Applications, Cape Town, South Africa* (pp. 336–339). DOI: 10.1109/ICEAA49419.2022.9900029
5. Sharma, P., Tiwari, R. N., Singh, P., Kumar, P., & Kanaujia, B. K. (2022). MIMO antennas: Design approaches, techniques and applications. *Sensors, 22*, 7813–7820. DOI: 10.3390/s22207813
6. Marzouk, H. M., Ahmed, M. I., & Shaalan, A. A. (2019). A novel dual-band 28/38 GHz slotted microstip MIMO antenna for 5G mobile applications. In *IEEE International Symposium on Antennas and Propagation and USNC-URSI Radio Science Meeting, Atlanta, GA, USA* (pp. 607–608). DOI: 10.1109/APUSNCURSINRSM.2019.8888799
7. Lamri, I. E., Mansoul, A., Nakmouche, M. F., & Belattar, M. (2021). Design of novel UWB 4-element MIMO microstrip patch antenna for sub-6 GHz 5G applications. In *International Conference on Radar, Antenna, Microwave, Electronics, and Telecommunications, Bandung, Indonesia* (pp. 7–11). DOI: 10.1109/ICRAMET53537.2021.9650462
8. Chu, S., Hasan, M. N., Yan, J., & Chu, C. C. (2018). Tri-band 2 × 2 5G MIMO antenna array. In *Asia-Pacific Microwave Conference* (pp. 1543–1545). DOI: 10.23919/APMC.2018.8617590
9. Khalid, H., Khalid, M., Fatima, A., & Khalid, N. (2019). 2×2 MIMO antenna with defected ground structure for mm-wave 5G applications. In *International Conference on Mathematics, Actuarial Science, Computer Science and Statistics* (pp. 1–6). DOI: 10.1109/MACS48846.2019.9024801
10. Nithya, S., & Seethalakshmi, V. (2022). MIMO antenna with isolation enrichment for 5G mobile information. *Mobile Information Systems*, 1–14.
11. Hediya, A. M., Ahmed, A., & Walid, S. (2022). 5G MIMO antenna system based on patched folded antenna with EBG substrate. *Progress In Electromagnetics Research M*, *109*, 149–161. DOI: 10.2528/PIERM22020101
12. Chen, H. N., Song, J. M., & Park, J. D. (2019). A compact circularly polarized MIMO dielectric resonator antenna over electromagnetic band-gap surface for 5G applications. *IEEE Access*, *7*, 140889–140898. DOI: 10.1109/ACCESS.2019.2943880
13. Jiang, W., Liu, B., Cui, Y., & Hu, W. (2019). High-isolation eight-element MIMO array for 5G smartphone applications. *IEEE Access, 7*, 34104–34112.
14. Guo, J., Cui, L., Li, C., & Sun, B. (2018). Side-edge frame printed eight-port dual-band antenna array for 5G smartphone applications. *IEEE Transactions on Antennas and Propagation, 66*, 7412–7417. DOI: 10. /TAP.2018.2872130

15. Tariq, S., Naqvi, S. I., Hussain, N., & Amin, Y. (2021). A metasurface-based MIMO antenna for 5G millimeter-wave applications. *IEEE Access, 9*, 51805–51817.
16. Supreeyatitikul, N., Phungasem, A., & Aeimopas, P. (2021). Design of wideband sub-6 GHz 5G MIMO antenna with isolation enhancement using an MTM-inspired resonators. In *Joint International Conference on Digital Arts, Media and Technology with ECTI Northern Section Conference on Electrical, Electronics, Computer and Telecommunication Engineering* (pp. 206–209). DOI: 10.1109/ECTIDAMTNCONc51128.2021.9425774
17. Tadesse, A. D., Acharya, O. P., & Sahu, S. (2022). A compact planar four-port MIMO antenna for 28/38 GHz millimeter-wave 5G Applications. *AEM, 11*, 16–25.
18. Ishfaq, M. K., Rahman, T. A., Yamada, Y., & Sakakibara, K. (2017). 8×8 phased series fed patch antenna array at 28 GHz for 5G mobile base station antennas. In *IEEE-APS Topical Conference on Antennas and* Propagation in Wireless Communications (pp. 160–162). DOI: 10.1109/APEC.2017.8062268

CHAPTER 8

ROAD AND POTHOLE SEGMENTATION USING AN ATTENTION-BASED COUPLED FRAMEWORK

C. V. LAKSHMI NARAYANA, VISWANADHAM VAMSI, VARUN RACHARLA, NARAPUREDDY SANDHYA, JANGITI UMA SANKAR, and GALISETTY SRAVANI

CSE Department, AITS, Rajampet, Andhra Pradesh, India

ABSTRACT

We present a unique linked attention-based paradigm for road and pothole segmentation. Many developing countries, as well as rural areas, lack well-defined and well-maintained drivable areas. In such cases, an advance driver assistance system (ADAS) is required to assess the drivable area and warn of impending potholes to ensure vehicle safety. The goal of this research is to create a standard for pothole classification to help support a decision-making system for pothole repair. In this study, we first examine the existing criteria. Following that, we gather photographs of potholes via video recording on a moving vehicle, propose a pothole categorization guideline, and then apply the guideline to the obtained images and pothole 2D DB. Using three types of training data, we compare GAL-DeepLabv3+, our best-performing implementation, to SoTA CNN using RGB images, disparity images, and modified disparity images. Furthermore, these data can be used in structured systems to analyze and maintain road health. We show how a few-shot learning strategy can improve accuracy even with

Cyber-Physical Systems: Applications, Challenges, and Research Directions. Rajanikanth Aluvalu, Samarjeet Borah, & Lalit Garg (Eds.)

DOI: 10.1201/9781779643421-8

fewer training data. We present comprehensive experimental results for road segmentation using the KITTI and Indian driving set (IDD) datasets. On IDD, we also present pothole segmentation.

8.1 INTRODUCTION

Computerized inspection and evaluation of both the physical and functional states of road enhance both road maintenance and vehicle safety. Researchers have become more interested in self-driving cars and sophisticated systems for driver assistance in recent years. A critical responsibility in any ADS/advance driver assistance system (ADAS) is road segmentation or defining the drivable territory. In general, well-maintained roads with distinct lane markers perform best for camera-based road segmentation solutions. In unstructured situations, at which roads are potholed or improperly marked, it is difficult to divide the drivable path. On the road surface, potholes represent structural faults. Regular road pothole inspection and repair are essential for ensuring traffic safety. To locate potholes both effectively and objectively, automated pothole detecting technologies are required. If the motorist in an ADS/ADAS scenario can be warned about the pothole beforehand, he or she can make an educated choice. Single-image field of vision and pothole segmentation is difficult in nonstructured road surroundings due to lighting fluctuations, occlusion caused by variable entity or image visual appeal, and poor road conditions.

Within this analysis, we present a completely new linked architecture for segmenting roads and potholes. The framework presupposes that the camera is mounted on the car and is constantly scanning the surroundings. The following are the chapter's key contributions:

1. In especially unstructured situations, we provide a coupled framework for segmenting roads and potholes.
2. We add attention-based refinement and feature fusion to DeepLabv3+ architecture to enhance segmentation.
3. According to the study referred, it is the initial study to evaluate some of the few learning strategies of potholes recognition on the Indian driving set (IDD) [1] datasets in order to achieve maximum high accuracy also with less supervised learning.

The structure of the chapter is indicated below. The following section contains a list of the relevant works. Section 8.3 goes over the suggested approach in detail. Section 8.4 discusses the experimental analysis and results, which are followed by a conclusion.

8.2 RELEVANT WORK

This section can be broken into:

1. Road segmentation.
2. Road profiling or classification.
3. Pothole detection methods.

8.2.1 ROAD SEGMENTATION

To prevent the development of self-driving cars and driver aid systems from going off course, one of the essential tasks is the division of roads. Recent approaches to road segmentation include Deeplabv3+ [1] and BiSeNet [2]. The authors of Ref. [3] established a framework for regular road scenario classification utilizing RGB images. Based on traditional convolutional neural networks, this framework uses semantic segmentation techniques. Mostly on network's development side, they created a representation between the categories and the frames, which allows for precise road label prediction. Reference [4] looked at the concerns with class imbalance around road labeling in the satellite and aerial images. To precisely calculate the extent of the path markings and avoid over and under concerns, they suggested a guided RANSAC-based technique. Fully convolutional neural networks were suggested in Ref. [5] for road segmentation using satellite pictures. A real-time curb detection approach using 3D point cloud data from a Lidar sensor was reported by the authors in Ref. [6]. Additionally, it offers a method for differentiating between off-road and on-road locations. To segment unmarked roads, the writers in Ref. [7] applied deep learning approaches with color line designs refined using a continuous randomized data refinement approach.

Image is at the left, road segmentation of pothole identification output is on the right. Blue-colored sections represent potholes found on the drivable path, while pink-colored parts show the drivable path. 1×1 convolution is shown in the blue and pink box, while Atrous convolutions with separation rate are shown in the orange box. Improved road and pothole boundary extraction is taken into consideration.

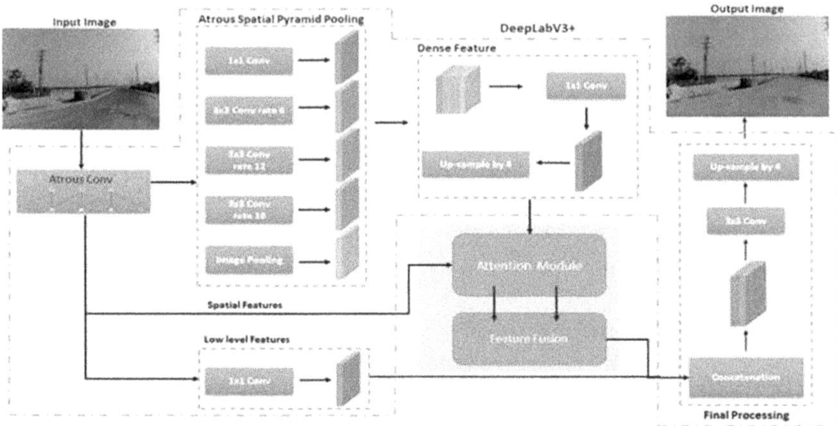

FIGURE 8.1 Framework for the suggested road and pothole detecting system.

The authors of Ref. [8] took the road regions out of the dense disparity maps. With the use of projection CNNs, the writers in Ref. [9] took advantage of the idea inter-learning approach that is personal self. They used unsupervised approaches to detect roads before fine-tuning it with personal self, CNN while obtaining depth data into consideration.

8.2.2 ROAD PROFILING OR CLASSIFICATION

A significant aspect that may cause vehicle instability is the road profile. The authors of Ref. [10] created a digital sensing with complete control that calculates the height of the path by simulating it, which is represented as a continuous impulsivity and hyperactivity. The author of Ref. [11] uses independent component analysis (ICA) to solve an inverse problem on three different automobile model variations to estimate the anomalies in the road profile.

Using the system's known dynamic reactions, the path excitation concept is being taught to ICA. Data from sensor devices, such as operating point, tyre land forces, weight fraction, return journey, and route severity, are used by the authors in Refs. [12–14], to predict the road profile. For two-wheeled vehicles, the authors in Ref. [15] calculated road curviness and hilliness. Using 3D point cloud data, the author of Ref. [16] detects road markings and reflexive arrow marks. Pavement markings are estimated in Ref. [17] using data from MLS sensor point clouds. Utilizing radiometric correction and picture enhancement methods, the data is preprocessed.

8.2.3 POTHOLE DETECTION

Pothole identification has been a major focus of active study for quite some time. Pothole detection has been addressed using image features [18, 19] acquired from 2D or 3D surveillance systems. There are several techniques to evaluate pictorial content, spanning from algorithms using deep learning [18] to the smart split of pictures into primary and secondary components [20]. As an alternative, [21] proposes using acoustical sensors to track noise quality around the vehicle in order to detect road damage. In Ref. [22], high-sensor frequencies were employed to capture acceleration (in three dimensions) and position (GPS). A reliable detection system is built by the authors in Ref. [23] using a variety of sensors, including a two-dimensional camera, a laser sensor, and a three-axis accelerometer.

The final projections are the product of the combined choices of these several sensors. Most of these solutions rely on a vertical acceleration threshold. Just a few approaches employ in-vehicle sensors, but one of them, [24], outlines a detection system that attempts to measure the road bank and inclination angle to anticipate predicted acceleration levels. A pothole is proclaimed at a location if the observed accelerations diverge significantly from the predicted ones. The strategy was assessed using simulated data; however, it is unclear whether the same performance can be obtained using actual sensor data. To determine if a picture has a pothole or not, the writers in Ref. [25] created a categorization issue. CNN with sigmoid activation is used in the architecture. The collection is mostly made up of pothole images, some of which were hand gathered from Bangalore roads and others that were taken from YouTube videos. Pothole segmentation was carried out in Ref. [26] without the use of any deep learning techniques. Using a computer technique that incorporates both two-dimensional and three-dimensional road data, they created a color differential model based on the thickness of 3D models and segmented 2D pothole images. The approach generates a segmented image with a backdrop and potholes.

The dataset created by the authors consists of 67 images taken at various resolutions, each of which depicts a segment of street with a single small yet severe pothole. In conducting research on pothole detection, we used pictures of various sizes of potholes as well as images of other objects such as vehicles, road barriers, buildings, and so on.

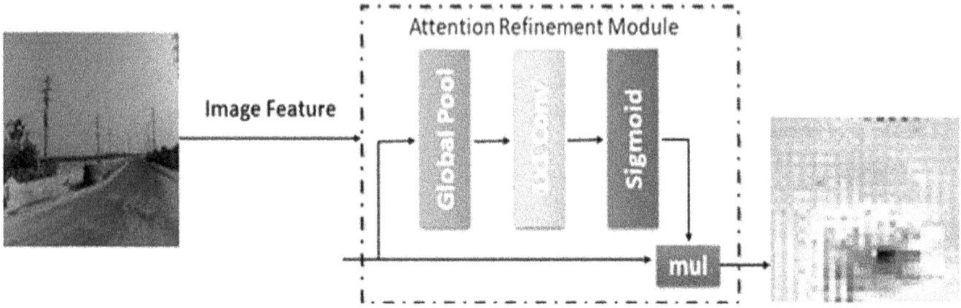

(a) Right: Matching attention map, left: Attention based refinement module.

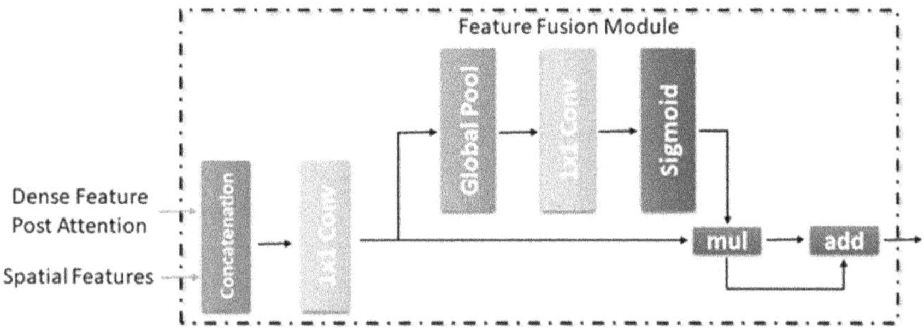

(b) Module for feature fusion.

FIGURE 8.2 Feature fusion and attention-based refinement.

8.3 SUGGESTED APPROACH

Unlike driverless vehicles in structured operating settings, road scene interpretation frequently exhibits internal and multi-variation in the scale, form, and material of objects or the look of the backdrop. Traditional CNNs contain local features, but they do not encode long-range contextual data, which affects segmentation efficiency. In order to solve this issue, we first adopt a multiscale technique similar to that of DeepLabv3+, we gather rich high accuracy and dynamic context features and then feed them into a consideration domain adaptation module, which directs feature learning.

DeepLabV3+ improves accuracy by collecting pits of varying widths and reducing misclassification.

In Figure 8.1, the suggested architecture for pothole and road identification is shown in broad strokes. Three steps make up the suggested method: Drivable route segmentation, The three approaches are Deeplabv3+ for effective road and pothole border extraction, consideration refined for efficient street and pothole border filtration, and so few shot training for pothole detection. The idea behind this pipeline is to provide a general framework that can accommodate both organized (such as smooth, drivable roads in wealthy nations) and unstructured (such as unpaved, unlanded roads with significant potholes in underdeveloped countries) road conditions.

8.3.1 SEMANTIC SEGMENTATION

Pothole identification and drivable path segmentation are both two-step semantic segmentation problems that define the road area in front of the automobile as the drivable path. Road segmentation, or defining the drivable region, is what we do in the first stage. The next step is to help the driver by alerting them of upcoming potholes so they may take the necessary action, such as slow down, detour, or apply the brakes.

We demonstrate the solution utilizing DeepLabv3+ [1] as the fundamental foundation. Rich geographical and contextual characteristics are captured by the DeepLabv3+ architecture. Atrous convolution, also known as dilated convolution, and atrous spatial pyramid pooling are the two main elements of the network (atrous spatial pyramid pooling (ASPP)). The following describes how atrous convolution works with an input feature map (x):

$$y(i) = \sum_{k} x(i + r * k)w(k) \tag{8.1}$$

where i is a point in the output feature map y, w is the convolutional filter, r is the atrous rate that controls the stride with which the input is sampled, and w is the atrous rate. The ASPP method uses parallel input image processing to extract extremely dense characteristics at multiple sizes. This aids in minimizing the spatial information loss brought on by earlier convolutions with stride operations. As a result, the spatial context characteristics evaluated at each layer contribute to a smaller number of parameters and a lower computing cost. This enables expanding the field of vision.

To make better use of the deeper data types, the convolution on low level features is used to decrease the number of channels. The features from the ASPP module are concatenated with the equivalent limited characteristics after being up sampled by a factor of 4. Following concatenation, convolution is used to sharpen the features, and then up sampling by a factor of 4 is used to get the input image's resolution. MobileNetV2 served as the backbone network for our evaluation of DeepLabv3+ [27]. For an ADAS/ADS-like use-case, this approach is recognized to have reduced computing complexity, which is a crucial need.

8.3.2 ATTENTION-BASED REFINEMENT WITH FEATURE FUSION (AFF)

Multiscale characteristics have proven to be helpful in the context of semantic segmentation. Atrous convolution encapsulates detailed spatial information in DeepLabv3+, whereas ASPP provides a capability for aggregating global to local context that has been learnt at various scales. To improve the features at each level, we suggest using an attention-guided feature learning module. As shown in Figure 2a, the focus method includes average global pooling to increase the spatial and international context information in order to construct a concentration matrix, resulting in a more potent representation. While ASPP dense context data are high level, arous convolution characteristics vary in small amount. As a result, we are unable to simply concatenate these characteristics. For the purpose of fusing both the traits, we provide the feature fusion module (Figure 2b). Prior to batch normalization, which corrects the scale of the features, the specified postattention features are first concatenated. Then, to create a feature vector and calculate weights, we apply global pooling to the concatenated features. The learned weights can help with feature combination and selection.

8.4 EXPERIMENTAL RESULTS

This section contains the experiments and analyses that were conducted to assess the suggested framework. The test was run on a Tesla K80 GPU computing 3.7, which has 2496 CUDA cores and 12 GB GDDR5 VRAM.

8.4.1 DATASETS

In the past few years, a number of datasets for autonomous navigation have become available. These datasets are generally concerned with organized driving settings, such as lanes and other clearly defined infrastructure, a small number of strongly outlined transportation participant groups, and rigorous observance to traffic regulations. There are 289 training pictures and 290 test images in the KITTI Visual Benchmark Suite1 [28, 29].

The IDD2 [1] is a reference set for driver assistance systems under irregular road situations. The collection comprises 10,000 highly annotated photos from 182 driving sequences on Indian highways, totalling 34 classes. It has 981 validation photos and 6993 training images. In our trials, we solely use annotations for roads. IDD provides pictures of potholes and other signs of bad road conditions, but there are no commentaries to be seen. Therefore, using LabelMe toolbox3 [30], we labeled 237 photos from the IDD dataset with varied pothole sizes for our studies on pothole identification.

8.4.2 NETWORK PARAMETER SETTINGS

Nadam, with a learning rate of 0.00001, is the optimizer that we use. Just 2.4 million parameters make up DeepLabv3backbone+'s network, MobileNetV2. With the goal of extracting denser features, toward the end of the design, we increased the ASPP component's production stride to 8 and included an attrition rate of 0.5. The concentration components were incorporated, and our final result included almost 4 million parameters. For the suggested framework's training, we used correlation damage like an error function. Jaccard Loss function convergence was discovered empirically.

1. http://www.cvlibs.net/datasets/kitti/eval_road.php
2. https://idd.insaan.iiit.ac.in/
3. http://labelme2.csail.mit.edu/Release3.0/

Compared to binary cross entropy, quicker with higher performance, this shows that Correlation Losing is better equipped to deal with issues of class inequality. In this situation, 70% of the region was used to create the backdrop, whereas the remaining pixels were used to represent the road or potholes. The epoch was set to 100 for all experiments.

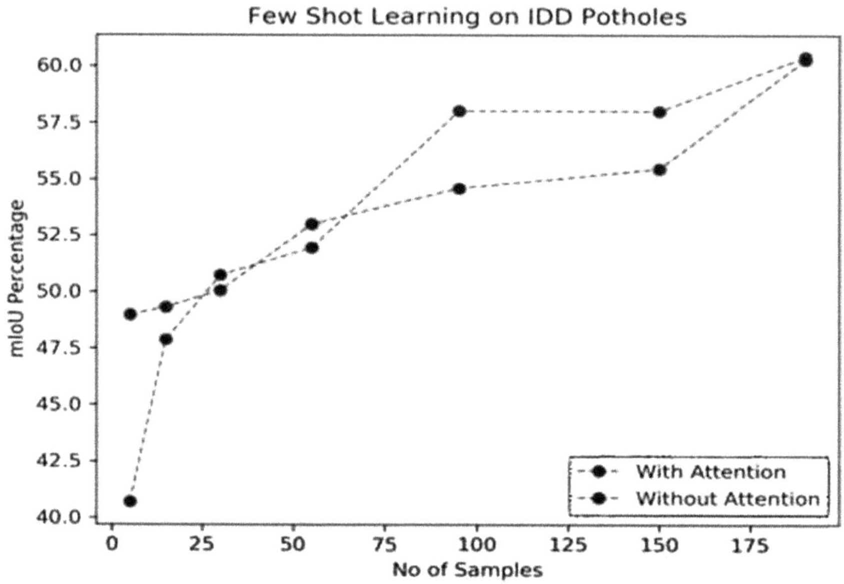

FIGURE 8.3 The graph depicts the variation of the mIoU value as the number of training samples increases. Potholes can be identified using a few short learnings on the IDD dataset (Green- To Attn, Pink- W/o Attn).

8.5 PERFORMANCE METRICS FOR SEGMENTATION MODELS

A model should ideally be assessed across a range of criteria, including quantitative accuracy, inference speed, and model quality. The typical measure used to judge the segmentation algorithm's accuracy is summarized here. The average IoU or Jaccard Index across all the classes is referred to as mean IoU. It is determined by dividing the area of union between the predicted and the ground truth by the area of intersection between the predicted segmentation areas and the ground truth:

$$mIoU = mean\left(\frac{A \cap B}{A \cup B}\right) \tag{8.2}$$

Semantic image segmentation accuracy algorithms are frequently assessed using the metrics precision/recall/F1 score. Each class's definition of recall and precision may be found here:

$$Precision = \frac{TP}{TP + FP} \tag{8.3}$$

$$Recall = \frac{TP}{TP + FN}$$

where FN stands for the false-negative proportion, FP for false-positive, and TP for true-positive. The symmetrical mean of recollection and precision is represented by the F1 score rate.

$$F1score = \frac{2 \times Precision \times Recall}{Precision + Recall} \tag{8.4}$$

The suggested framework is assessed using the precision, recall, and F1 score on the KITTI dataset.

8.5.1 QUANTITATIVE ANALYSIS

The experiments and analyses are split into two categories: Pothole detection and road segmentation.

Networks Initialized:

[Network GAL Net] Total number of parameters: 55.031M

Epoch best test loss: 0.027, ss: 0.059

Epoch best glob acc: 0.992, pre: 0,855, recall: 0.897, F-score: 0.875, IoU: 0.778

FIGURE 8.4 IDD outcomes of road segmentation. (a) Without paying attention, DeepLabv3+ for street classification, DeepLabv3+ classification results, (e–h) Road naming based on reality as opposed to context (i–l) (pink: road, red: background).

8.5.1.1 ROAD SEGMENTATION

Using a DeepLabv3+ variation and MobileNetV2 as the backbone network, the first experiment for road segmentation was carried out. The initialization of the network and subsequent network fine-tuning on IDD both employed cityscape weights. IDD's initial picture size was 1920 × 1080; using the antialiasing filter, it was down sampled to 512 × 512. We present the indicated baseline findings for comparison. Dilated Residual Networks (DRN-D-38) on the road class label in Ref. [1]. The module mIoU is 92% while developed and evaluated on IDD, compared to 83% when trained and tested on Berkeley Deep Drive (BD) [33]. Because the Berkeley Deep Drive Dataset, as opposed to IDD, only takes into account organized road conditions, the trials exhibit weak generalization. CRF-based postprocessing has been widely employed by researchers to increase the semantic correctness of FCN results. As a result, we compare the outcomes of BiSeNet [2] with and without CRF. This framework's main flaw is the requirement for separate training of FCN and CRF. We add an attention module to DeepLabV3+ to enhance the segmentation performance to provide an all-encompassing solution. The suggested framework's classification technique without or with attentiveness is 97.3% and 98.32%, respectively. Analyze the quantitative lane classification performance on IDD.

On the basis of the KITTI road segmentation dataset, we assess the suggested framework, and it contains a summary of the findings. Modern techniques for road segmentation include [31] employing both picture and LiDAR data and [32] using a single, specialized image. The approach suggested in this study focuses mostly on the unstructured Indian road dataset and solely makes use of the monocular picture obtained by a camera mounted on the vehicle. Our segmentation results are comparable, and our approach requires less computing effort. Compared to the state-of-the-art run-time, we see an 81% decrease in run-time.

8.5.1.2 POTHOLE DETECTION

We used the same network for pothole detection trials, providing a connected end-to-end road and pothole segmentation framework. There is no annotated data for potholes for unstructured road conditions. Using the LabelMe toolset, we added pothole annotations to 237 pictures from the IDD. Formulating a pothole detecting system is difficult on two fronts. First, there were just a few photos of roads with potholes, and second, the ground-truth image had a higher-class imbalance and more backdrop.

FIGURE 8.5 Results from the KITTI street classification dataset. Road segmentation results from DeepLabv3+ (a) through (d) without attention (e) through (h) with attention roads' ground-truth labelling versus backdrop (i–l) (pink: lane, red: BG).

Therefore, to handle learning from a small dataset, we use the few-shot learning paradigm. Second, to increase pothole accuracy regardless of background, we use class-weights during training to enable the network to give more weight to pothole areas and substitute category accuracy with per class accuracy of potholes. In the case of road areas with a weight of one, the category for potholes is fixed at 5000. These figures were derived experimentally. Hyperparameter tweaking was seldom employed. We use the Adam optimizer with a fixed learning rate of 104. The few-shot learning studies are depicted in Figure 8.3. A conceptual episode comprises n samples from two classes for the reference set and n instances that belong to two categories for the query set, as illustrated by the graph below, which shows the mIoU value obtained at varied n can be 5, 30, 90, 150, and 190 for $k = 2$. The trials using a small number of shots demonstrate that we can attain the same accuracy using attention methods while using fewer samples. With attention, we were able to reach a maximum mIoU of 73.88%; without it, it was only 70.18%.

FIGURE 8.6 A few examples of road segmentation issues.

FIGURE 8.7 Pothole identification results on IDD demonstrating variation in size, depth, and shadow.

8.6 QUALITATIVE ANALYSIS

We demonstrate a few example outcomes from the proposed methodology on the IDD dataset in this section. Without paying attention to IDD, DeepLabv3+ produces the findings shown in Figures 4a through 4d, whereas paying attention produces Figures 4e through 4h. With increased focus, the road segmentation is seen to be more in line with its bounds. In Figure 4i, the ground-truth labeling is displayed for Ref. [1]. Similar findings for road segmentation on the KITTI dataset are shown in Figure 8.5.

In Figure 8.6, we illustrate a few instances in which a sidewalk and a piece of a car are covered by road annotations. We notice a similar tendency in prediction since the model is refined using IDD data. The pink overlay shows the anticipated road segmentation, while the green pixels indicate the actual ground reality. Figure 8.7 displays some pothole segmentation findings. The areas in green are the actual potholes, while the areas in blue are where the potholes are expected to be. We also see that the road sections are covered by the pothole annotations (green regions). The example results photographs clearly show that DeepLabv3+ enhances pothole segmentation by lowering the false-negative rate.

8.7 CONCLUSION

The chapter's major contribution is a unique linked attention-based paradigm for segmenting roads and potholes. In terms of road segmentation, the framework works well in both organized and unstructured driving contexts. In order to identify potholes with decent accuracy using fewer labeled samples, we investigate few-shot learning.

The experiments reveal that the suggested attention framework's overall mIoU for road segmentation on the IDD dataset is 98.42%, while for pothole segmentation it is 73.83%. In compared to state-of-the-art, the road segmentation on the KITTI dataset has an F1-score of 95.21% and a run-time reduction of 81%. The goal of this work's future expansion is to improve the model's ability to handle photos of roads in various lighting (such as at night for assessing road condition) and weather (such as rain, fog, and snow) circumstances. This can be accomplished by reducing the impacts of light scattering and making use of a priori visual meteorological information. When there is a large difference between the distributions of the unlabeled and labeled data, a semi-supervised learning strategy to improve classification performance would be another intriguing topic to investigate.

KEYWORDS

- **attention-based coupled framework**
- **ADAS**
- **GAL-DeepLabv3+**
- **IDD**
- **KITTI**
- **road and pothole segmentation**

REFERENCES

1. Varma, G., Subramanian, A., Namboodiri, A., Chandraker, M., & Jawahar, C. V. (2019, January). IDD: A dataset for exploring problems of autonomous navigation

in unconstrained environments. In *2019 IEEE Winter Conference on Applications of Computer Vision (WACV)* (pp. 1743–1751). IEEE.

2. Yu, C., Wang, J., Peng, C., Gao, C., Yu, G., & Sang, N. (2018). Bisenet: Bilateral segmentation network for real-time semantic segmentation. In *Proceedings of the European Conference on Computer Vision (ECCV)* (pp. 325–341).
3. Oliveira, G. L., Burgard, W., & Brox, T. (2016, October). Efficient deep models for monocular road segmentation. In *2016 IEEE/RSJ International Conference on Intelligent Robots and Systems (IROS)* (pp. 4885–4891). IEEE.
4. Yuan, M., Liu, Z., Wang, F., & Jin, F. (2019). Rethinking labelling in road segmentation. *International Journal of Remote Sensing*, *40*(22), 8359–8378.
5. Henry, C., Azimi, S. M., & Merkle, N. (2018). Road segmentation in SAR satellite images with deep fully convolutional neural networks. *IEEE Geoscience and Remote Sensing Letters*, *15*(12), 1867–1871.
6. Zhang, Y., Wang, J., Wang, X., & Dolan, J. M. (2018). Road-segmentation-based curb detection method for self-driving via a 3D-LiDAR sensor. *IEEE Transactions on Intelligent Transportation Systems*, *19*(12), 3981–3991.
7. Yadav, S., Patra, S., Arora, C., & Banerjee, S. (2017, September). Deep CNN with color lines model for unmarked road segmentation. In *2017 IEEE International Conference on Image Processing (ICIP)* (pp. 585–589). IEEE.
8. Fan, R., Bocus, M. J., & Dahnoun, N. (2018). A novel disparity transformation algorithm for road segmentation. *Information Processing Letters*, *140*, 18–24.
9. Wang, W., Wang, N., Wu, X., You, S., & Neumann, U. (2017, May). Self-paced cross-modality transfer learning for efficient road segmentation. In *2017 IEEE International Conference on Robotics and Automation (ICRA)* (pp. 1394–1401). IEEE.
10. Doumiati, M., Martinez, J., Sename, O., Dugard, L., & Lechner, D. (2017). Road profile estimation using an adaptive Youla–Kučera parametric observer: Comparison to real profilers. *Control Engineering Practice*, *61*, 270–278.
11. Hassen, D. B., Miladi, M., Abbes, M. S., Baslamisli, S. C., Chaari, F., & Haddar, M. (2019). Road profile estimation using the dynamic responses of the full vehicle model. *Applied Acoustics*, *147*, 87–99.
12. Vazquez, A. G. A., Vaseur, C., Correa-Victorino, A., & Charara, A. (2019, June). Road profile and suspension state estimation boosted with vehicle dynamics conjectures. In *2019 IEEE Intelligent Vehicles Symposium (IV)* (pp. 1809–1815). IEEE.
13. Qin, Y., Langari, R., Wang, Z., Xiang, C., & Dong, M. (2017, May). Road profile estimation for semi-active suspension using an adaptive Kalman filter and an adaptive super-twisting observer. In *2017 American Control Conference (ACC)* (pp. 973–978). IEEE.
14. Li, Z., Kolmanovsky, I. V., Atkins, E. M., Lu, J., Filev, D. P., & Bai, Y. (2016). Road disturbance estimation and cloud-aided comfort-based route planning. *IEEE Transactions on Cybernetics*, *47*(11), 3879–3891.
15. Gorges, C., Öztürk, K., & Liebich, R. (2018). Road classification for two-wheeled vehicles. *Vehicle System Dynamics*, *56*(8), 1289–1314.
16. Soilán, M., Riveiro, B., Martínez-Sánchez, J., & Arias, P. (2017). Segmentation and classification of road markings using MLS data. *ISPRS Journal of Photogrammetry and Remote Sensing*, *123*, 94–103.

17. Cheng, M., Zhang, H., Wang, C., & Li, J. (2016). Extraction and classification of road markings using mobile laser scanning point clouds. *IEEE Journal of Selected Topics in Applied Earth Observations and Remote Sensing*, *10*(3), 1182–1196.
18. Eisenbach, M., Stricker, R., Debes, K., & Gross, H. M. (2017). Crack detection with an interactive and adaptive video inspection system. *Arbeitsgruppentagung Infrastrukturmanagement*, *94*, 94–103.
19. Seichter, D., Eisenbach, M., Stricker, R., & Gross, H. M. (2018, August). How to improve deep learning based pavement distress detection while minimizing human effort. In *2018 IEEE 14th International Conference on Automation Science and Engineering (CASE)* (pp. 63–70). IEEE.
20. Yebes, J. J., Montero, D., & Arriola, I. (2020). Learning to automatically catch potholes in worldwide road scene images. *IEEE Intelligent Transportation Systems Magazine*, *13*(3), 192–205.
21. Mednis, A., Strazdins, G., Liepins, M., Gordjusins, A., & Selavo, L. (2010). RoadMic: Road surface monitoring using vehicular sensor networks with microphones. In *Networked Digital Technologies: Second International Conference, NDT 2010, Prague, Czech Republic, July 7-9, 2010. Proceedings, Part II 2* (pp. 417–429). Springer Berlin Heidelberg.
22. Ghadge, M., Pandey, D., & Kalbande, D. (2015, October). Machine learning approach for predicting bumps on road. In *2015 International Conference on Applied and Theoretical Computing and Communication Technology (iCATccT)* (pp. 481–485). IEEE.
23. Hsu, Y. W., Perng, J. W., & Wu, Z. H. (2016, July). Design and implementation of an intelligent road detection system with multisensor integration. In *2016 International Conference on Machine Learning and Cybernetics (ICMLC)* (Vol. 1, pp. 219–225). IEEE.
24. Fox, A., Kumar, B. V., Chen, J., & Bai, F. (2017). Multi-lane pothole detection from crowdsourced undersampled vehicle sensor data. *IEEE Transactions on Mobile Computing*, *16*(12), 3417–3430.
25. Pereira, V., Tamura, S., Hayamizu, S., & Fukai, H. (2018, July). A deep learning-based approach for road pothole detection in timor leste. In *2018 IEEE International Conference on Service Operations and Logistics, and Informatics (SOLI)* (pp. 279–284). IEEE.
26. Fan, R., Ozgunalp, U., Hosking, B., Liu, M., & Pitas, I. (2019). Pothole detection based on disparity transformation and road surface modeling. *IEEE Transactions on Image Processing*, *29*, 897–908.
27. Howard, A., Zhmoginov, A., Chen, L. C., Sandler, M., & Zhu, M. (2018). Inverted residuals and linear bottlenecks: Mobile networks for classification, detection and segmentation.
28. Fritsch, J., Kuehnl, T., & Geiger, A. (2013, October). A new performance measure and evaluation benchmark for road detection algorithms. In *16th International IEEE Conference on Intelligent Transportation Systems (ITSC 2013)* (pp. 1693–1700). IEEE.
29. Geiger, A., Lenz, P., & Urtasun, R. (2012, June). Are we ready for autonomous driving? the kitti vision benchmark suite. In *2012 IEEE Conference on Computer Vision and Pattern Recognition* (pp. 3354–3361). IEEE.

30. Russell, B. C., Torralba, A., Murphy, K. P., & Freeman, W. T. (2008). LabelMe: A database and web-based tool for image annotation. *International Journal of Computer Vision, 77*, 157–173.
31. Chen, Z., Zhang, J., & Tao, D. (2019). Progressive lidar adaptation for road detection. *IEEE/CAA Journal of Automatica Sinica, 6*(3), 693–702.
32. Sun, J. Y., Kim, S. W., Lee, S. W., Kim, Y. W., & Ko, S. J. (2019). Reverse and boundary attention network for road segmentation. In *Proceedings of the IEEE/CVF International Conference on Computer Vision Workshops* (pp. 0–0).
33. Yu, F., Chen, H., Wang, X., Xian, W., Chen, Y., Liu, F., ... & Darrell, T. (2020). Bdd100k: A diverse driving dataset for heterogeneous multitask learning. In *Proceedings of the IEEE/CVF Conference on Computer Vision and Pattern Recognition* (pp. 2636–2645).

CHAPTER 9

ENHANCING ELECTORAL INTEGRITY THROUGH CYBER-PHYSICAL SYSTEMS: A DECENTRALIZED APPROACH FOR SECURE AND TRANSPARENT VOTING

MALLELLU SAI PRASHANTH[1], RAMESH KARNATI[2], MUNI SEKHAR VELPURU[3], and H. VENKATESHWARA REDDY[2]

[1]Symbiosis Institute of Technology, Hyderabad Campus, Symbiosis International University, Pune, India

[2]Department of Computer Science and Engineering, Vardhaman College of Engineering, Hyderabad, India

[3]Department of Information Technology, Vardhaman College of Engineering, Hyderabad, India

ABSTRACT

With the increasing prevalence of cyber-physical systems (CPS) in various domains, it is crucial to explore their potential for revolutionizing the electoral process. This chapter presents a novel approach that combines blockchain technology with CPS to create a decentralized voting system, ensuring enhanced security, transparency, and verifiability in elections. By utilizing the novel attributes of CPS, such as the integration of physical devices with computational elements, the proposed solution addresses the challenges of traditional voting systems and mitigates potential vulnerabilities in the digital realm. Voting is a crucial process

Cyber-Physical Systems: Applications, Challenges, and Research Directions. Rajanikanth Aluvalu, Samarjeet Borah, & Lalit Garg (Eds.)

DOI: 10.1201/9781779643421-9

in any democratic society. However, traditional voting systems are often centralized, and the results can be manipulated, leading to fraud and mistrust. The emergence of blockchain provides an answer to these issues by enabling a decentralized and transparent voting system. This research chapter proposes a decentralized smart contract-based voting mechanism on the Ethereum blockchain. The suggested solution creates a smart contract that controls the voting process using the Solidity programming language. The smart contract defines two structs, Candidate and Voter, that represent candidates and voters respectively. It also includes functions that allow the owner of the contract to add candidates, authorize voters, and record votes. The proposed system leverages the transparency and immutability of the Ethereum blockchain to ensure the security and integrity of the voting process. Once a voter casts a vote, the vote is recorded on the blockchain, making it immutable and transparent. This approach eliminates the possibility of fraudulent activities and boosts the voting process's credibility. The chapter gives a thorough analysis of the proposed system, including the design, implementation, and evaluation. The system is evaluated using various metrics, such as security, scalability, and performance. The outcomes demonstrate that the suggested system is secure, scalable, and efficient. In conclusion, the proposed decentralized smart contracts-based voting method on the Ethereum blockchain provides an effective solution to the challenges of traditional voting systems. The system is transparent, immutable, and secure, making it an ideal solution for democratic societies that value transparency, accountability, and trust.

9.1 INTRODUCTION

Cyber-physical systems, which seamlessly integrate computing, communication, and physical elements, have shown tremendous potential in transforming various industries. In the realm of elections, the traditional voting systems often face concerns connected to security, trust, and integrity. The advent using blockchain technology has introduced promising solutions to these challenges, but the incorporation of cyber-physical machines can further enhance the integrity of the voting process.

9.1.1 CYBER-PHYSICAL MACHINES IN THE VOTING SYSTEM

In the proposed voting system based on blockchain, cyber-physical machines play an important role in bridging the gap between the physical and digital realms. These machines encompass a range of devices such as voting machines, biometric scanners, and networked sensors, which interact with the blockchain network to facilitate secure and transparent voting. By integrating physical hardware components with cryptographic algorithms and decentralized consensus protocols, the system guarantees the accuracy and immutability of the recorded votes. The deployment of cyber-physical machines enables several key advantages. First, it reduces the reliance on traditional paper-based voting, minimizing the potential for human errors and manipulation. Second, the machines provide real-time data collection and analysis, allowing election officials to promptly identify and address any irregularities. Third, the integration of cryptographic techniques in these machines ensures end-to-end security, protecting the confidentiality and integrity of the voting process. It is a fact that the majority of countries in the globe have adopted the right to vote. It gives people a way to select their representatives, voice their opinions, and exercise their right to take part in decision-making. The voting system utilized in a specific country can vary greatly depending on a range of criteria, including its history, culture, and political system [1]. Nonetheless, a few fundamental ideas guide most voting procedures. First off, a means for casting and tallying votes is usually included in voting systems. Paper ballots, computerized voting machines, or a combination of both may be used for this. Also, there are other voting procedures such as proportional representation, ranked choice voting, and plurality. These techniques may significantly affect election results and produce various patterns of governmental representation [2].

Second, the layout of the voting process might affect voter participation and turnout. For instance, several nations have laws requiring citizens to cast ballots in elections. Others have put into place methods to make voting simpler for citizens, such as early voting, mail-in voting, or online voting. Moreover, voter outreach and education programs can make sure that voters are informed about the electoral process and feel empowered to participate in elections [3]. Last but not least, securing fair and valid election results depends on the voting system's transparency and integrity. This can entail actions like maintaining impartial control of the election procedure, auditing voting equipment and ballots, and enforcing laws that prohibit voter fraud or intimidation. Also, the implementation of open and secure voting procedures can foster public confidence in the electoral process and encourage belief in the veracity of election results [4].

A digital platform called a "blockchain-based voting system" utilizes blockchain technology to provide secure and open voting procedures. All vote information is recorded and stored on the blockchain, a decentralized and tamper-proof ledger, creating a transparent and auditable voting system. A system like this would do away with the necessity for a centralized authority to oversee the sometimes manipulated and fraudulent electoral process [5].

Among the primary benefits of an Ethereum-based voting system is its heightened security. By using blockchain technology, all voting-related information is protected with encryption and kept in a secure ledger. This basically eliminates any chance of data manipulation or voting process interference by hackers or other hostile actors. The fact that the blockchain is decentralized further increases its security and resilience because there isn't a single point of failure in the system [6].

Transparency is yet another important advantage of a blockchain-based voting system. Since the blockchain ledger is open to the public, anyone may inspect the voting information and confirm its accuracy. This ensures that the voting process is fair and unbiased and makes it much simpler to spot any efforts at fraud or manipulation [7]. Additionally, the automated verification and tally of votes made possible by the usage of smart contracts within the blockchain lessens the possibility of mistakes or discrepancies [8].

Yet, there are inherent difficulties with voting systems based on blockchains. Making the system accessible and user-friendly for all voters, regardless of their level of technical ability or acquaintance with blockchain technology, is one of the primary issues [9]. There are also worries about the possibility of cyberattacks or other types of voting process meddling, which might jeopardize the system's integrity. The legal and legislative frameworks that would be necessary to support the adoption of a blockchain-based voting system also raise concerns, notably with regard to safeguarding the security and privacy of voter data [10].

9.2 RELATED WORK

9.2.1 BLOCKCHAIN

Blockchain is a decentralized, transparent, and secure digital ledger technology that keeps track of transactions and preserves them. Without a central authority or middleman, the technology enables numerous parties to access and validate the data kept in the ledger. Each transaction is stored as a block, which is then cryptographically connected to the block before it creates a chain of blocks, or blockchain [11].

The cryptographic hashing technique that is used to connect the blocks is the foundation for the security and immutability of blockchain. Each block contains a distinct code, or hash, that is produced by fusing the block's data with a cryptographic procedure. It is practically hard to tamper with the safe chain of transactions since each block's hash includes the block before its hash. Any effort to edit a block in the chain would cause a change in the block's hash, which would be immediately apparent to all other network users [12].

Blockchain technology offers accountability and transparency in addition to security. The ledger records every transaction, and when a block joins the chain, it can't be easily changed or removed. As a result, a clear and unchangeable record of all network transactions is produced. This capability is especially helpful in sectors like finance and supply chain management where accountability and transparency are essential [13].

Additionally, blockchain technology has the ability to open up new avenues for decentralized programs and services. Smart contracts, for instance, are contracts that execute on their own and have the terms of the contract encoded directly into the code. These contracts can be used to automate a wide range of processes, from supply chain management to financial derivatives. Blockchain networks can be used to build decentralized applications, or dApps, which provide users new decentralized services and applications [14].

In general, blockchain technology has the power to completely change how we handle, store, and move data. It is a potent tool for a variety of businesses and applications thanks to its security, transparency, and decentralization. The blockchain technology is still a very young and sophisticated one, and there are still numerous issues and restrictions that need to be resolved before it can be extensively used [15].

9.2.2 SMART CONTRACTS

A self-executing smart contract program that is kept on a blockchain network, such as Ethereum. When specific criteria are met, smart contracts are intended to automatically carry out the terms of a contract between two or more parties [16]. The smart contract's code contains these requirements, and when they are satisfied, the smart contract automatically carries out the predetermined activities [17].

Intelligent contracts provide a secure and transparent means to record and validate votes; hence, they can also be used in voting systems. The conditions of the agreement would specify the election's specifics, including the candidates and voting procedures. The smart contract would automatically count the votes and declare the winner after the election was over [18].

In short, smart contracts automate the conditions of a contract between two or more parties by self-executing software. They are an excellent instrument for automating and enforcing agreements without the need for middlemen because they are tamper-proof and unchangeable [19].

9.3 PROPOSED WORK

9.3.1 ARCHITECTURE

The architecture of the model is comprised of a smart contract named VotingSystem. This contains several components, including data structures, functions, and events.

The data structures in the code include two structs: (a) Candidate and (b) Voter. Candidate represents a candidate in the voting system and contains three fields: (a) an unsigned integer id to identify the candidate, (b) a string name to store the candidate's name, and (c) an unsigned integer voteCount to keep track of votes cast received by the candidate. Voter represents a voter in the voting system and contains three fields: (a) a Boolean voted to indicate whether the voter has cast a vote, (b) a uint voteIndex to store the index of the candidate voted for, and (c) a uint weight to represent the weight of the voter's vote.

The address variable owner is also declared in the code and is used to store the address of the owner of the voting system. The mapping voters is used to map addresses to Voter structs and is used to monitor which electors have projected votes and how much weight each voter's vote carries. The dynamic array candidates are used to store all the candidates in the voting system.

The events in the code are used to emit information about the voting system to the blockchain. The NewCandidate event is emitted when a new candidate is added to the candidates' array and contains the uint candidateId and the string candidateName. The NewVote event is emitted at the point when a citizen makes a choice and contains the uint candidateId and the uint voteCount.

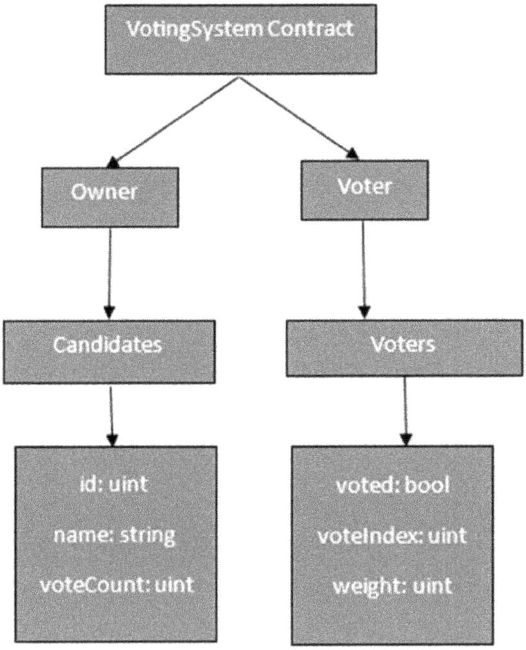

FIGURE 9.1 The architecture of a voting system based on blockchain.

9.3.1.1 ADDCANDIDATE

The addCandidate function is utilized to add another competitor to the voting system. It takes a single argument, name, which is a string representing the name of the candidate. The function is defined as a public function, which means it can be called from outside the contract by anyone with an Ethereum account.

Before adding a new candidate, the function checks if the msg.sender (i.e., the location of the account that called the function) is the owner of the contract. This is done using the require keyword, which throws an exception and stops the execution of the function if the condition in the parentheses is not true. In this case, if the msg.sender is not the proprietor of the agreement, the function will stop executing and return an error message to the caller. If the msg.sender is the proprietor of the agreement, the function makes another Competitor struct using the name argument and the length of the candidates array as the id of the candidate. The voteCount of the candidate is initialized to 0.

The function then pushes the new Candidate struct to the end of the candidates array using the push() method, and assigns the new length of the array minus 1 as the candidateld. This ensures that each candidate has a unique id that corresponds to their index in the candidate's array.

Finally, the function emits a NewCandidate event, passing in the candidateld and name of the new candidate as arguments. This allows external systems or users to be notified when a new candidate is added to the framework. Overall, the addCandidate function is a simple and straightforward function that allows the owner of the contract to add new candidates to the voting system. The use of the require keyword ensures that only the owner can add new candidates, adding a layer of security to the system. The use of events provides a way to notify external systems or users when a new candidate is added, allowing for easy monitoring and analysis of the voting system.

9.3.1.2 AUTHORIZE

The authorize function in the VotingSystem contract is used to grant voting permission to a specific address. The function takes in one parameter, voter, which is the address of the voter being authorized to vote.

The first line of the function uses the require statement to ensure that only the owner of the contract can authorize voters. If the msg.sender (i.e., the address of the caller) is not the owner, the function will throw an exception and revert the changes made to the state of the contract. The next line uses another require statement to ensure that the voter has not already voted. If the voters[voter].voted boolean flag is set to true, it means that the voter has already cast a vote, and is thus not eligible to vote again. In this case, the function will throw an exception and revert the changes.

Assuming both checks pass, the function assigns a weight of 1 to the voter by setting voters[voter].weight to 1. This weight will be used later to ensure that the vote count for each candidate is incremented by the correct amount based on the weight of the voter. It's important to note that the authorize function does not directly modify the state of the contract in terms of vote counts or candidate data. Instead, it only modifies the weight of the voter, which is used in the vote function when the voter casts their vote. This separation of concerns ensures that each function in the contract has a clear and specific responsibility and reduces the risk of unexpected behavior or errors.

9.3.1.3 VOTE

The vote function is responsible for registering a vote for a particular candidate. It takes in a single parameter voteIndex, which is the index of the candidate in the candidate's array that the voter wishes to vote for.

The first step of the function is to check whether the sender of the transaction has already voted by accessing the voted variable of the corresponding Voter struct. If the voter has already voted, the function throws an error with the message "Already voted" and terminates. If the voter has not already voted, the function sets the voted variable of the corresponding Voter struct to true, indicating that the voter has now cast a vote. It also sets the voteIndex variable of the Voter struct to the voteIndex parameter passed to the function, indicating which candidate the voter has voted for.

Next, the function updates the voteCount variable of the corresponding candidate by adding the voter's weight to it. The weight is stored in the Voter struct and is set to 1 by default when a voter is authorized. This ensures that each voter has equal voting power, and prevents any single voter from having an undue influence on the outcome of the election. After updating the candidate's voteCount, the function emits a NewVote event with the voteIndex of the candidate and the new voteCount. This allows external systems or users to monitor the progress of the election in real-time and keep track of the vote counts for each candidate. Finally, the function terminates, and control is returned to the caller.

Overall, the vote function is relatively simple and straightforward and performs the necessary operations to register a vote for a candidate and update the corresponding data structures. The use of the require keyword to check whether the voter has already voted ensures that the integrity of the election is maintained and prevents any attempts at double voting or other forms of fraud.

9.3.1.4 WINNINGCANDIDATE

The purpose of the winningCandidate function is to decide the candidate with the most votes in the voting system. The function iterates through the candidate's array and compares the voteCount of each candidate to determine the one with the highest number of votes.

The function starts by declaring a local variable winningVoteCount of type uint and initializes it to 0. This variable will store the current highest number of votes that a candidate has received. Next, the function enters a for loop that iterates over all the candidates in the candidates array. The loop uses a variable I of type uint to keep track of the current index of the candidate being evaluated. The loop continues until I is less than the length of the candidates array. Within the loop, the function checks if the voteCount of the candidate at index I is greater than the current winningVoteCount. If it is, the function updates the value of winningVoteCount to the value of the candidate's voteCount and assigns the candidate's id to a local variable candidateId. This process continues until the loop has evaluated all the candidates in the candidates' array.

Once the loop is complete, the function returns the value of candidateId. This value represents the identity of the candidate who received the most votes in the voting system. It's important to note that if multiple candidates have the same number of votes, the function will return the ID of the candidate with the lowest index in the candidates array. This is because the function only updates the value of candidateId when it finds a candidate with a higher number of votes than the current highest value. In a scenario where two candidates have the same highest number of votes, the function will not update the value of candidateId, and it will remain equal to the ID of the candidate with the lowest index.

In summary, the winningCandidate function determines the candidate with the most votes in the voting system by iterating through the candidate's array and comparing the voteCount of each candidate. It returns the identity of the candidate who received the most votes, or the ID of the candidate with the lowest index if multiple candidates have the same highest number of votes. Overall, the architecture of the code follows a simple model-view-controller (MVC) pattern. The Candidate and Voter structs act as the model, storing data about the candidates and voters respectively. The functions addCandidate, authorize, and vote act as the controller, providing methods for modifying the model data. The winningCandidate function acts as a helper function to retrieve data from the model. The events in the code provide a mechanism for notifying external systems or users of changes to the model data, and allow for easy monitoring and analysis of the voting system. Utilizing the require keyword in the functions provides a way to guarantee that only approved users can alter the model data and prevents the occurrence of unexpected behavior or errors in the system.

9.3.2 PROCESS MODEL

The process model consists of three main processes: (a) Candidate Management, (b) Voter Management, and (c) Voting.

9.3.2.1 CANDIDATE MANAGEMENT

The Candidate Management process in the VotingSystem contract is responsible for allowing the owner to manage candidates participating in the voting process.

The process consists of a single function called addCandidate, which allows the owner to add new candidates to the candidate's array. The addCandidate function takes one argument, name, which is a string representing the name of the candidate being added. To add a new candidate, the owner must call the addCandidate function and pass in the candidate's name. The function first checks that the caller is the owner by using the require statement. If the caller is not the owner, the function will throw an exception and stop executing.

If the caller is the owner, the function creates a new candidateId by getting the length of the candidate's array. The candidateId is used to uniquely identify the new candidate being added. The function then creates a new Candidate struct using the candidateId and name arguments, and sets the voteCount to 0, indicating that the candidate has not received any votes yet. The new Candidate struct is then pushed onto the end of the candidates array, and the function emits a NewCandidate event with the candidateId and name of the new candidate. Once a candidate has been added, their information will be stored in the candidates array and can be accessed by calling the array with the index of the candidate. For example, candidates[0] would return the first candidate added to the array.

Overall, the Candidate Management process provides the owner with the ability to add candidates to the voting process, which is an essential step in the overall voting process.

9.3.2.2 VOTER MANAGEMENT

In the VotingSystem contract, the Voter struct is used to keep track of voters' information. The voters mapping is used to map the address of a voter to their corresponding Voter struct.

The Voter struct has three fields:

1. voted: A boolean value that shows if a voter has actually cast their ballot or not.
2. voteIndex: An integer esteem that shows the index of the candidate that the voter voted for in the candidate's array.
3. Weight: A whole number worth that addresses the number of votes the voter can cast.

The authorize function is used to authorize a voter. Just the agreement proprietor can authorize voters. The function takes an address as input, which represents the address of the voter to be authorized. The function first checks if the caller of the function is the contract owner. If the caller is not the owner, the function reverts with an error message. If the caller is the owner, the function checks assuming the elector has proactively made their choice by checking the voted field of their corresponding Voter struct in the voters mapping. assuming the elector has proactively made their choice, the function reverts with an error message, assuming the elector has proactively did not make their choice, the function sets the voter's weight to 1.

The weight field is used to ensure that each voter can project their only once. Once a voter has cast their vote, their weight is set to 0, which ensures that they cannot cast any more votes. The weight field is also used to monitor the quantity of votes a voter can cast. By default, the weight of a voter is 0. When the authorize function is called for a voter, their weight is set to 1, which means they can cast one vote. If a voter is authorized multiple times, their weight will be incremented by 1 each time, which means they can cast more than one vote.

Overall, the Voter Management process ensures that every elector would be able cast their vote only once, and their vote is counted only once. It also allows the contract owner to authorize voters and keep track of their voting rights.

9.3.2.3 VOTING

In the VotingSystem smart contract, a voter can cast their vote by calling the vote function, which takes a voteIndex parameter representing the index of the candidate they wish to vote for in the candidate's array. Here's a step-by-step breakdown of how the vote function works:

The function first retrieves the Voter struct associated with the caller's address from the voters mapping using the msg.sender keyword. It then checks whether the voter has already cast their vote by verifying that the voted field of the Voter struct is false. If the voter has already voted, the function throws an error and exits. If the voter has not yet voted, the function sets the voted field of the Voter struct to true, indicating that the voter has now cast their vote. The function then assigns the voteIndex parameter to the voteIndex field of the Voter struct, indicating which candidate the voter has voted for.

Finally, the function updates the voteCount of the candidate at the voteIndex position in the candidate's array by adding the weight of the voter. The weight of the voter is stored in the weight field of the Voter struct and is set to 1 by default in the authorize function, which is called by the owner to authorize voters. The weight field ensures that each voter's vote is counted only once, even if they attempt to cast multiple votes.

After updating the voteCount of the candidate, the function emits a NewVote event, which includes the voteIndex and the updated voteCount of the candidate. Once all voters have cast their votes, the winningCandidate function can be called to determine the candidate with the most votes. The winningCandidate function iterates over the candidate's array, compares the voteCount of each candidate to the current highest winningVoteCount, and updates winningVoteCount and candidateId accordingly. The winningCandidate function returns the candidateId of the candidate with the most votes.

Overall, the VotingSystem smart contract provides a simple and secure way for voters to cast their votes and for the owner to manage candidates and voters.

9.3.3 ALGORITHM

The algorithm for the proposed model is as follows:

Step 1. Define two structs, Candidate and Voter, that represent candidates and voters respectively.

Step 2. Declare a public address variable owner, a dynamic array Candidate [ ] named candidates, and a mapping voter that maps addresses to Voter structs.

Step 3. Define an event NewCandidate that emits a uint candidateld and a string candidateName, and an event NewVote that emits a uint candidateld and a uint voteCount.

Step 4. Define a constructor that sets the owner variable to the address of the contract deployer.

Step 5. Write a function addCandidate that adds a candidate to the candidate's array. It requires that the caller is the owner. It assigns a unique uint candidateld to the new candidate and initializes the Candidate struct with the candidateld, name, and voteCount. It emits a NewCandidate event.

Step 6. Write a function authorize that authorizes a voter to make a choice to vote. It needs that the caller is the owner, the voter has not already voted, and assigns a uint weight of 1 to the voter's weight field in the voters mapping.

Step 7. Write a function vote that allows a voter to make a choice to vote for a candidate. It needs that the voter has not already voted. It assigns true to the voter's voted field in the voters mapping, assigns the voteIndex to the voter's voteIndex field in the voters mapping, increments the voteCount of the candidate at the voteIndex position in the candidate's array by the voter's weight, and emits a NewVote event.

Step 8. Write a function winningCandidate that determines the candidate with the most votes. It iterates over the candidate's array, compares the voteCount of each candidate to the current highest winningVoteCount, and updates winningVoteCount and candidateld accordingly.

Step 9. End of algorithm.

9.4 RESULTS

The blockchain-based voting typically describes the design, implementation, and evaluation of a voting system that uses blockchain innovation to guarantee the integrity, transparency, and security of the democratic interaction.

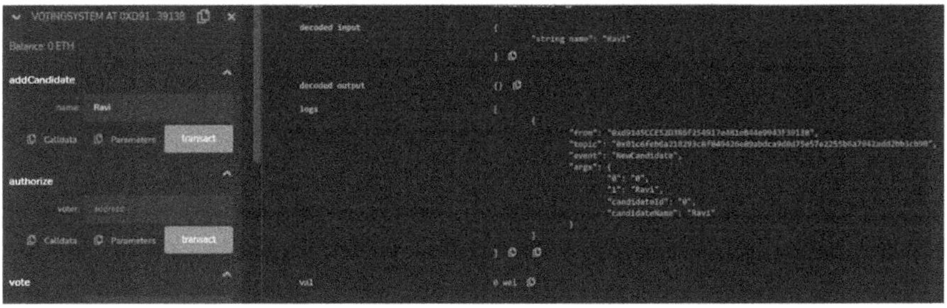

FIGURE 9.2 User interface for adding a candidate.

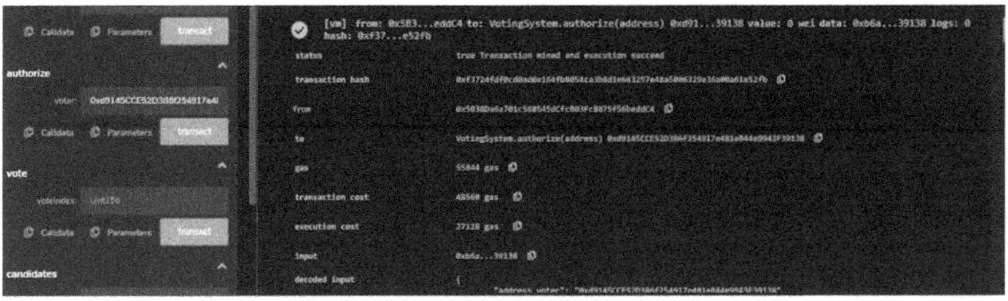

FIGURE 9.3 User interface for authorizing a candidate.

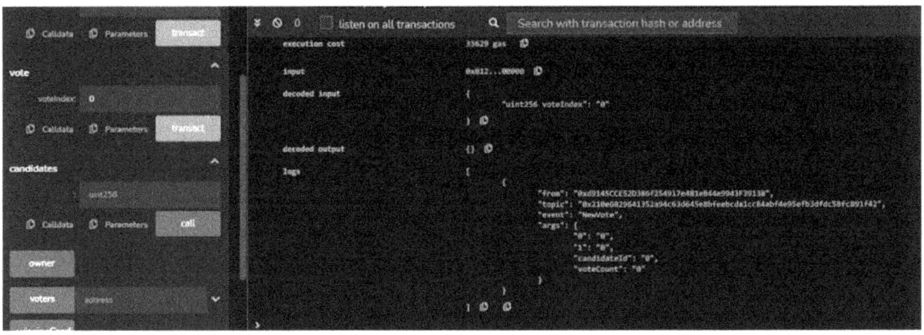

FIGURE 9.4 The user interface for voting a candidate.

FIGURE 9.5 The user interface shows the winner.

9.5 DISCUSSION

The table provides a comparison of the Solidity voting system with traditional paper ballot and electronic voting methods using various metrics. Here's a brief explanation of the trends observed in Table 9.1.

TABLE 9.1 Comparison of Proposed Model with Existing Model.

Metric	Solidity Voting System	Traditional Paper Ballot	Electronic Voting
Accuracy	High	High	High
Transparency	High	Medium	Low
Security	High	Medium	Low
Scalability	Limited	Limited	High
Cost	Low	Medium	High
Accessibility	Limited	High	Medium
Efficiency	Medium	Low	High

9.5.1 ACCURACY

All three methods have high accuracy since they are designed to record and count votes accurately.

9.5.2 TRANSPARENCY

The Solidity voting system has high transparency since all exchanges on the Ethereum blockchain are freely apparent. Traditional paper ballots may have medium transparency since they can be subject to human error or tampering, but they are generally publicly counted and audited. Electronic voting may have low transparency since it may rely on closed-source software or proprietary systems that are difficult to audit.

9.5.3 SECURITY

The Solidity voting system has high security since it uses the Ethereum blockchain's decentralized and carefully designed nature to forestall unapproved access or tampering. Traditional paper ballots may have medium security since they can be vulnerable to physical tampering or loss, but they can also be audited and verified. Electronic voting may have low security since it can be vulnerable to cyber-attacks, hacking, or software bugs.

9.5.4 SCALABILITY

The Solidity voting system may have limited scalability since it relies on the Ethereum blockchain, which may have restrictions as far as processing speed and gas costs. Traditional paper ballots may also have limited scalability since they require physical ballot boxes and manual counting, which can be slow for large elections. Electronic voting may have high scalability since it can be designed to process large numbers of votes quickly and efficiently.

9.5.5 COST

The Solidity voting system may have a low cost since it only requires gas fees to execute on the Ethereum blockchain. Traditional paper ballots may have medium cost since they require printing, distribution, and storage of physical ballots, as well as personnel to count and verify the votes. Electronic voting may have high cost since it requires specialized hardware and software, as well as maintenance and security costs.

9.5.6 ACCESSIBILITY

The Solidity voting system may have limited accessibility since it requires voters to have access to an Ethereum wallet and be familiar with the

Ethereum blockchain. Traditional paper ballots may have high accessibility since they are widely used and understood. Electronic voting may have medium accessibility since it requires voters to have access to specialized hardware and software, which may not be available to all voters.

9.5.7 *EFFICIENCY*

The Solidity voting system may have medium efficiency since it relies on the Ethereum blockchain's processing speed and the winningCandidate function may be slow for large arrays. Traditional paper ballots may have low efficiency since they require manual counting and verification. Electronic voting may have high efficiency since it can be designed to quickly tabulate results and provide real-time updates.

9.6 CONCLUSION

In conclusion, the integration of cyber-physical machines with blockchain technology presents a decentralized solution that addresses the security and transparency challenges of traditional voting systems. By combining the strengths of both domains, the proposed approach enhances electoral integrity and fosters trust among voters. Future research in this area should focus on standardizing the implementation of cyber-physical machines in the voting process and conducting rigorous testing to ensure their reliability and resilience against potential cyber threats. Ultimately, this innovative combination holds the potential to transform the landscape of elections also, and prepare for secure and transparent democratic practices. It can be concluded that the proposed model is an implementation of a voting system in Solidity. The smart contract allows an owner to add candidates, authorize voters, and record votes. The code is written in a straightforward and easy-to-understand manner, making it accessible to developers with a basic understanding of Solidity. Overall, the code is a good starting point for developing more complex voting systems on the Ethereum blockchain. In any case, taking note of that is significant because the current implementation is susceptible to various security vulnerabilities, such as the possibility of double voting and the lack of anonymity for voters. Future work can focus on addressing these issues and improving the overall security and usability of the system. In conclusion, the code demonstrates the potential of smart contracts to create decentralized and transparent voting systems. With further development and refinement, these systems can help facilitate fair and trustworthy elections in a variety of contexts, from political elections to corporate shareholder votes.

9.7 FUTURE SCOPE

Furthermore, the use of cyber-physical machines enables advanced verification mechanisms. For instance, biometric scanners can authenticate voters' identities, ensuring that main qualified people partake in the electoral process. Also, networked sensors can detect any tampering attempts or unauthorized access to the machines, triggering immediate alerts and safeguarding the sanctity of the system. With further development and refinement, these systems can help facilitate fair and trustworthy elections in a variety of contexts, from political elections to corporate shareholder votes.

KEYWORDS

- **blockchain**
- **ethereum**
- **security**
- **transparency**
- **voting**

REFERENCES

1. Hjálmarsson, F. Þ., Hreiðarsson, G. K., Hamdaqa, M., & Hjálmtýsson, G. (2018, July). Blockchain-based e-voting system. In *2018 IEEE 11th International Conference on Cloud Computing (CLOUD) San Francisco, CA, USA* (pp. 983–986). IEEE.
2. Echchaoui, H., Roumaissa, B., & Boudour, R. (2022, November). A proposal of blockchain and NFC-based electronic voting system. In M. Hatti (Eds.), *Advanced computational techniques for renewable energy systems. IC-AIRES* (Vol. 591, pp. 66–75). Cham: Springer International Publishing.

3. Adiputra, C. K., Hjort, R., & Sato, H. (2018, October). A proposal of blockchain-based electronic voting system. In *2018 Second World Conference on Smart Trends in Systems, Security and Sustainability (WorldS4)* (pp. 22–27). IEEE.
4. Kirillov, D., Korkhov, V., Petrunin, V., Makarov, M., Khamitov, I. M., & Dostov, V. (2019, July 1–4). Implementation of an e-voting scheme using hyperledger fabric permissioned blockchain. In *Computational Science and Its Applications–ICCSA 2019: 19th International Conference, Saint Petersburg, Russia, Proceedings, Part II 19* (pp. 509–521). Springer International Publishing.
5. Garg, K., Saraswat, P., Bisht, S., Aggarwal, S. K., Kothuri, S. K., & Gupta, S. (2019, April). A comparitive analysis on e-voting system using blockchain. In *2019 4th International Conference on Internet of Things: Smart Innovation and Usages (IoT-SIU)* (pp. 1–4). IEEE.
6. Wang, B., Sun, J., He, Y., Pang, D., & Lu, N. (2018). Large-scale election based on blockchain. *Procedia Computer Science*, *129*, 234–237.
7. Hanifatunnisa, R., & Rahardjo, B. (2017, October). Blockchain based e-voting recording system design. In *2017 11th International Conference on Telecommunication Systems Services and Applications (TSSA)* (pp. 1–6). IEEE.
8. Moura, T., & Gomes, A. (2017, June). Blockchain voting and its effects on election transparency and voter confidence. In *Proceedings of the 18th Annual International Conference on Digital Government Research* (pp. 574–575).
9. Xiao, S., Wang, X. A., Wang, W., & Wang, H. (2020). Survey on blockchain-based electronic voting. In *Advances in Intelligent Networking and Collaborative Systems: The 11th International Conference on Intelligent Networking and Collaborative Systems (INCoS-2019)* (pp. 559–567). Springer International Publishing.
10. Vivek, S. K., Yashank, R. S., Prashanth, Y., Yashas, N., & Namratha, M. (2020, July). E-voting systems using blockchain: An exploratory literature survey. In *2020 Second International Conference on Inventive Research in Computing Applications (ICIRCA)* (pp. 890–895). IEEE.
11. Li, Y., Susilo, W., Yang, G., Yu, Y., Liu, D., Du, X., & Guizani, M. (2020). A blockchain-based self-tallying voting protocol in decentralized IoT. *IEEE Transactions on Dependable and Secure Computing*, *19*(1), 119–130.
12. Vivek, S. K., Yashank, R. S., Prashanth, Y., Yashas, N., & Namratha, M. (2020, July). E-voting systems using blockchain: An exploratory literature survey. In *2020 Second International Conference on Inventive Research in Computing Applications (ICIRCA)* (pp. 890–895). IEEE.
13. Patidar, K., & Jain, S. (2019, July). Decentralized e-voting portal using blockchain. In *2019 10th International Conference on Computing, Communication and Networking Technologies (ICCCNT)* (pp. 1–4). IEEE.
14. Hardwick, F. S., Gioulis, A., Akram, R. N., & Markantonakis, K. (2018, July). E-voting with blockchain: An e-voting protocol with decentralisation and voter privacy. In *2018 IEEE International Conference on Internet of Things (iThings) and IEEE Green Computing and Communications (GreenCom) and IEEE Cyber, Physical and Social Computing (CPSCom) and IEEE Smart Data (SmartData)* (pp. 1561–1567). IEEE.
15. Zhang, Y., Li, Y., Fang, L., Chen, P., & Dong, X. (2019, December). Privacy-protected electronic voting system based on blockchin and trusted execution environment. In

2019 IEEE 5th International Conference on Computer and Communications (ICCC) (pp. 1252–1257). IEEE.
16. Ayed, A. B. (2017). A conceptual secure blockchain-based electronic voting system. *International Journal of Network Security and Its Applications*, *9*(3), 1–9.
17. Roopak, T. M., & Sumathi, R. (2020, March). Electronic voting based on virtual id of Aadhar using blockchain technology. In *2020 2nd International Conference on Innovative Mechanisms for Industry Applications (ICIMIA)* (pp. 71–75). IEEE.
18. Liu, Y., & Wang, Q. (2017). An e-voting protocol based on blockchain. *IACR Cryptology ePrint Archive*, *2017*(1043).
19. Abuidris, Y., Hassan, A., Hadabi, A., & Elfadul, I. (2019, December). Risks and opportunities of blockchain based on e-voting systems. In *2019 16th International Computer Conference on Wavelet Active Media Technology and Information Processing* (pp. 365–368). IEEE.

CHAPTER 10

UNVEILING MEDIA INSIGHTS IN INTELLIGENT CYBER-PHYSICAL SYSTEMS: TRANSCRIBING AND SUMMARIZING FOR GENERATION AND COMPARATIVE ANALYSIS USING NATURAL LANGUAGE PROCESSING

KISHOR KUMAR REDDY C.[1], RAMANA MURTHY B. V.[1], MARLIA M. HANAFIAH[2,3], and AREENA MAHEK[4]

[1]*Department of Computer Science and Engineering, Stanley College of Engineering and Technology for Women, India*

[2]*Department of Earth Sciences and Environment, Faculty of Science and Technology, Universiti Kebangsaan Malaysia (UKM), Bangi, Selangor, Malaysia*

[3]*Centre for Tropical Climate Change System, Institute of Climate Change, Universiti Kebangsaan Malaysia (UKM), Bangi, Selangor, Malaysia*

[4]*School of Computing and Digital Media, Depaul University, Chicago, IL, USA*

ABSTRACT

This chapter mainly focuses on dealing with big data where media, that is, video and audio, is converted into text using MoviePy and Google Speech

Cyber-Physical Systems: Applications, Challenges, and Research Directions. Rajanikanth Aluvalu, Samarjeet Borah, & Lalit Garg (Eds.)

DOI: 10.1201/9781779643421-9

Recognition. The text is summarized in different ways that include Extractive and Abstractive approaches. The comparison is done for Extractive approaches that include SpaCy, Natural Language ToolKit (NLTK), Gensim, Term Frequency and Inverse Document Frequency (TF-IDF), and TextRank; and Abstractive approaches that use Hugging Face Transformer models like Text-to-Text Transfer Transformer (T5), Generative Pre-trained Transformer2 (GPT-2), Bidirectional Auto-Regressive Transformers (BART), DistalBART, and Transformer Pipeline Model. The dataset used is BBC News Summary with news and reference summary articles to compare the accuracy of computer-generated summary with the reference summary provided. The accuracy is computed using the Recall-Oriented Understudy for Gisting Evaluation (ROUGE) metric. The earlier researchers have not considered the comparison of approaches; hence, this study has illustrated the comparison to determine a close-to-an accurate method which has been illustrated through scatter plots, line graphs and bar, and graphs for comparing the F-scores for multiple news articles and their summaries. Thus, this study efficiently can convert any video or audio file to text with automatic subtitle generation and summarizes its text along with defining the accurate summarization approach that can be adopted for all summarization tasks.

10.1 INTRODUCTION

The existing systems used for summarization were human-based that increased the cost of labor to summarize long documents and generate abstracts. This was time-consuming and was dependent on labor work. Many of the online summarizers also generate summaries but are not accurate enough and only extract the sentences from the long text that denotes an extractive summarization approach to depict the content from the original input text [1].

This study is based on the Video-to-Audio conversion using the MoviePy Python library and then the audio is recognized using Google Speech Recognition to convert the audio into a textual format. The converted text data can be summarized using different summarization approaches that include Extractive and Abstractive. The summarization approach that merely extracts the sentences from the original input text and generates the summary using the same sentences is named Extractive

Summarization. This type of summarization is faster to execute and more efficient to be working with. The other is the Abstractive summarization approach which paraphrases the text sentences to keep the meaning intact but to give out a type of human-generated summary. The text's original meaning is retained by forming new and understandable sentences. Nowadays, there is no pure abstractive summary [2].

The Abstractive summarization approaches use the Deep Learning approaches to feed the data into a model and train it for an accurate summary generation. The summary generated can be compared with respect to the human-generated summary equivalent to the given text to analyze how accurate is the system summary when compared with the human summary. The textual data for summarization can be copied directly onto the platform or can be web-scraped through a URL provided. Web Scraping is used to extract and handle large amounts of data and use it as input for summarization. This text can be extracted based on the criteria of only extracting the paragraphs or headings as well. This is up to the user to retrieve the text and utilize it in the way needed [3].

After the generation of the summary, the key phrases from the summary that are only the noun phrases extracted through POS tagging is a method that depicts the various parts of speech used in the summary. In the case of generating keywords, POS tagging is done for Noun Phrases to extract the nouns that can represent the important words in the summary. The analysis of accuracy can be determined using the ROUGE metric that specifies the recall value which is the significance of the amount of the text from the reference summary that has been captured by the system abstract and the precision value that determines the amount or the quantity of the system summary that is relevant. Thus, from these values, an F-score can be computed that determines the actual accuracy of the system summary. The above research is useful to generate as well as compare the various extractive and abstractive summarization approaches [4].

Apart from the generation of the summary using the text provided, extensive research is conducted using the BBC News summary dataset to extract text from it, generate a summary, and compare the summary with a human-generated summary of that text in order to determine how accurate the summarization approach is with respect to the human-based abstract. The comparison between the extractive and abstractive approaches explains how each of the methods works and how well these perform with the texts and how accurate they are with respect to the text provided as well as the given summary [5].

The study was conducted to tackle many issues. The grasping of facts from long informative videos can be difficult and time-consuming, if the video has to be paused at every interval. Transcription generates the entire text at once, which is time-saving and helps to fit larger amounts of text in a smaller area. Summarization of text helps to grasp concepts in less time with a better understanding for any layman and helps to take better decisions in less time regarding any paper or book choice. This shortening of texts is also beneficial to read news in a shorter and more understandable manner instead of repetitive sentences and long paragraphs which is out of our scope to understand [6].

The objectives of the proposed study are as follows:

1. Video transcribing using the open-source Speech Recognition API, as video/audio-to-text is manually done and is a very time-consuming process.
2. Generated transcribed text from the media file (video/audio) is summarized.
3. Summarization using both extractions of sentences and paraphrasing of text.
4. Comparison of different extractive and abstractive summaries based on the F-score accuracies.
5. Generating keywords for the summary that highlight important concepts covered.
6. Determining the accuracy of the different summaries to justify which works better.
7. The study helps to generate a user-friendly platform that does not need any understanding in-depth for its functioning.

This research chapter has been organized in the following way: Section 10.2 presents an extensive review of the existing methodologies and technologies applied to the earlier works along with the parameters involved. Section 10.3 presents our proposed work along with the data being used. A general framework of the proposed architecture is presented in this section to illustrate the algorithm of study to better understand the design flow. This is then followed by Section 10.4 that depicts all the results for various approaches and the comparison outcomes and in Section 10.5, we conclude the chapter.

10.2 RELEVANT WORKS

Ángel Hernández-Castañeda, René Arnulfo García-Hernández, Yulia Ledeneva, and Christian Eduardo Millán-Hernández presented a proposed methodology, where an automatic extractive summarization task deals with the clustering approach supported by the Genetic Algorithm. This approach is better than any other previous approach as it is platform and domain-independent because it does not require any apriori information. CNN/Daily Mail dataset is used to evaluate the performance of the supervised and unsupervised approaches. Proximity measures are used to evaluate the proximity of different objects in the partition using Euclidean distance and cosine distance measures. This approach outperforms all the other approaches but gives competitive results for CNN/Daily Mail dataset. LDA cannot capture correlations and the number of topics (K value) has to be shown in advance [5].

Reem A. K. Aljorani and Boshra F. ZoponAl_bayaty presented an article that includes the audio features extraction using the Microsoft Azure Cognitive Services which helps to generate transcripts in the form of text. The various filtering options and pre-processing functions are applied to the text transcripts. Audio Extraction from the videos is carried out using MoviePy python library and using the Azure Cognitive Services API, the audio transcripts are generated. Preprocessing of carried out in order to remove the unwanted/stop words and each word is assigned weight using the TF-IDF technique. This chapter illustrates the process of conversion of Arabic audio to Arabic text which is a time-consuming task and cost expensive to keep the API running through regular payments [17].

Chalothon Chootong, Timothy K. Shih, AnkhtuyaOchirbat, Worapot-Sommool, and Yung-Yu Zhuang presented a paper that explains the summarization process of the human-provided subtitle documents using CNN and Bi-directional LSTM in order to capture the important information which can be used to determine the sentence significance. These are performed on both Text documents as well as subtitle documents. This summarization is based on the multi-attention approach for the Extractive Summarization that consists of three main components—Pretraining Process that defines the vector generation using GloVe from the sentences with words and forming sentence matrices where each sentence is given a score. Sentence Feature Extraction is done using the WACNN-based model that indicates how much attention has to be given to each word based on its contribution to the meaning of the sentence. The dataset utilized involves subtitles from TED.com and YouTube channels. The text is used from DUC 2002 or CNN/Daily Mail. The proposed system gives better results as compared to the three baseline architectures: (a) CNN, (b) WACNN, and (c) CNN-bi-LSTM [7].

The previous research had a few limitations that will be looked into, in the proposed work. These include the Speech Recognition API that is used in the previous studies and is the one that needs regular payment, but Google Speech Recognition is much more convenient as it is highly accurate and is open-source. The previous studies transcribed only the subtitle documents instead of the manually generated files as they contain noise and inaccuracy, which will be tackled in the proposed system. No explanation was provided regarding the accurate methods that can work with all datasets and ROUGE score specified as ROUGE-1, ROUGE-2, and ROUGE-L that does not explain much about the accuracy rate of that approach for the text, which will be solved through this proposed study [8].

10.3 PROPOSED METHODOLOGY

The process flow of the video transcription and text summarization tool is that it takes the input as a media file that includes both videos and audio files which is converted into a textual format using the Python libraries and API which can then be used to apply various summarization approaches to generate a coherent and an abstract version of the input text as a summary. The proposed architecture has been illustrated as follows:

The study is carried out on different datasets to analyze the best and the most accurate text summarization approach and illustrate the generation of a summarized version of the video or audio text. The following shows the various sections and steps that are to be carried out in order to generate a summarized textual format of the media content.

***Algorithm:* The step-by-step algorithm of the proposed research is as follows:**

Step 1. Choose a media file for text conversion from the User Interface.

Step 2. Audio extraction and Speech Recognition in case of the video file and Speech Recognition in case of an audio file uploaded from which Video or Audio text will be generated.

TABLE 10.1 Comparative Study on the Previous Approaches and the Datasets Used.

S. No.	Title of Paper	Approach	Advantage	Limitation
1.	**Language-independent extractive automatic text summarization based on automatic keyword extraction** Elsevier, Computer Speech and Language, Volume 71, January 2022	Clustering using Genetic Algorithm representation and evaluation of relevance using LDA [10].	This proposed system is platform and domain-independent because it needs no apriori knowledge.	Gives competitive results when using large-size dataset and number of topics should be shown ahead of time.
2.	**An attention-enhanced sentence feature network for subtitle extraction and summarization** Elsevier Ltd, 23 March 2021	Extractive Summarization on Subtitle and Text Documents using Attention Mechanism [11].	Works well on the subtitle document. Closely similar to the human-generated summaries.	Works only on the subtitle document provided as the auto-generated subtitles have inaccurate data and contain noise.
3.	**Summary and Keyword Extraction from YouTube Video Transcript** International Research Journal of Modernization in Engineering Technology and Science Volume:03, Issue:06, June-2021	Extractive Summarization, Abstractive Summarization, Keyword Extraction.	Generates summary using both the techniques of summarization.	Only performs summarization of video transcripts, but not of audio or simple text.
4.	**A Study on FFmpeg Multimedia Framework** International Journal of Trend in Scientific Research and Development (IJTSRD), Volume 5, Issue 4, May-June 2021	Image and Video editing using FFmpeg [12].	FFmpeg supports many programming languages to and can handle basic to complex editing of the videos.	FFmpeg is not user friendly as everything is based on the writing of commands.

TABLE 10.1 *(Continued)*

S. No.	Title of Paper	Approach	Advantage	Limitation
5.	**A Deep Look into Extractive Text Summarization** Journal of Computer and Communications, 9, 24-37, 2021	Extractive Text Summarization using TF-IDF method [13].	Consumes fewer memory resources and can be applied to single as well as multiple documents.	Laborious process with satisfactory results. Only applied to the text in Spanish.
6.	**Abstractive Text Summarization Techniques** International Journal of Engineering Science and Computing (IJESC), Volume 10 Issue No.7, July 2020	Abstractive Summarization using LSTM model building [14].	Produces an exact, relevant and a less redundant summary, close to the human-based summary.	This is difficult to implement and it is a challenge to achieve its results as it uses the non-uniform semantics of the text.
7.	**Extractive Automatic Text Summarization Based on Lexical-Semantic Keywords** IEEE, Volume 8, March 11, 2020	Sentence Tokenization, vector generation, clustering, selection of key sentences using LDA [15].	This approach is domain and language independent. Outperforms other methods and gives better results through ROUGE.	Constraint of using Genetic Algorithm is that maximum words that can be generated for the summary are 100.
8.	**A Novel Web Scraping Approach Using the Additional Information Obtained from Web Pages** IEEE, Volume 8, 2020	Web Scraping using UzunExt, DOM tree creation [16].	DOM-based method is 60 times slower than the UzunExt method when extracting text.	DOM tree construction increases the cost and the time of the extraction of data based on the structure.

Step 3. Choose a summarization method either Extractive or Abstractive Summarization (as per choice).
Step 4. Summary is generated along with the reading time in minutes.
Step 5. Keywords are extracted from the summary that highlight key concepts.
Step 6. Accuracy of the summary is computed in the form of F-score with respect to the recall and precision values using the ROUGE metric.
Step 7. Comparison based on F-score can be done to determine the most accurate summary.

10.3.1 VIDEO TRANSCRIPTION

Audio Extraction from Video files

MoviePy is an open-source, python library that helps in video editing. MoviePy.editor is used to extract audio and store it with a filename and extension. It is capable of reading all audio and video formats.

Speech Recognition

This is an API that is used with the speech recognition function to convert audio to textual format. It uses Pydub for working with wav files. Sound level under 14dBFs is considered to be silence and then based on the silence the chunks are divided if the silence minimum length is 500 ms. The speech recognizer extracts text for each chunk, respectively [17].

10.3.2 TEXT SUMMARIZATION

10.3.2.1 EXTRACTIVE SUMMARIZATION

Spacy, NLTK

Spacy and NLTK are used for Extractive Summarization and are based on the word frequency technique. Each of these NLP libraries has its own corpus of stop words, method of preprocessing, and speed through which they differ. The process (shown in Figures 10.1 and 10.2) illustrates the steps that are to be followed for a summary generation.

Gensim, TextRank

Gensim is a summarization library and TextRank is a summarization approach based on TextRank and PageRank algorithms (Figure 10.1). This involves the computing of vector for sentences and checking the similarity between sentences using Cosine Similarity or BM25 Ranking Function. A Similarity matrix is generated to determine the probabilities of each sentence for a summary [18].

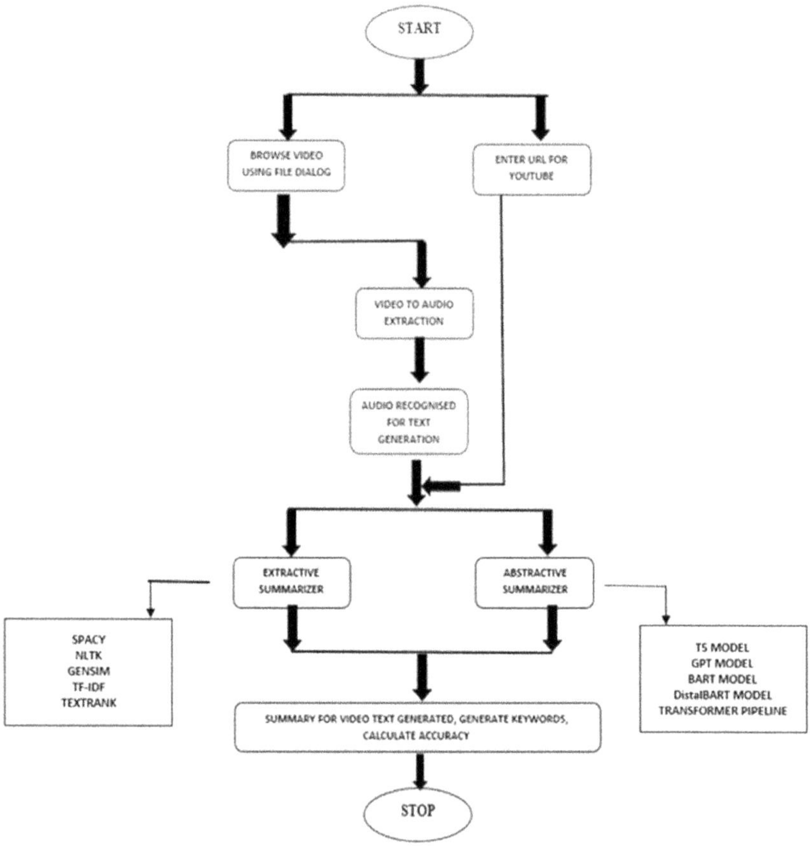

FIGURE 10.1 Process flow of the proposed system.

TF-IDF

This summarization involves the use of the Term Frequency (TF) matrix and Inverse Document Frequency (IDF) matrix. It uses word frequency and its relevance in the text based on the number of sentences in which the words are occurring. It involves scoring the sentences based on the product of TF and IDF values for each of the sentences in the text to generate the summary.

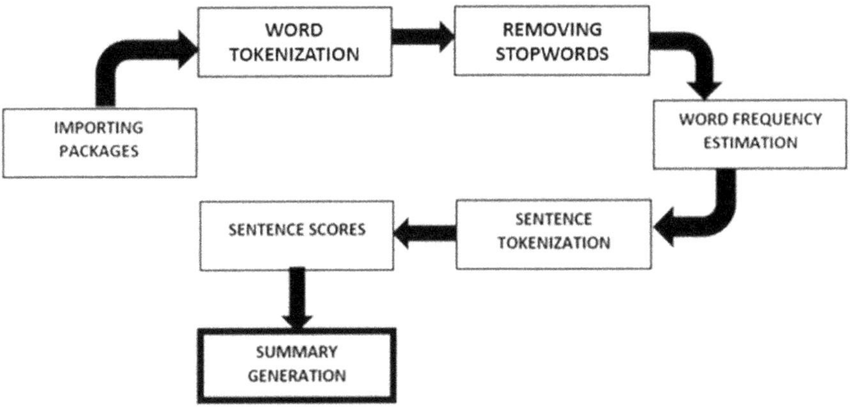

FIGURE 10.2 The working of SpaCy and NLTK libraries.

The process of summary generation is shown in Figure 10.3.

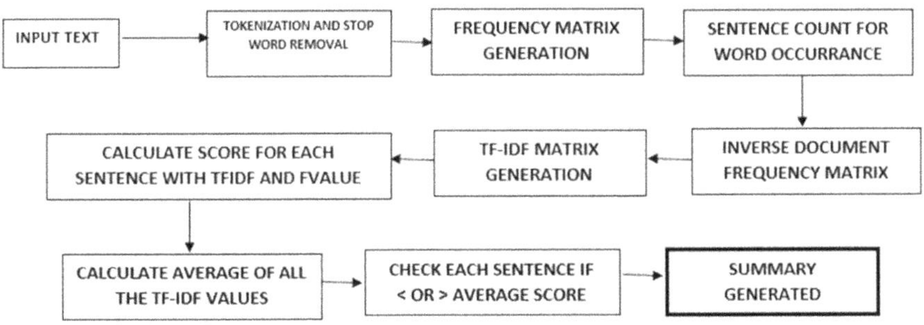

FIGURE 10.3 Working of TF-IDF for summary generation.

10.3.2.2 ABSTRACTIVE SUMMARIZATION

T5 Model

It represents a Text-To-Text Transfer Transformer, a transformer model by Google that takes input as text and modifies it to a text output. It is used to perform any task that gives varied outputs only in text format. It is based on the pre-trained model that uses T5Configurations and T5Tokenizer to perform the summarization. The advantage of this model is that it can use the same model, functionsm], and hyperparameters for any task.

BART

BART stands for Bidirectional Auto-Regressive Transformer. It is used to train on noisy or corrupt data to retrieve the original data. BART is based on the usage of BERT for its bi-directional encoder and GPT for text generation (depicted in Figure). For summary generation, it uses the pre-trained BARTConfig and BARTTokenizer.

DistalBART

DistalBART is the distilled version of BART which compresses the size of the neural network without much loss of accuracy and fewer parameters needed. Distilled versions of models have better accuracy as compared to the originally trained counterparts.

Transformer Pipeline

This is a great and easy way to use models for inference. This pipeline is easy-to-use, and avoids redundancy of code and enhanced reusability due to shortened code. It is dedicated to several tasks, such as Sentiment Analysis, Feature Extraction, and Summarization. This can work with any model when supplied in the model's name. If no is model provided, it uses DistalBART model by default for abstractive summarization.

Keyword Extraction Using POS Tagging

The summary generated is a coherent version of the original text and it's important to highlight the main topics in the summary in order to understand its context faster and in a better way. The keywords are extracted using POS tagging that is used to highlight only the Noun Phrases in the summary and the words whose frequency is greater than equal to 2 are represented as important words from the summary. The illustration can be done using a word cloud that shows the important words in order of their frequency of occurrence in the summary.

Accuracy Computation Using ROUGE

Rouge is a metric that is used to analyze the accuracy with respect to the system summary with the reference summary or the text to depict the fact that how relevant is the summary generated and the amount of text that is captured by the summary. This is an important aspect to determine the usefulness of a particular approach in a summarization task.

1. Recall: Determines: How much of the text is captured by the summary.
2. Precision: Determines: How much of the summary is relevant?
3. F-score: Combination of the recall and precision values to determine accuracy.

10.4 RESULTS AND DISCUSSION

Dataset Description

1. Video/Audio
   1) Videos/Audios from Local Storage.
   2) URL from https://www.youtube.com/.
2. Text
   1) Text of your choice.
   2) Web scraping through any URL.
3. Dataset
   1) BBC News Summary (Containing 2225 Text and Summary Files each).

   Total: 4450 Files.

   [Source:https://www.kaggle.com/datasets/pariza/bbc-news-summary].

10.4.1 COMPARISON OF EXTRACTIVE SUMMARIZATION APPROACHES

Figure 10.4 depicts the comparison of the summarization algorithms with the online summary that can be used to compare the accuracies and choose the best algorithm that is nearest to the input text provided. This represents how relevant the summary is and how much of the input text is captured by the system summary.

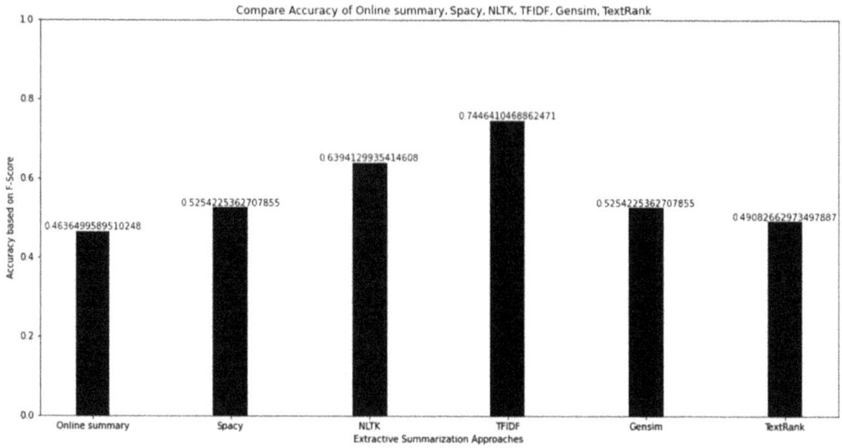

FIGURE 10.4 Comparing online summary, SpaCy, NLTK, TF-IDF, Gensim, and TextRank against an input text.

The BBC News Summary is used as a dataset with news articles and a reference summary to be compared against the system summary that is generated. The news article is extracted from the dataset and is used as input for the summarizer to generate an abstract version of the text. The summary is compared against the reference summary which is a human-based summary from the dataset to calculate the accuracy of the algorithm and check how accurate a summary it generates. Based on the accuracy range of 0–1, the appropriate algorithm is decided and can be used for future summarizations.

A number of summaries are generated for different news articles from the dataset and the recall and precision values for these summaries are added to a list. Then these values are retrieved altogether for each of the extractive summarization approaches. A line graph is plotted against each change of the recall value on the x-axis and precision on the y-axis. The changes are depicted as follows (depicted in Figure 10.5). SpaCy, NLTK, and Gensim are seen to have the highest values for recall and precision among all the extractive approaches under consideration.

The recall and precision values from the above graph are extracted and are used to compute the F-score for these summaries with respect to the reference summary. The average of all the F-scores is computed for each summarization approach and is plotted as shown in Figures 10.5 and 10.6, respectively.

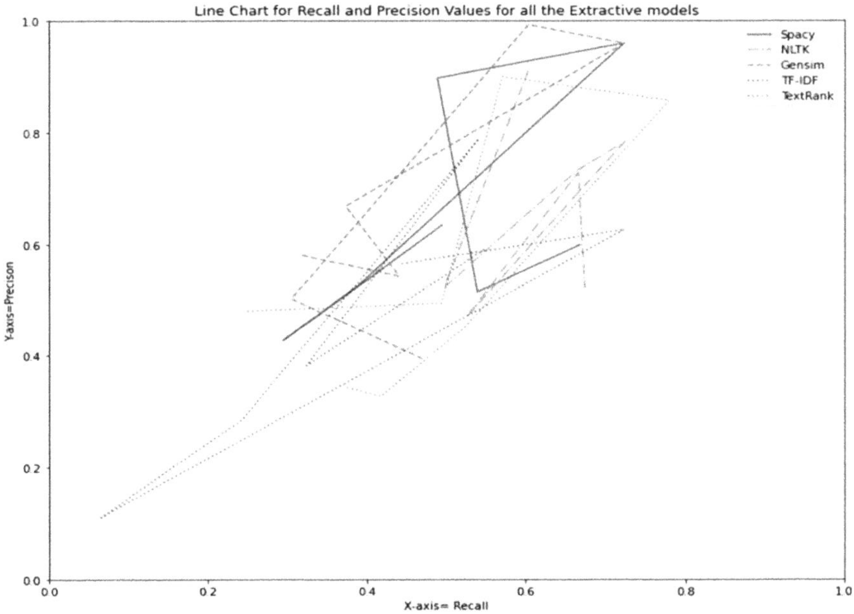

FIGURE 10.5 Distribution of recall and precision values for the extractive summarization approaches.

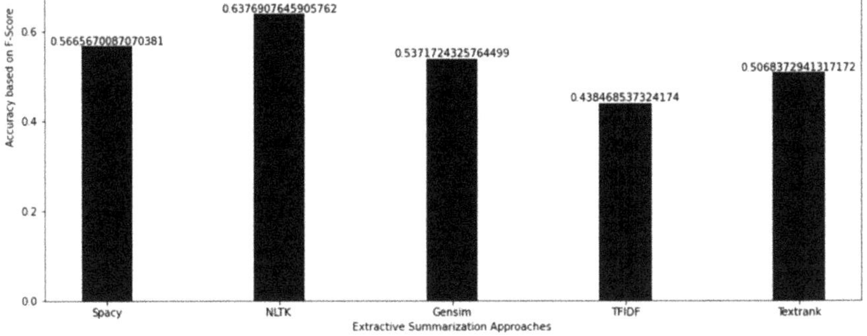

FIGURE 10.6 Average of all the F-scores for different news article summaries.

TABLE 10.2 Average Accuracy of Extractive Approaches wrt to Reference Summary for Multiple Texts.

S. No.	Algorithm	Accuracy wrt Reference Summary	Accuracy
1.	SpaCy	0.5666567008707038l	0.57
2.	**NLTK**	**0.6376907645905762**	**0.64**
3.	Gensim	0.5371724325764499	0.54
4.	TFIDF	0.438468537324174	0.44
5.	TextRank	0.5068372941317172	0.51

10.4.2 COMPARISON OF ABSTRACTIVE SUMMARIZATION APPROACHES

The below graphs (Figure 10.7) illustrate the Abstractive Summarization approaches that are compared against the input text. BART consists of the GPT model thus they are compared with each other. The transformer pipeline uses the DistalBART model as default thus they are compared together. DistalBART is the distilled version of the BART model, hence these are compared. The accuracy of each model is determined and an appropriate model can be chosen with respect to the input text by the user.

The BBC News Summary with the news article and summaries is used as an input dataset for the summarization algorithms. The abstractive summaries are generated from the news article and the summaries are compared against the reference summary to compute an F-score value that determines the accuracy of the abstractive approach. Thus, a suitable summary can be determined and used in the future [9].

The below scatter plot (Figure 10.8) depicts the recall and precision values for a single news article summary and its reference summary. The recall and precision values of the abstractive approaches are determined in order to determine the accurate approach for a multiple news article.

The bar graph below (Figure 10.9), illustrates the average of the F-scores derived from recall and precision values for a number of news articles. GPT shows the highest F-score whereas T5 shows the least for six news articles.

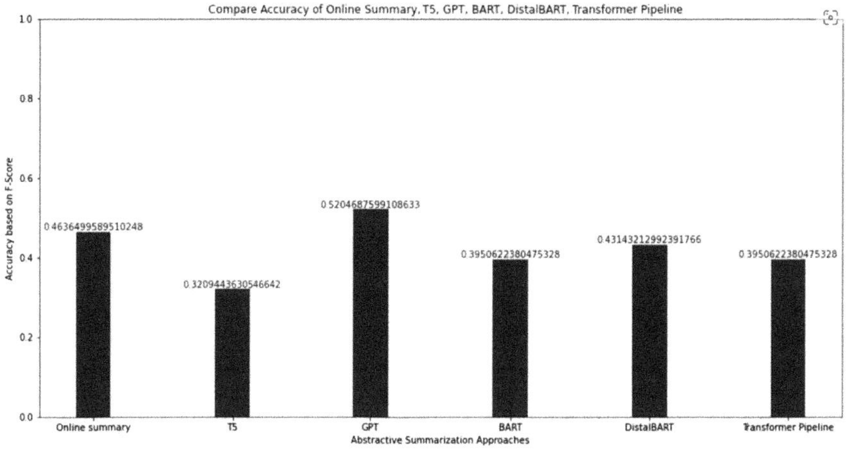

FIGURE 10.7 Comparing online Summary, T5, BART, GPT, DistalBART, and Pipeline models against an input text.

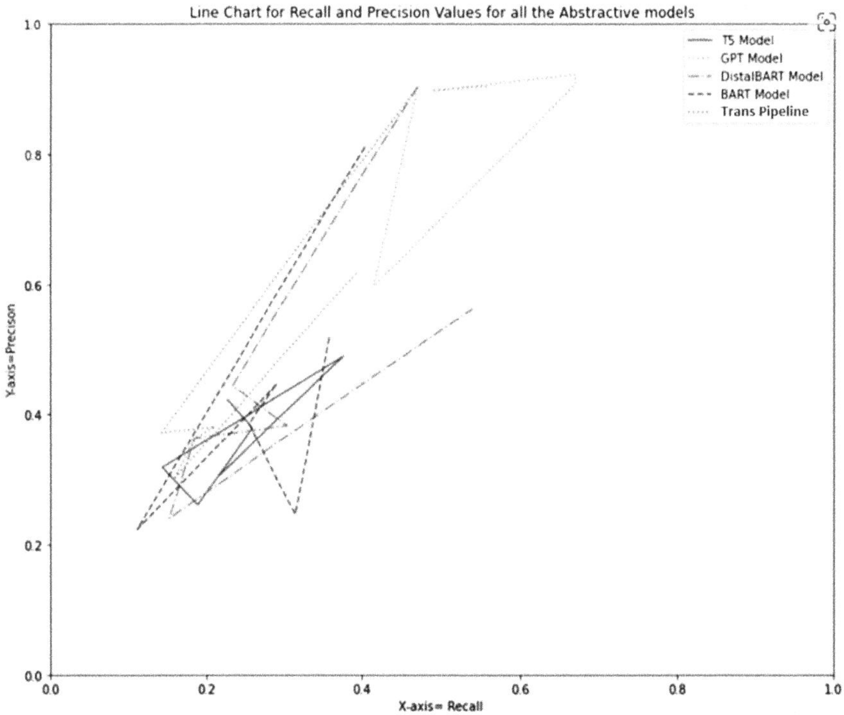

FIGURE 10.8 Line chart for changes in recall and precision for multiple news articles.

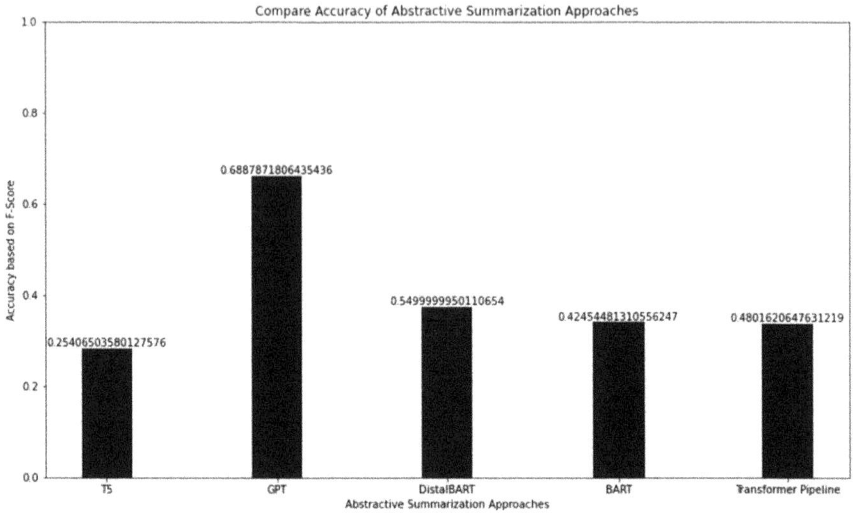

FIGURE 10.9 Averaged F-score values for the abstractive summarization approaches.

TABLE 10.3 Average Accuracy of Abstractive Approaches wrt to Reference Summary for Multiple Texts.

S. No.	Algorithm	Accuracy wrt Reference Summary	Accuracy
1.	T5	0.25406503580127576	0.57
2.	**GPT**	**0.6887871806435436**	**0.64**
3.	DistalBART	0.5499999950110654	0.54
4.	BART	0.42454481310556247	0.44
5.	Transformer Pipeline	0.480162064761219	0.51

10.5 CONCLUSION

This chapter compares the different extractive and abstractive summarization methods that are applied to video text/audio text or web-scraped data. Among the extractive approaches, NLTK has shown the greatest accuracy score when compared to the human-generated summary. This represents that the summary generated by NLTK is the closest to our human-generated summaries that were provided by the users. Among the abstractive approaches, the GPT model has shown the greatest and the best results with respect to the human-generated summary. The limitations of this study are that the abstractive summarization approaches have slower processing time and take a long time to process and generate a summary. These methods also consume a lot of runtime memory, which can be a disadvantage for users with slow processor systems and less memory storage.

This study can be enhanced by comparing the other extractive and abstractive models available along with training and then using our own tokenizers and models, while comparing them against the pre-trained models, to determine which of the models is accurate enough to meet the standards. The BLEU metric can also be used as a metric of comparison along with ROUGE to find the best summarization approach.

KEYWORDS

- **BLEU**
- **intelligent cyber-physical systems**
- **media insights**
- **natural language processing**
- **summarize**

REFERENCES

1. Hernández-Castañeda, Á., García-Hernández, R. A., Ledeneva, Y., & Millán-Hernández, C. E. (2020). Extractive automatic text summarization based on lexical-semantic keywords. *IEEE Access*, *8*, 49896–49907.
2. Aiswarya, K. R. (2020). Automatic multiple language subtitle generation for videos. *International Research Journal of Engineering and Technology (IRJET)*, *7*(5).
3. Huang, Y., Yu, Z., Guo, J., Yu, Z., & Xian, Y. (2020). Legal public opinion news abstractive summarization by incorporating topic information. *International Journal of Machine Learning and Cybernetics*, *11*(9), 2039–2050.
4. Bharadwaj, R. K. (2020). Automatic text summarizer application using extractive text summarization. *International Journal of Creative Research Thoughts (IJCRT), 8*(7).
5. Hernández-Castañeda, Á., García-Hernández, R. A., Ledeneva, Y., & Millán-Hernández, C. E. (2022). Language-independent extractive automatic text summarization based on automatic keyword extraction. *Computer Speech and Language*, *71*(C), 15.
6. Sumesh Singha, H., & Bhuvana, J. (2021). A study on FFmpeg multimedia framework. *International Journal of Trend in Scientific Research and Development (IJTSRD)*, *5*(4), 580–584.

7. Uzun, E. (2020). A novel web scraping approach using the additional information obtained from web pages. *IEEE Access*, *8*, 61726–61740.
8. Dantis, A., Fernandes, R., & Rodrigues, A. P. (2019). An extractive summarization technique for text documents, *International Journal of Engineering and Advanced Technology (IJEAT)*, *8*(6).
9. Khandare, P., Gaikwad, S., Kukade, A., Panicker, R., & Thamke, S. (2019). Audio data summarization system using natural language processing, *International Research Journal of Engineering and Technology (IRJET)*, *6*(9).
10. Kumar, S., Kar, A. K., & Ilavarasan, P. V. (2021). Applications of text mining in services management: A systematic literature review. *International Journal of Information Management Data Insights*, *1*(1), 100008.
11. Quillo-Espino, J., Romero-González, R. M., & Herrera-Navarro, A. M. (2021). A deep look into extractive text summarization. *Journal of Computer and Communications*, *9*(6), 24–37.
12. Al-Sabahi, K., Zuping, Z., & Nadher, M. (2018). A hierarchical structured self-attentive model for extractive document summarization (HSSAS). *IEEE Access*, *6*, 24205–24212.
13. Yadav, S., Behra, A. K., Sahu, C. S., & Chandrakar, N. (2021). Summary and keyword extraction from YouTube video transcript. *International Research Journal of Modernization in Engineering Technology and Science, 3*(6).
14. Qaiser, S., & Ali, R. (2018). Text mining: Use of TF-IDF to examine the relevance of words to documents. *International Journal of Computer Applications*, *181*(1), 25–29.
15. Jose, D. V. (2019). Speech to text conversion and summarization for effective understanding and documentation. *International Journal of Electrical and Computer Engineering (2088–8708)*, *9*(5).
16. Magar, Y. Automatic text Summarization for Abstract Generation.
17. Aljorani, R. A. (2021). Encapsulation video classification and retrieval based on Arabic text. *Diyala Journal for Pure Science*, *17*(4).
18. Kishor Kumar Reddy, C., Anisha, P. R., Shastry, R., & Ramana Murthy, B. V. (2020). Comparative study on internet of things: Enablers and constraints. In *Data Engineering and Communication Technology: Proceedings of 3rd ICDECT-2K19* (pp. 677–684). Singapore: Springer Singapore.

CHAPTER 11

INDIVIDUAL PLAYER ACTION RECOGNITION USING KEYPOINT-BASED UNWEIGHTED GRAPHICAL APPROACH IN SPORTS

KANIMOZHI S.[1], DEEPIKA ROSELIND J.[2], and SAI RAMESH L.[1]

[1]*School of Computer Science Engineering, VIT, Chennai, India*

[2]*Department of Computer Science and Engineering, St. Joseph's Institute of Technology, Chennai, India*

ABSTRACT

Sports involve a variety of interactions between people and objects with little change in posture. Visual classification makes an effort to comprehend and recognize these actions. There are two main processes in the identification of human–object interaction (HOI). The initial step in HOI is localizing the subject and the object. The second element of it involves the identification of complicated relationships between them. Hence, the proposed work focused on detection on the player actions individually by considering their body gesture across the entire video. To attain this, a novel approach is proposed which localizes the joint points with normalized values using KDCnet architecture. Using the plotted points an unweighted graph-based approach is applied to determine the resultant posture labels. At last, the action class is defined by using the body part labels along with posture combination LSTM. The proposed keypoint mapping and action

Cyber-Physical Systems: Applications, Challenges, and Research Directions. Rajanikanth Aluvalu, Samarjeet Borah, & Lalit Garg (Eds.)

DOI: 10.1201/9781779643421-11

recognition (KM-AR) framework provides a meaningful classification result as the graph follows single source traversal along with minimal path coverage across the plotted points. Experiments were conducted using two action datasets such as Sports Video Wild (SVW) and MPII for both keypoint plotting and action detection. The performance of the proposed work is evaluated by comparing with other state-of-the-art methods used for action recognition in the field of sports.

11.1 INTRODUCTION

There is a significant need in the modern world to understand and predict the player's next move in the field of sports. Understanding how players move and how other players will react in order to score is crucial for sports training. It is necessary to evaluate the variations involved in player posture to execute a single action in order to predict their next move. Consequently, locating and plotting the keypoints that identify the player's body joints is a difficult task. As the sports video consists of sequence of motion, the spatio point over body joints varies in relation to the temporal changes.

As a result, it is essential to monitor the player's posture during the entire film and combine all of the joint scores earned. Sports involve both single and many participants; it is difficult to keep track of everyone and identify their shared points throughout the full film. While still in the development stage, several technologies, such as convolutional neural networks and deep neural networks, aid in calculating the poses. To solve these problems, a network model-based method (KM-AR) was created, which connected the joint locations and created the player's skeletal structure. This proposed approach offers a solution in two ways:

1. first, by increasing the posture's level of visibility; and
2. second, by precisely identifying the posture labels.

Despite the fact that the skeleton structure offers geometric proof of human posture, relational logic and normalized joint points for a single player's posture throughout the video cannot be located [1]. The Keypoint Deeper Cut network (KDCnet), which plots the joint points based on higher probability values obtained for a player action, is utilized in the proposed work to address this issue. The obtained skeleton structure is placed in the 3D space for a better understanding of reasonable construction of player posture [4, 5]. For determining the posture class, the unweighted minimum spanning tree method is combined with the single source shortest (SSS) path algorithm, which follows a path without looping and traverses from a single source, such as the head. Finally, LSTM is used to define the player action class [3].

11.2 RELATED WORKS

Classification of Sports Action

Detecting human action in the field of sports is a tedious process as there is change in movement on a timely basis. To recognize the playfield actions a neural-based method [14] was proposed, which focuses on frame level action recognition. Transfer learning based on convolution neural network (CNN) along with the VGG-model is used to fine tune the recognized actions. The proposed work is focused only on the actions performed on the entire frame [7] and not based on individual player. This approach results in detection of action patterns only for the trained frames and not for any newer sports. Next, to monitor the player performance, gesture-based individual player action recognition is proposed [15]. In this transformer networks were used to recognize the gesture along with 3D CNN. As they use bounding boxes over the player to determine the actions, results can cause occlusion by overlapping one player with another.

So, to overcome this issue, spatial attention at the channel level [16] is proposed. Residual convolution neural network (CNN) [11] is used to address the specific action class by mapping the feature graph with its relevant temporal action category. A spatial attention module is created, which is an attention mechanism that employs a local cross-channel interaction method to offer greater relevance scores to spatial regions and video frames that are more pertinent to the action category without dimensionality reduction.

Since the suggested approach concentrated on the highlighted spatial locations, it neglected to classify the actions of the entire player body. An unweighted graph-based technique that considers the entire body structure to categorize the individual player's posture is provided to address the problems with action classification. In order to traverse and fix the action class based on player posture, spatial points are also plotted over the body joints.

11.3 PROPOSED WORK

The proposed work targets to define the player's action through their recognized posture pattern. Continuous changes in the movement among the players make it challenging for the system to understand and finalize the action performed by them [20]. Fixing keypoints in the human body parts helps to overcome the problem of recognizing the movement pattern.

Mostly, keypoints have been placed at the joint portions of the human body in the view to make tracking of movements easier. It is located through spatial values in the body parts using dots or numbering systems [9] and is connected to provide a skeleton structure. However, there may be variations in fixation of points when there is a continuous movement. This results in a rough calculation about estimating values for the pattern identification process. Actions are highly dependent on posture movement, thereby facilitating researchers to find various modes to fit the keypoints in the human body.

Among the various human body parts, wrist, elbows, knees, and ankles are the major areas through which final valuation is done with respect to the spatial values. Instead of considering only the X and Y coordinates for each joint, it would be better to choose the X, Y, and Z coordinates to get better results. The RGB image is used for plotting the points due to its additive and nontransformation property. Point values are joined together to get a skeletal structure that makes the process of estimation clear and intelligible. The overall workflow with regard to the process of player posture identification is portrayed in Figure 11.1.

11.3.1 PLAYER POSTURE ESTIMATION

Initially, the player and other functional objects are marked as Region of Interest (ROI) as in [10]. The pose pattern of that player is identified by fixing the keypoints in their human body parts. Owing to the better results obtained through the usage of X, Y, and Z coordinates, 3D-based landmark of limb points is plotted over the body of the player [8]. The keypoints were plotted from head to foot with the index value starting from 0 and end in the foot with value as 21. The estimator accepts the input (RGB) with a dimension of $254 \times 254 \times 3$. The Z-value is based on the distance between the hip of the considered player to that of the keypoints. Then finally, the landmark keypoints are plotted in the 3D space using the player pose estimation algorithm, as described in Algorithm 1.

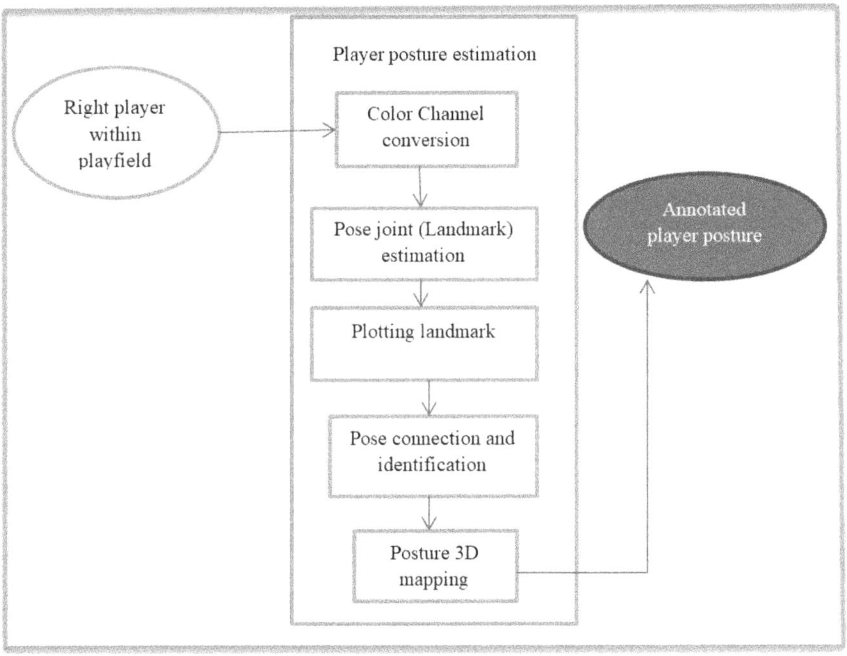

FIGURE 11.1 Workflow diagram for player posture identification.

Algorithm 1 for 3D Pose estimation over player's body

Input: Frames containing player in close contact with functional object (O_i)

Output: Skeletal keypoints of player in 3D space

1. Read the input
2. If (static == true)
3. then set det_confidence = 0.3 && mdl_complexity = 2
4. Else if (static == false)
5. then set det_ confidence = 0.5 && mdl_complexity = 1
6. While (cap.isOpened()) : #Read the video in terms of frames
7. Success, img = cap.read()
8. If not success:
9. Print ("ignoring empty frame")

```
10. End if
11. det_pose= process(input frame, conv_BGRtoRGB)
12. While det_pose_plotpoint < 33 do
13. For i = 0 to 32
14. # Player body parts are identified and corresponding spatial point values are noted
    Print (MP_pose. det_pose_plotpoint(i).label & MP_pose.det_pose_plotpoint(i).
    value)
15. End while
16. # keypoints plotted over body parts of player
    Print(MP_pose. det_pose_plotpoint(i).value *img_w) & Print(MP_pose.
    det_ pose_plotpoint(i).value *img_h) & Print(MP_pose.
    det_pose_plotpoint(i).value *img_d)
17. # Draw the detected pose plotpoint and pose connection in 3D space
    mp_draw(det_pose_plotpoint, pose_connection)
18. End if
```

Initially, the input is verified as whether it is a frame or a video using static function (step-2). When the static mode of the image is true, it is verified to be a frame and the det_confidence () function is used to check the possibility of player in the detected frame. When mode value holds false, it signifies that the input is a video (step 4). In that case, the availability of player is checked with the detection score set to 0.5. Further, the model complexity to locate the keypoints is set to 1 in videos in the view to balance the inference latency values of key points (step 5). Then, once the frames are read using step 6, the set of frames are converted from BGR to RGB for better localization of pose points as in Figure 11.2(a).

Based on the Region of Interest (ROI), the player's body parts are identified using MP_pose function (step-14). Each point projects the spatial value of the equivalent joint points (step-16) in x, y, and z axes as shown in Figure 11.2(b). The z-value represents the distance between the hip to that of its corresponding pose joint. Subsequently, the keypoint landmark is plotted in the 3D space (step-17) as shown in Figure 11.2(c) using the player pose estimation algorithm as described in Algorithm 1.

The keypoint values obtained from each input frame like that in Figure 11.3 with identified right player are stored in a CSV file for calculating angle heuristics. From the obtained pose joint point values as in Figure 11.2(d), normalization is taken place using the proposed Keypoint Deeper Cut network (KDCnet).

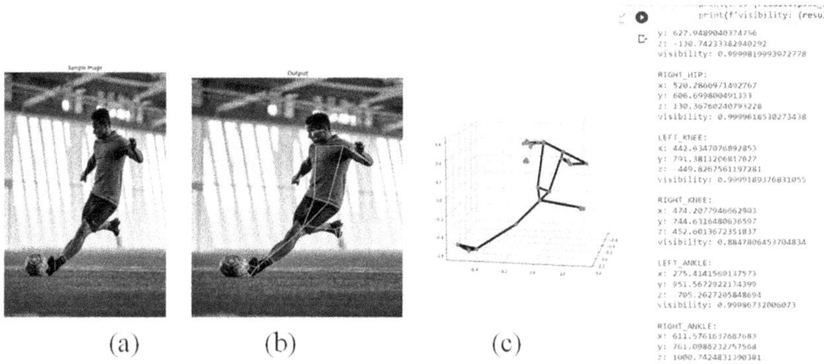

(a) (b) (c) (d)

FIGURE 11.2 Output of identified player posture in 3D space: (a) input frame, (b) plotted landmark and pose connection, (c) posture in 3D map, and (d) actual 3D keypoint values.

1	left_elbow	left_elbow	left_elbow	right_elbo	right_elbo	right_elbo	left_wrist:	left_wrist:	left_wrist:	right_wris	right_wris	right_wris	left_pinky	left_pinky	left_pinky	right_pink	right_pink	right_pink	left_index	left
2	921.6582	506.42	442.5624	907.6373	515.6384	-426.512	946.9891	557.7266	416.4862	961.5859	510.5866	-247.373	958.1801	571.421	437.6941	982.186	499.7892	-288.088	959.7766	5
3	493.3339	219.7332	-316.86	480.4427	141.2051	-142.185	393.6702	223.355	-314.644	404.7617	176.25	-160.828	368.0142	231.6422	-348.367	381.4427	181.5924	-185.74	370.0065	21
4	489.9745	255.5952	222.4752	534.042	279.0573	-25.4737	556.1385	250.9356	183.8297	564.4133	253.7951	122.5753	574.7305	245.3099	184.1093	578.9605	247.3749	116.5481	575.3886	24
5	522.5741	233.6716	-165.021	428.2612	230.9243	-532.987	499.5489	218.7998	-468.026	501.0058	220.1932	-787.553	502.5031	222.6506	-523.304	513.5831	231.0691	-855.934	486.2365	21
6	457.2122	233.3121	-39.4575	394.5865	226.7686	-132.063	413.5073	236.2851	-151.956	416.6815	225.2166	-275.994	402.3911	239.8776	-179.928	420.0326	223.9762	-309.128	400.7748	23
7	122.4784	379.1717	-82.3933	72.99736	397.6407	-133.526	137.1878	350.8951	-177.013	112.5748	407.6633	-98.2005	141.2573	343.1227	-194.224	121.9018	412.807	-107.686	141.4496	33
8	381.4012	175.3597	44.74435	545.1759	89.559	187.2539	344.7337	188.8466	29.74826	560.4362	49.8004	278.3169	335.5201	195.2615	5.46442	568.1852	41.30838	290.9692	335.6049	1
9	329.619	215.3426	245.8767	367.5418	229.8151	-79.3359	378.694	197.9822	266.7306	391.5578	210.259	28.24437	391.3159	192.2708	276.0327	399.7595	205.1788	20.53168	391.1446	18
10	202.909	239.9344	188.3445	246.2457	239.9902	-38.9446	235.652	217.0836	177.5346	248.0228	213.6846	35.66395	245.1216	209.9931	175.2821	252.9446	206.261	30.42777	243.2466	20
11	338.5347	429.1911	-491.357	231.4866	367.6738	58.19622	279.717	376.4323	-687.039	166.3715	358.8711	13.96754	262.9648	364.0444	-759.775	146.4871	353.1367	15.4674	269.8765	34
12	246.3438	192.1114	-150.028	211.0964	199.4026	81.5755	206.8162	150.3321	-99.5007	191.4627	153.3494	-91.0754	196.0528	141.8658	-117.398	185.0078	142.6544	-116.539	199.7736	13
13	549.2923	198.4379	-286.042	466.625	115.2663	48.07659	551.726	235.4237	-376.822	453.9103	103.222	155.3715	555.5708	243.604	-419.727	451.9316	96.59544	155.0062	543.7069	2
14	247.1119	270.8055	301.22	355.0554	249.5269	-206.315	260.2491	203.7638	429.7757	299.9746	163.6678	-68.6036	268.2799	183.5505	451.6785	295.3569	144.8244	-80.208	259.8462	18
15	456.1955	215.8826	-92.6918	599.9834	196.5065	184.6054	438.7882	269.4502	17.13284	609.7323	230.6928	188.0147	435.2808	282.8267	12.56436	613.4015	241.2657	199.1967	431.2298	28

FIGURE 11.3 3D keypoint values obtained from the identified right players of Football Sport.

11.3.2 *KEYPOINT DEEPER CUT NETWORK (KDCNET)*

The body part of the player is identified by its fully connected graphical structure whereas the upper and lower parts of the body are differentiated by its disconnected organization of plotted keypoints [17]. The labels and the coordinate keypoint values in the CSV file help to finalize the body parts which help to determine the actions. The time complexity used for traveling the graphical structure of a single player in the 3D space is $O(|V|+|U_E^D|)$.

The coordinate value of the identified keypoints and deeper-cut network (KDCnet) is used to map the class labels of the player body parts. The model works on three distinct properties such as uniqueness, consistency, and transitivity with Integer Liner Program (ILP) to optimize the obtained float values to single integer. This approach helps with easy mapping of keypoint value (K) to its corresponding class labels (L) of player (P) in each considered frame as in Equation 11.1.

$$P(K,K',L,L') = x(K,L)x(K',L')y(K,K') \tag{11.1}$$

Here, y is used to predict whether the parts belong to the same person or not and K, K', L, L' represent the right and left sides of the body part. After applying the ILP, if $y(K, K') = 1$ means that the obtained keypoints belong to a single person. Similarly if $x(K, L) = 1$ means the keypoint value matches its corresponding body part label. Finally, $P(K,K',L,L') = 1$ is used to relate the keypoint (K, K') and the class label (L, L') all belong to single player.

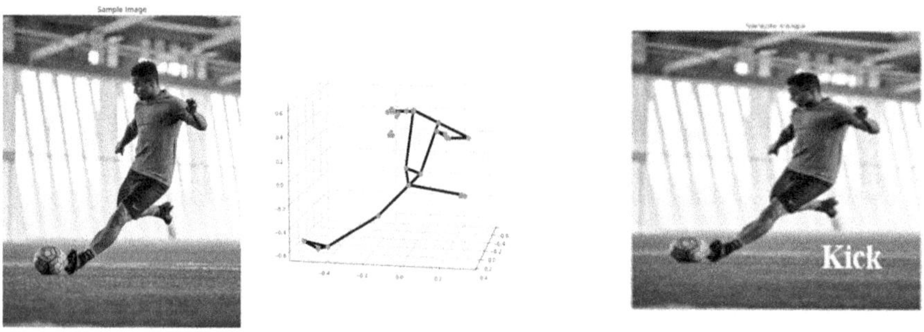

(a) (b) (c)

FIGURE 11.4 Output of the identified player posture: (a) input frame, (b) plotted pose keypoints 3D space, and (c) identified action label.

The obtained outcome pertaining to identification of posture is projected in the 3D space as in Figure 11.4(b). Likewise, to conclude a single posture nearly 5–10 frame sequences are taken into account. Thus, a posture label is derived by investigating a sequence of consecutive frames based on KDCnet. Finally, an action is concluded with the help of the derived posture labels, using the unweighted SSS algorithm. In this experiment, the traversal is conducted among nodes, which are joint points with specified index values and are not associated with any weights, except for body joint points represented by coordinate values. Therefore, an unweighted approach is employed for this research.

11.3.3 DETECTION OF POSTURE LABELS USING PROPOSED UNWEIGHTED SSS ALGORITHM

As the mapped points in the 3D space (Figure 11.3(b)) provide a graphical structure, the minimum spanning of the unweighted graph-based approach is implemented. A set of posture labels are determined using this unweighted minimum spanning tree approach. To achieve this single source shortest (SSS) path algorithm is implemented. Initially, the undirected edges (UE) are considered two parallel edges (E) which can traverse in both the directions. The traversal of the search algorithm starts from the root (R) node and in the 3D space the head of the player is considered as root element. Next, the index (I) previous to current index (C_I) is considered as parent ($C_I - I$). Thus, each vertex $\left(V(\frac{v-1}{2})\right)$ in the graph will have exactly one parent. The landmark keypoints were used to define traversal path and the search ends when there is no child node for the parent.

$$\text{Posture} = \left(U_E\left(R(V(\frac{v-1}{2})\right),(C_I - I)\right) \tag{11.2}$$

As majority of collections in the datasets (SVW and MPII) characterize closeup camera views, the coordinates of hip, knee, and ankle are used to determine the posture of 1 KneeBend (1KB). On an average about eleven postures such as 1 LegStretch (1LS), 1 FootUp (1FU), 1 KneeBend (1KB), 1 FootDown (1FD), 2 FootDown (2FD), 2 LegStretch (2LS), 1 HandMid (1HM), 1 HandUp (1HU), 1 HandDown (1HD), 2 HandUp (2HU), and 2 HandMid (2HM) are identified in football.

11.3.4 RECOGNITION OF INDIVIDUAL PLAYER ACTION

Once the player posture is mapped, then the player action is to be classified using the Long Short-Term Memory (LSTM) in Figure 11.4(c). The postures for a single player vary in every keyframe and hence the time-based variation is correctly captured by LSTM [12]. Likewise for every player in the video the sequence of posture input is fetched from the SSS unweighted MST approach and feed through the input gate. ReLU activation function is used in the conv layer and the processed output is sent bi-conv layer which consists of 64 filters with the size as 3. In the end the softmax layer confined the action class based on the probability value. The learning rate is assigned as 0.0025 and the iteration is fixed to 300 based on repetitive training. Next, the batch size is fixed with 1500 as suggested in [2].

11.4 RESULTS AND DISCUSSION

11.4.1 DATASET

The dataset used for the proposed work is SVW (Sports Video Wild) which is the standard dataset which consists of more than 85,000 videos of 44 sports. Among them the most popular sport such as football is considered for analysis. Around 4100 videos of football sport are taken among which 2280 videos are used for training and 1820 used for testing. Similarly, the MPII dataset consists of 25K images among which 5343 images with football annotated body joints were used for training and 3503 images used for testing. Initially, the videos are converted into frames at the rate of 35fps as instructed in [6].

11.4.2 TRAINING

For the proposed pose estimation, the i7 Processor (16 GB/1 TB SSD/ Windows 11 Home/NVIDIA ® GeForce RTX ¢ 4060) PHN16-71 is used. As each video available in the SVW dataset is less than 3 minutes, the frame conversion count is fixed at the rate of 35fps to capture the posture in a precise manner. The color channels are converted from RGB to BGR due to the reading ability of OpenCV on each converted frame. Next the converted frames were fed into the 3D pose estimation algorithm as in Table 11.1 which results in keypoints marked over player body in the form of skeletal in 3D space. From the obtained skeleton structure, the body parts were mapped using Keypoint-Deeper-cut model as explained in Section 11.3.2. At last, the posture labels are determined using the proposed SSS path algorithm with an unweighted minimum spanning tree approach. The novelty of the proposed work depends on decision-making strategy by concentrating only on highlighted body part labels (either upper or lower body) using the disconnected graph method.

For the football sport the possible set of poses taken in account were: 2 HandMid (2HM), 1 HandMid (1HM), 1 HandDown(1HD), 1 KneeBend

(1KB), 1 LegStretch (1LS), 2 LegStretch (2LS), 1 FootUp (1FU), 1 FootDown (1FD), 2 FootDown (2FD). At last, the action classes are determined by consolidating the identified postures from various frames using heuristics and define a single action class which the player performed in that scene. Similarly, actions such as Kick, Rolls, Run, Block, Stand, Hit, and Foul were detected from the obtained poses.

11.4.3 EVALUATION METRICS

Standard metrics such as precision, recall, f-measure, and accuracy is used to evaluate the efficiency of the entire proposed model. Initially, the SVW and MDPII dataset is used for the posture estimation process whereas player action detection is deployed in the SVW dataset. The MDPII dataset primarily comprises image data and is not well suited for action sequence detection. As the SVW dataset consists of labeled action sequence, a comparative study is carried out between the ground truth and other related models in detecting player actions. Obtained results were converted in terms of percentage and our proposed method scores higher precision and accuracy value than other models.

Also, F-measure is used to evaluate the prediction rate of the proposed model as compared with other popular models used for the action detection purpose. The obtained results are plotted as a graph and shown in Figure 11.5. As each model focuses on either upper or full-body approach, the proposed heuristic model makes the decision based on highlighted poses of player body part to distinct the work from others.

11.4.4 EVALUATION METRIC FOR POSE ESTIMATION

Two standard metrics such as Object Keypoint Similarity (OKS) and Probability of Corrected Keypoints (PCK) as in [13] were used to justify the performance of proposed work. The metrics were applied over SVW, MPII human pose dataset with fixed resized frames of 254 × 254 and applied joint lines over detected keypoints to attain the skeleton structure instead of boxes over the detected players.

$$OKS = \frac{\sum_j \exp(-E_j^2 / 2O_S^2 f_j^2)\delta(Vf_j > 0)}{\sum_j \delta(Vf_j > 0)} \tag{11.3}$$

In Equation 11.3 for every player's j^{th} keypoint, the Euclidean distance (E) is calculated between the ground-truth and identified joint points, as well as a flag (f) if set to define the visibility level of ground-truth, O_S is used to define the object scaling whereas Vf_i deals with the falloff with the ranges as 0, 1, 2. Next the normalized probability corrected keypoints (PCK) are calculated as in Equation 11.4, to fix a single point over the detected player body part irrespective of player movements.

$$\frac{\left\| X_i - \overrightarrow{X_i} \right\| 2}{\left\| \left(X_{L_{hp}} - X_{L_{sh}} \right)\left(X_{R_{hp}} - X_{R_{sh}} \right) \right\| 2} \leq t_h \tag{11.4}$$

where X_i is the ground-truth spatio point and $\overrightarrow{X_i}$ is the detected spatio point. Similarly, the torso distance is calculated by measuring the distance between the right side and the left side of the player body. $X_{L_{hp}}$ is left hip point, $X_{L_{sh}}$ left shoulder, $X_{R_{hp}}$ right hip point, and $X_{R_{sh}}$ right shoulder where the key parameters are considered for the torso calculation. Also the threshold t_h is fixed between 0 and 1 so as to finalize the point location.

To fix a single torso size of a player four keypoints were used. As torso size helps in correcting the player posture which leads to action detection, more importance is given for its calculation. Once the posture labels are identified, next the player action is classified using the proposed single source shortest (SSS) unweighted minimum spanning tree (MST) approach. In the table the comparative study is carried out among various state-of-the-art methods used for action recognition in the field of football sport.

Figure 11.5 shows the performance outcome of the proposed KM-AR framework in detecting action class when compared to other state-of-the-art methods. Among all actions, rolls and hit scores have a lower F-measure due to the use of videos with less human interaction in the dataset (SVW and MPII). As the proposed work focused toward detection of action in centric with the player posture, the two actions (rolls and hit) score 92.351% when compared with other action recognition methods. Similarly, the proposed method outperforms in other player-centric actions such as kick, run, block, stand, and foul. Since the keypoints are localized more precisely over the player body using proposed KDCnet, it helps with proper posture labels results with more accurately classified actions in football sport.

TABLE 11.1 Comparison Results of Keypoint Fixing Using Object Keypoint Similarity (OKS) Metric.

Methods	Left Shoulder	Right Shoulder	Left Elbow	Right Elbow	Left Hip	Right Hip	Left Knee	Right Knee	Left Ankle	Right Ankle	Left Heel	Right Heel	Total
Ning et al. [18]	96.3	95.1	92.2	91.6	90.6	89.6	87.6	86.5	82.7	83.3	81.7	80.7	88.2
Luvizon et al. [19]	96.6	95.4	92.0	91.5	90.6	89.6	88.0	87.3	82.7	83.2	82.0	81.7	88.4
HRNet-W32 [13]	96.3	95.0	92.0	91.3	90.3	89.7	88.1	87.6	84.1	83.1	81.7	81.9	88.5
KDCNet-ours	98.2	97.1	92.7	91.8	90.8	90.4	88.3	88.2	83.9	83.5	81.9	81.6	89.0

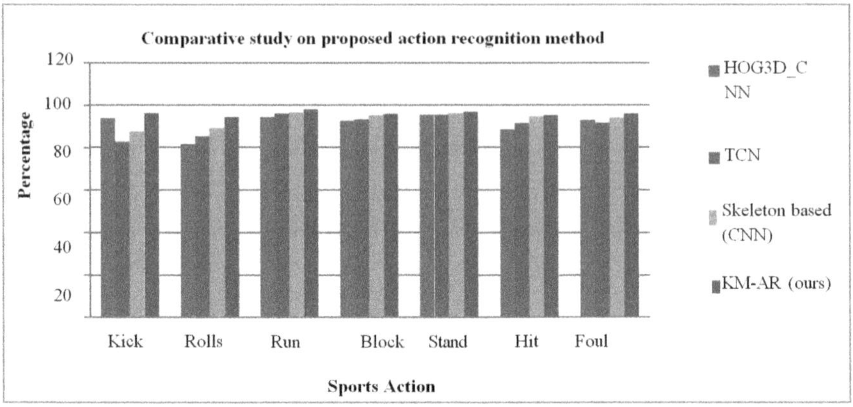

FIGURE 11.5 Comparative study on various state-of-the-art methods in recognizing player actions.

11.5 CONCLUSION

In the detection of sports action, it is essential to localize the player and their current posture from long temporal sequences such as videos. Due to change in position of the player across the entire game, it is difficult to localize the keypoints and derive the posture pattern. Hence, a single source shortest (SSS) path method is proposed which follows the unweighted graph technique along with minimum spanning tree approach in deciding the action class of every player. The novelty of this framework is the proposed Keypoint Deeper Cut network (KDCnet), which concentrates on mapping the keypoints location along with its body part besides temporal changes [21]. The probability of matching keypoints over player body is verified using the keypoint evaluation metrics against other relevant methods. The performance of the proposed framework is assessed in terms of significant methods that are used for the player action recognition in sports.

The proposed keypoint mapping method can be enhanced by incorporating occlusion culling techniques, which can improve the scene visibility level irrespective of motion blur. Rather than football, the proposed Keypoint Mapping and Action Recognition (KM-AR) framework can be applied on live stream data for multi-sport concepts.

KEYWORDS

- **keypoints**
- **posture**
- **skeletal**
- **spatio point**
- **unweighted**

REFERENCES

1. Ahad, M. A. R., Ahmed, M., Antar, A. D., Makihara, Y., & Yagi, Y. (2021). Action recognition using kinematics posture feature on 3D skeleton joint locations. *Pattern Recognition Letters*, *145*, 216–224.
2. Karunratanakul, K., Spurr, A., Fan, Z., Hilliges, O., & Tang, S. (2021, December). A skeleton-driven neural occupancy representation for articulated hands. In *2021 International Conference on 3D Vision (3DV)* (pp. 11–21). IEEE.
3. Chen, J., Samuel, R. D. J., & Poovendran, P. (2021). LSTM with bio inspired algorithm for action recognition in sports videos. *Image and Vision Computing*, *112*, 104214.
4. Brau, E., & Jiang, H. (2016, October). 3d human pose estimation via deep learning from 2d annotations. In *2016 Fourth International Conference on 3D Vision (3DV)* (pp. 582-591). IEEE.
5. Rayat Imtiaz Hossain, M., & Little, J. J. (2017). Exploiting temporal information for 3D pose estimation. *arXiv e-prints*, arXiv-1711.
6. Luvizon, D. C., Picard, D., & Tabia, H. (2018). 2d/3d pose estimation and action recognition using multitask deep learning. In *Proceedings of the IEEE Conference on Computer Vision and Pattern Recognition* (pp. 5137–5146).
7. Tekin, B., Rozantsev, A., Lepetit, V., & Fua, P. (2016). Direct prediction of 3d body poses from motion compensated sequences. In *Proceedings of the IEEE Conference on Computer Vision and Pattern Recognition* (pp. 991–1000).
8. Park, S., Hwang, J., & Kwak, N. (2016). 3D human pose estimation using convolutional neural networks with 2D pose information. In *Computer Vision–ECCV 2016 Workshops: Amsterdam, The Netherlands, October 8-10 and 15-16, 2016, Proceedings, Part III 14* (pp. 156–169). Springer International Publishing.
9. Fang, H. S., Xu, Y., Wang, W., Liu, X., & Zhu, S. C. (2018, April). Learning pose grammar to encode human body configuration for 3d pose estimation. In *Proceedings of the AAAI Conference on Artificial Intelligence* (Vol. 32, No. 1).
10. Lin, T. Y., Dollár, P., Girshick, R., He, K., Hariharan, B., & Belongie, S. (2017). Feature pyramid networks for object detection. In *Proceedings of the IEEE Conference on Computer Vision and Pattern Recognition* (pp. 2117–2125).

11. Bogo, F., Kanazawa, A., Lassner, C., Gehler, P., Romero, J., & Black, M. J. (2016). Keep it SMPL: Automatic estimation of 3D human pose and shape from a single image. In *Computer Vision–ECCV 2016: 14th European Conference, Amsterdam, The Netherlands, October 11-14, 2016, Proceedings, Part V 14* (pp. 561–578). Springer International Publishing.
12. Malik, N. U. R., Abu-Bakar, S. A. R., Sheikh, U. U., Channa, A., & Popescu, N. (2023). Cascading pose features with CNN-LSTM for multiview human action recognition. *Signals*, *4*(1), 40–55.
13. Wang, R., Huang, C., & Wang, X. (2020). Global relation reasoning graph convolutional networks for human pose estimation. *IEEE Access*, *8*, 38472–38480.
14. Wu, F., Wang, Q., Bian, J., Ding, N., Lu, F., Cheng, J., Dou, D., & Xiong, H. (2022). A survey on video action recognition in sports: datasets, methods and applications. *IEEE Transactions on Multimedia*, *25*, 7943–7966.
15. Martin, P. E., Calandre, J., Mansencal, B., Benois-Pineau, J., Péteri, R., Mascarilla, L., & Morlier, J. (2021). Sports video: fine-grained action detection and classification of table tennis strokes from videos for mediaeval 2021. *arXiv preprint arXiv:2112.11384*.
16. Rangasamy, K., As'ari, M. A., Rahmad, N. A., & Ghazali, N. F. (2020). Hockey activity recognition using pre-trained deep learning model. *ICT Express*, *6*(3), 170–174.
17. Chen, B., Meng, F., Tang, H., & Tong, G. (2023). Two-level attention module based on spurious-3d residual networks for human action recognition. *Sensors*, *23*(3), 1707.
18. Ning, G., Zhang, Z., & He, Z. (2017). Knowledge-guided deep fractal neural networks for human pose estimation. *IEEE Transactions on Multimedia*, *20*(5), 1246–1259.
19. Luvizon, D. C., Tabia, H., & Picard, D. (2019). Human pose regression by combining indirect part detection and contextual information. *Computers & Graphics*, *85*, 15–22.
20. Wu, F., Wang, Q., Bian, J., Ding, N., Lu, F., Cheng, J., Dou, D., & Xiong, H. (2022). A survey on video action recognition in sports: Datasets, methods and applications. *IEEE Transactions on Multimedia*, *25*, 7943–7966.
21. Xiang, W., Li, C., Wang, B., Wei, X., Hua, X. S., & Zhang, L. (2022). Spatiotemporal self-attention modeling with temporal patch shift for action recognition. In *European Conference on Computer Vision* (pp. 627–644). Springer.

CHAPTER 12

VIRTUAL PIANO: SIMULATING A REAL-LIFE INSTRUMENT FOR INTERACTIVE USER EXPERIENCE

LOKITA VARMA, PRATHAM BHOIR, KUSH MANIAR, URMI DEDHIA, and SINDHU NAIR

Department of Computer Engineering, Dwarkadas J. Sanghvi College of Engineering, Mumbai, India

ABSTRACT

With the rapid advancement of technology in the domain of machine learning, virtual musical instruments have gained considerable popularity, enabling individuals to engage in musical expression without the need for physical instruments. This evolution has spurred from the rampant virtualization of all technology to increase user convenience and usability. Our work focuses on designing and integrating a software to emulate a virtual piano that utilizes computer vision capabilities provided by OpenCV and the remarkable learning capabilities of deep learning models. By employing image processing techniques and convolutional neural networks (CNNs), we propose a system that can detect and track the movements of a user's fingers on a flat surface, thereby enabling them to play the piano virtually. We discuss the various stages involved in the development process, including hand detection, gesture recognition, and sound modulation and frequency additive synthesis. Experimental results demonstrate the effectiveness and accuracy of the proposed system, offering promising prospects for the future of virtual musical instruments.

Cyber-Physical Systems: Applications, Challenges, and Research Directions. Rajanikanth Aluvalu, Samarjeet Borah, & Lalit Garg (Eds.)

DOI: 10.1201/9781779643421-12

12.1 INTRODUCTION

In 2020, the entire world turned upside down with the onset of the Covid-19 pandemic. New norms were soon adapted to get things started. The concept of work from home, distance learning, online learning, and so on became increasingly popular [15]. With a lot of time on their hands, we saw people exploring their creative side and encouraging their hobbies more and more [13]. We also conducted a field survey and according to the responses we received, we saw that a large majority of the public (Figure 12.1) was keen on learning musical instruments but were often deterred by the price or other factors like maintenance and expenses. This response led us to research more on virtual instrument technologies. We started off with the piano since that was the instrument that got the most responses in our survey (Figure 12.2).

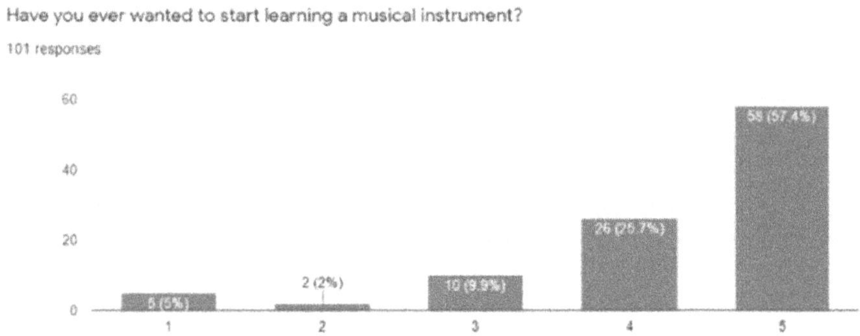

FIGURE 12.1 Majority people want to learn instruments.

Although there are other applications that claim to play music virtually, none of them give the real feel of the instrument which makes transitioning into playing the actual instrument difficult. We aimed to ease this transition by displaying the actual keys as they would be on a real piano on the screen and when the user's fingers encounter the edge of the key (as if they were to touch the keys on an actual piano), the notes are played. To implement this, we are making use of various technologies like hand recognition and tracking, matching the finger landmarks to the respective key of the piano, and sound generation. For hand recognition and tracking we are using mediapipe [11]. Mediapipe is a library developed by Google which lets one track the hand movements by mapping 21 hand landmarks. For synthetic audio generation we are using a python module named pygame mixer which allows the seamless playing of sound notes.

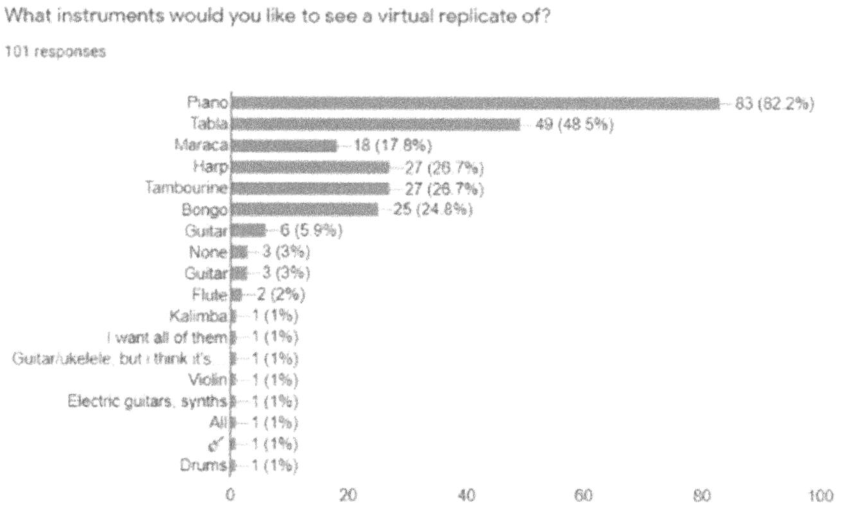

FIGURE 12.2 Piano was the most liked instrument.

12.2 RELATED WORKS

In another conference paper (Qiu et al.), the authors create and put into use a system for playing the piano based on the recognition of human posture in photos and videos. This system first identifies the human posture and then finds and locates the virtual key location. Intelligent simulation performance undertakes work to convert moves to sheet music by fusing human body recognition with the outcomes of key location. They trained a deep learning AlexNet model and achieved an accuracy of 98.36% on the testing dataset [3].

Among the various preexisting findings of real-time hand gesture detection, the approaches discussed by Oudah et al. give a broad overview of the limitations, merits and performance, as well as similarities and differences between various hand segmentation techniques [1, 8]. The paper also discusses the application of certain gesture detection algorithms based on their advantages and disadvantages. The authors grouped numerous approaches under the category of Computer Vision Approach, like Color-Based recognition. This simplistic method involves masking using OpenCV. Another interesting approach is deep learning-based recognition involving Convolutional Neural Networks like ADCNN [16], AlexNet, and VGGNet. To sacrifice real-time speed and improve performance, certain Deep Learning techniques involve two-shot detection, consisting of a hand segmentation and a detection phase followed by a gesture classification phase. These approaches significantly reduce the speed and are a dealbreaker for real-time applications.

Similarly, Zhang et al. have created an ML pipeline consisting of a one-shot BlazePalm detector for bounding boxes of the palm [9]. The bounding box landmark prediction of the previous frame is input to the next frame, which reduces the computational cost to apply BlazePalm model at each frame; this makes the approach suitable for real-time applications. The subsequent Hand Landmark model is used for precise landmark localization inside the detected bounding box region via regression.

In an alternate approach, Sung et al. discuss the hand pose detection algorithm defining two virtual key points, to describe the hand center, scale and rotation angle. Calculating the alignment vector based on these key points helps by eliminating extrinsic features like rotation, scale, and translation, thus preserving only the intrinsic features. The proposed single-shot heuristic-based classifier [10] first derives a small set of angles based on 2D key points, and then replaces those features with the ones in 3D world metric space [2, 19].

Using the Python OpenCV real-time video data processing capabilities, Ramadhan et al. have designed and built a hand gesture pattern detection module for finger segments [4, 8]. This hand-tracking module was implemented as a virtual painter, but the purpose of using Python real-time video data processing libraries like OpenCV and MediaPipe can be extrapolated to our use case.

Shriram et al. propose a solution using a real-time landmark detection library known as Mediapipe [5, 7], and create a Virtual mouse that boosts human–computer interaction by substituting gestures for physical devices. It was concluded in the study [6] that gesture control is the most frequently used alternative communication tool, along with being a more cost-effective and reliable HCI-based system. Employing human hand gesture-activated piano keystrokes enhances user interactivity making it more natural.

In another paper, Yashwanth et al. propose the use of hand segmentation and gesture recognition for finger tapping. The hand segmentation is done using the RANSAC algorithm [12, 14] and then fit in a 3D plane. For hand tracking a set of regression trees are used which give 7 points or joints of the hand. Although this method is good, it has its own limitations like not being robust enough to capture the different hand shape variations. It also gives only 7 landmarks as opposed to Mediapipe which gives 21. This method does not give the accurate position of the thumb when it is hidden under the palm. Depth sensors are also a limitation.

12.3 PROPOSED METHODOLOGY

Our proposed solution consists of mainly three phases. Identifying the user's hands and finger landmarks using MediaPipe, extracting and projecting those landmarks on the display where the user can visualize their finger movements on the piano in real-time, and a digital piano is emulated on the user's screen to interact with and play the piano notes. Lastly, playing the piano sounds that are mapped to the particular keys, and synthesizing the sound of the notes. This flow can be visualized in Figure 12.3.

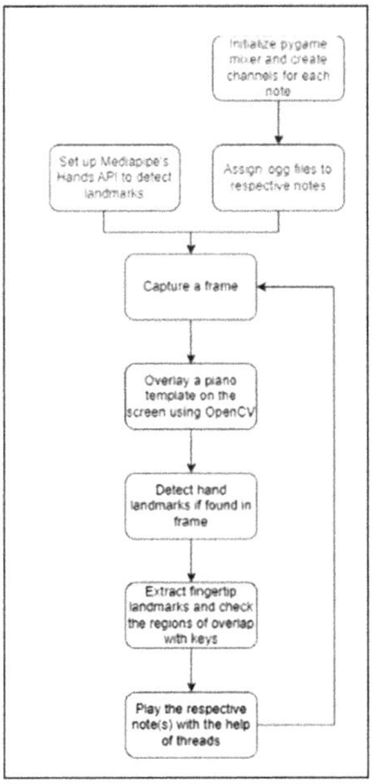

FIGURE 12.3 Proposed methodology.

12.3.1 FINGER LANDMARK IDENTIFICATION

Extracting the user's fingertip position in space in real time is challenging as it requires not only segmenting the video camera input but also comparing and analyzing frame by frame for changes in finger movements to detect keystrokes being played. We have achieved this by utilizing the MediaPipe library. Using the MediaPipe pretrained machine learning models for hand tracking, we can detect and track the user's hands, and extract the landmarks in real time [20]. These precise landmark coordinates corresponding to the fingertips, knuckles, and other critical points on the user's hands will be segmented and used to compute whether the user has played a keystroke or not. One of the key and crucial problems in computer vision is the topic of hand gesture recognition. Modern information and media technology has made it possible for human–computer interaction (HCI) systems to do hand-processing tasks including hand detection and hand gesture identification [17]. It is designed with the help of 2 ML models: the Palm Detection Model and the Hand Landmark Model. The API performs a key point localization of 21 3D hand-knuckle coordinates inside the detected hand regions via regression, that is, direct coordinate prediction.

With the help of Mediapipe's Hands API as seen in Figure 12.4, we can detect the 21-point landmark and thus we will be able to compute the distance between the fingertip landmarks and the piano key.

12.3.2 INTERACTIVE VIRTUAL PIANO VISUALIZATION

To create a simulated experience of playing the piano for the user, a virtual piano visualization phase is implemented. The piano interface is created using the OpenCV Python library. Rectangular outlines of black and white are used to represent keys of the piano, ranging from the note C4 to D5. Each note in the range is assigned to its respective key; for example, the first note (C4) was assigned to the first white rectangle, the next note (C4) was assigned to the first black rectangle that followed C4, and so on. The finger landmark coordinates obtained from MediaPipe are then reflected onto the user's display, overlaying them onto the virtual piano interface [18]. Based on the coordinates of each—those of the fingers fetched with the help of Mediapipe and those of the keys fetched with the help of the rectangles drawn using OpenCV—overlaps between any of the user's fingers and the keys are detected.

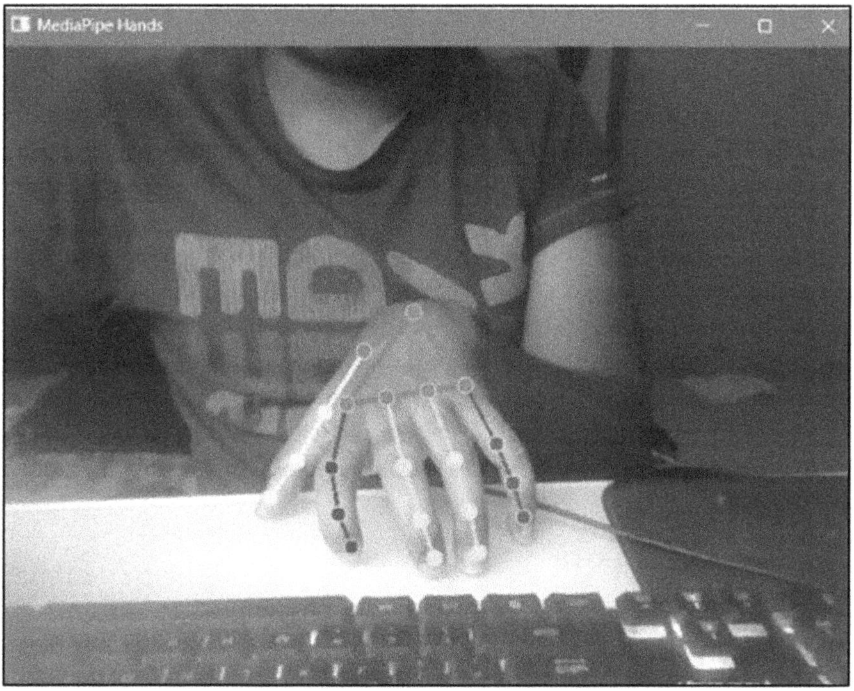

FIGURE 12.4 Hand detection view.

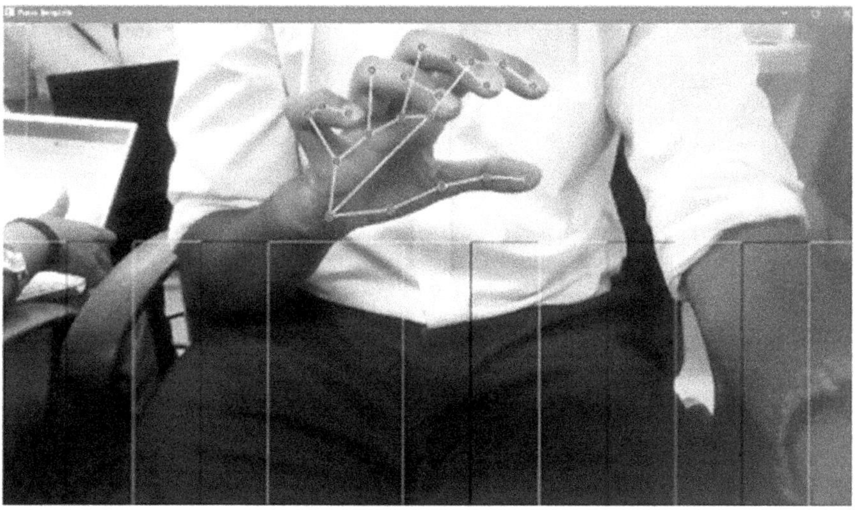

FIGURE 12.5 User interacting with virtual piano.

12.3.3 SOUND SYNTHESIS AND MODULATION

A pygame mixer is initialized to enable the use of channels. Channels are initialized to allow sustaining notes when a key is held, handling simultaneous key presses. The sounds for each note are played using prerecorded OGG files. The other approach is to synthesize the sound from the sine wave itself. However, this approach results in an increased amount of white noise. Prerecorded OGG files allow us to retain the authenticity of the piano notes. The OGG format is preferred over MP3 because of its features such as smaller file size, higher bitrate, and overall audio quality. When the finger overlap over a key is detected, the pygame mixer plays the sound in a separate thread and assigns it to a channel. This allows the complete note to play and thus creates more immersive music by overlapping several notes played simultaneously in separate threads.

12.4 REQUIREMENTS

The software mandates specific system requirements, which comprise an operating system of Windows 11, a central processing unit (CPU) equipped with an Intel(R) Core(TM) i7-10750H processor working at a frequency of 2.60 GHz, a minimum of 24 GB of random access memory (RAM), and a graphics processing unit (GPU) with a minimum specification of an NVIDIA GeForce GTX 1650 Ti. In terms of software requirements, the best performance and compatibility of the system depend on the installation and correct functioning of Python 3.9.5, Mediapipe version 0.8.9.1, OpenCV-Python version 4.5.5.62, and Pygame version 2.4.0.

12.5 RESULTS AND DISCUSSION

12.5.1 DELAY OF VIRTUAL SOUND SYNTHESIS

This section presents the research evaluation results, specifically focusing on the measured sound playback delay. The main goal was to evaluate the effectiveness of the virtual piano and identify any possible sound production delays, which are crucial for user experience in interactive musical applications. The Python programming language's time module was utilized to measure the duration between the key press event and the subsequent sound generation event in the virtual piano. The sound playback delay was determined by measuring the time interval between the key press and the subsequent auditory perception of the corresponding sound. Each note played in a single key scenario was measured. Each key was subjected to the same process, and the average delay was determined for each type of musical sequence. The study's findings indicate that the virtual piano showed minimal sound playback delays for single-key sequences. The mean delay for individual key presses was found to be 0.013806 seconds, with a standard deviation of 0.004076 seconds. The results suggest that the virtual piano effectively produces sound in a timely manner, allowing users to have a musical experience that closely resembles real-time. It is crucial to recognize the study's limitations. The evaluation was conducted on a specific computer system with specific hardware and software configurations. The results may vary across systems with differing performance capabilities. Furthermore, the participant sample size was limited, suggesting the need for future studies with a larger and more diverse pool of participants to improve the generalization of the results. In summary, the Python-based virtual piano exhibited favorable outcomes regarding sound playback latency. The measured delays for both single key and chord sequences were within acceptable limits, resulting in a satisfactory user experience. Future research could focus on optimizing the performance of the virtual piano on different hardware configurations and expanding the evaluation to include a larger sample of participants.

12.5.2 ANALYSIS OF MULTIPLE CHORDS

We sought the assistance of ten accomplished pianists who voluntarily offered to test the proposed system. Five different chords—A major, C major, D major, E major, F major, and G major—were required to be played by each pianist. These chords are made up of three keys being played concurrently. These chords were chosen based on how frequently they were played by pianists. We documented and examined the outcomes of these testing sessions in order to evaluate the system's performance. Each trial's outcome was labeled as either a "success" or a "failure." A trial was deemed successful if all of the chord's keys were played almost instantly. A trial was considered a failure, however, if one or more chord keys were either not played at all or played with a pronounced delay that could be heard by listening to the auditory output.

The comprehensive findings of this experiment are shown in Table 12.1, which also shows the number of both successful and failed trials for each chord. An average accuracy of 88.30% was discovered after the evaluation of data, incorporating every experiment. This indicates that the pianists used the approach to play the chords successfully almost 88.30% of the time. This shows how well the system works in identifying and recording common piano chords, with a high success rate in precisely identifying the simultaneous key presses made by skilled pianists.

TABLE 12.1 Time Delay in Playing Keys.

Key	Time Delay (Seconds)
C4	0.017991
C#4	0.016033
D4	0.017042
D#4	0.012507
E4	0.014133
F4	0.023046
F#4	0.012923
G4	0.011564
G#4	0.013032
A4	0.010513
A#4	0.009965
B4	0.006927
Average delay	0.013806
Standard deviation	0.004076

12.6 LIMITATIONS

As one can see from the results section, we have some limitations with respect to the system. Currently, we tested our system on only a Windows PC and are getting the respective results. Testing with respect to other operating systems like MacOS and Linux may lead to different results. Another significant limitation faced during this research was that the results were concluded with a small participant population. This can be thoroughly improved in future studies. The MediaPipe library used for landmark detection for the hands also was not reliable in every scenario. As the whole system is a 2D-camera input-based system, some angles where the image is captured fail to correctly detect the landmarks for the hands and therefore, cause discrepancies.

TABLE 12.2 Accuracy Analysis of Playing Different Chords.

Sr. No.	Chord	Keys	Success	Failure	Accuracy
1.	A Major (A)	A - C# - E	8	2	0.80
2.	C Major (C)	C - E - G	10	0	1.00
3.	D Major (D)	D - F# - A	9	1	0.90
4.	E Major (E)	E - G# - B	7	3	0.70
5.	F Major (F)	F - A - C	10	0	1.00
6.	G Major (G)	G - B - D	9	1	0.90
–	–	–	53	7	Average accuracy: 0.883

12.7 FUTURE SCOPE

The application has scope to support multiple users simultaneously and create a digital virtual jam room where different users can collaborate and play different virtual instruments. Advanced sound synthesis [22] techniques such as physical modeling synthesis [21] or sample-based synthesis will be explored in the future. This would allow for more realistic and nuanced piano sounds, improving the overall quality of the virtual piano experience. It would also allow the user to change the piano sounds by adjusting and tuning the frequencies of the notes as per the user's preference. For a more user-friendly interface and experience, the GUI (Graphic User Interface) can be improved significantly to make it more intuitive and improve user experience.

KEYWORDS

- **virtual instrument**
- **music synthesis**
- **computer vision**
- **hand detection**
- **fingertip segmentation**
- **MediaPipe**

REFERENCES

1. Zhang, F., Bazarevsky, V., Vakunov, A., Tkachenka, A., Sung, G., Chang, C. L., & Grundmann, M. (2020). Mediapipe hands: On-device real-time hand tracking. *arXiv preprint arXiv:2006.10214.*
2. Ismail, A. P., Abd Aziz, F. A., Kasim, N. M., & Daud, K. (2021, February). Hand gesture recognition on python and opencv. In *IOP Conference Series: Materials Science and Engineering* (Vol. 1045, No. 1, p. 012043). IOP Publishing.
3. Qiu, J., Wu, X., Chen, S., & Zhang, Y. (2020). Piano playing simulation system based on human posture recognition. In *2020 International Conference on Culture-Oriented Science & Technology (ICCST)* (pp. 39–44). IEEE.
4. Ramadhan, G., Wiratama, J., & Permana, A. A. (2022). Development of virtual painting method using OpenCV library with finger gesture on online learning platform. *JIKA (Journal Informatika)*, *6*(3), 332–341.
5. Shriram, S., Nagaraj, B., Jaya, J., Shankar, S., & Ajay, P. (2021). [Retracted] Deep learning-based real-time AI virtual mouse system using computer vision to avoid COVID-19 spread. *Journal of Healthcare Engineering*, *2021*(1), 8133076. https://doi.org/10.1155/2021/8133076
6. Katona, J. (2021). A review of human–computer interaction and virtual reality research fields in cognitive InfoCommunications. *Applied Sciences*, *11*(6), 2646. https://doi.org/10.3390/app11062646
7. Chavali, E. S. (2023). Virtual mouse using hand gesture. *International Journal of Scientific Research in Engineering and Management, 7*, https://doi.org/10.55041/IJSREM21501.
8. Oudah, M., Al-Naji, A., & Chahl, J. (2020). Hand gesture recognition based on computer vision: a review of techniques. *Journal of Imaging*, *6*(8), 73. https://doi.org/10.3390/jimaging6080073
9. Zhang, F., Bazarevsky, V., Vakunov, A., Tkachenka, A., Sung, G., Chang, C. L., & Grundmann, M. (2020). Mediapipe hands: On-device real-time hand tracking. *arXiv preprint arXiv:2006.10214.*
10. Sung, G., Sokal, K., Uboweja, E., Bazarevsky, V., Baccash, J., Bazavan, E. G., & Grundmann, M. (2021). On-device real-time hand gesture recognition. *arXiv preprint arXiv:2111.00038.*
11. Kavana, K. M., & Suma, N. R. (2022). Recognization of hand gestures using mediapipe hands. *International Research Journal of Modernization in Engineering Technology and Science*, *4*(06).
12. Thottathil, I. A., & Thivaharan, S. (2023). Virtual musical instruments with python and OpenCV. *Journal of Ubiquitous Computing and Communication Technologies*, *5*(1), 1–20.
13. Adhikari, A., Rao, T. S., Kar, K., & Dhruw, T. (2022). Computer vision based virtual musical instruments. *Mathematical Statistician and Engineering Applications*, *71*(4), 9600–9608.
14. Yashwanth, G., Khan, S., Reddy, T., & Varsha, A. (2020). The literature survey on virtual piano. international journal of engineering research and. V9. https://doi.org/10.17577/IJERTV9IS040405.

15. Ismail, M. J., Anuar, A. F., & Loo, F. C. (2022). From physical to virtual: A new learning norm in music education for gifted students. *The International Review of Research in Open and Distributed Learning, 23*(2), 44–62.
16. Yao, J., Wang, D., Hu, H., Xing, W., & Wang, L. (2022). ADCNN: Towards learning adaptive dilation for convolutional neural networks. *Pattern Recognition, 123*, 108369.
17. Lv, Z., Poiesi, F., Dong, Q., Lloret, J., & Song, H. (2022). Deep learning for intelligent human–computer interaction. *Applied Sciences, 12*(22), 11457. https://doi.org/10.3390/app122211457
18. Zhou, Z., & Lv, Y. (2023, May). The application of gesture recognition based on MediaPipe in virtual scene. In *Third International Conference on Machine Learning and Computer Application (ICMLCA 2022)* (Vol. 12636, pp. 403–409). SPIE. https://doi.org/10.1117/12.2675148.
19. Zanfir, A., Bazavan, E. G., Xu, H., Freeman, W. T., Sukthankar, R., & Sminchisescu, C. (2020). Weakly supervised 3d human pose and shape reconstruction with normalizing flows. In *Computer Vision–ECCV 2020: 16th European Conference, Glasgow, UK, August 23–28, 2020, Proceedings, Part VI 16* (pp. 465–481). Springer International Publishing.
20. Veluri, R. K., Sree, S. R., Vanathi, A., Aparna, G., & Vaidya, S. P. (2022, March). Hand gesture mapping using MediaPipe algorithm. In *Proceedings of Third International Conference on Communication, Computing and Electronics Systems: ICCCES 2021* (pp. 597–614). Springer Singapore.
21. Hawley, S. H., Chatziiannou, V., & Morrison, A. (2020). Synthesis of musical instrument sounds: Physics-based modeling or machine learning. *Phys. Today, 16*(1), 20–28.
22. Huzaifah, M., & Wyse, L. (2021). Deep generative models for musical audio synthesis. *Handbook of artificial intelligence for music: foundations, advanced approaches, and developments for creativity*, 639–678.

CHAPTER 13

EXPLORING GLOBAL PROSPECTS AND HURDLES IN INTEGRATING VEHICLE-TO-GRID (V2G) TECHNOLOGY

ANUBHAV AGRAWAL[1], PRADEEP KUMAR ARYA[2], RANBIR SINGH[1], and ANANYA AGGARWAL[1]

[1]*BML Munjal University, Gurugram, Haryana, India*

[2]*Symbiosis Institute of Technology, Pune, Maharashtra, India*

ABSTRACT

The rapid growth of electric vehicles has transformed the automotive industry and catalyzed a global transition toward sustainable transportation. The integration of electric vehicles with the power grid through vehicle-to-grid technology has become an emerging realm of research. Using the capabilities of the Internet of Things along with smart grid technology, electric vehicles are likely to transform into versatile portable energy storage units. This research article investigates the latest trends in electric vehicle technology, with a specific focus on vehicle-to-grid integration. Through an extensive review of literature and bibliometric analysis of Web of Science database, this study unveils the various prospects associated with V2G integration, such as grid stability enhancement and facilitation of renewable energy integration. Additionally, the paper highlights the hurdles that hinder seamless implementation of V2G technology, including energy management, regulatory complexities, and

Cyber-Physical Systems: Applications, Challenges, and Research Directions. Rajanikanth Aluvalu, Samarjeet Borah, & Lalit Garg (Eds.)

DOI: 10.1201/9781779643421-13

infrastructure requirements. By shedding light on these dimensions, the paper offers valuable insights for policymakers, industry stakeholders, and researchers in navigating the path toward effective V2G implementation on a global scale.

13.1 INTRODUCTION

Electric vehicle (EV) technology has witnessed significant advancements in recent years, transforming the automotive industry and promoting sustainable transportation solutions. Electric vehicle (EV) is among the most accepted technological solutions to deal with of climate change [1] and dependency on fossil fuels to promote sustainability in transportation [2] sector. One of the key advantages of electric vehicles is their potential to significantly reduce greenhouse gas emissions and air pollution from transportation sector [3]. The increasing contribution of renewable energy sources to the grid supply will further reduce the emissions to the environment. EVs are powered by electric motors driven by energy stored in rechargeable batteries. Advancements in battery technology [4] have played a significant role in adoption of electric vehicles across the globe. Battery technology is accepted across the transportation sectors, including commercial trucks, buses, cars, motorcycles, and even bicycles. Scalability of battery technology has enabled its wider acceptance [5].

Battery technology is core of electric vehicle development. Advances in battery chemistry are increasing the energy density with durability and performance of the batteries. 98% of high-energy density battery demand corresponds to EVs currently [6]. This is a large energy storage infrastructure available and will grow rapidly in the future at a high rate. The emergence of solid-state batteries has shown new path of progress. The vision of cleaner, greener, and sustainable energy storage systems can be achieved in a short time by using the available EV battery infrastructure. By lever-aging the advantages of electric vehicle technology and addressing the challenges, the transition in transportation sector can lead the growth in the sales of EVs exponentially. This exponential growth will then ensure availability of more energy storage infrastructure to assist grid to meet peak requirements and enable a sense of energy management.

Technologies related to EVs like battery technology, charging infrastructure, range optimization, vehicle-to-grid integration, autonomous driving, and connected vehicles need to be critically analyzed to identify the present trends, challenges, and future research directions. Critical analysis of the current state of research and development in these areas will help to provide insights of latest advancements and future directions in EV technology. This research article presents on such a critical analysis for vehicle-to-grid integration. Vehicle-to-grid (V2G) integration is the concept of bidirectional power flow between EV and the electrical grid, technically referred to as V2G technology as shown in Figure 13.1. Grid stabilization, demand response management, and renewable energy integration [7] are some technological dominating areas of V2G technology.

V2G systems offer the opportunity for EV owners to monetize their vehicle's energy storage capabilities [8]. Researchers and experts in addition to government initiatives need to address the challenges related to grid infrastructure, regulatory frameworks, and user participation [9]. V2G integration ensures grid stability and peak load management through the power exchange between EVs and the grid [10] offering capabilities of frequency regulation, voltage support, and grid balancing in addition to voltage regulation and reactive power support [11]. Figure 13.2 demonstrates word cloud of V2G, developed by bibliometric analysis of articles from 1993 to 2023 from WoS.

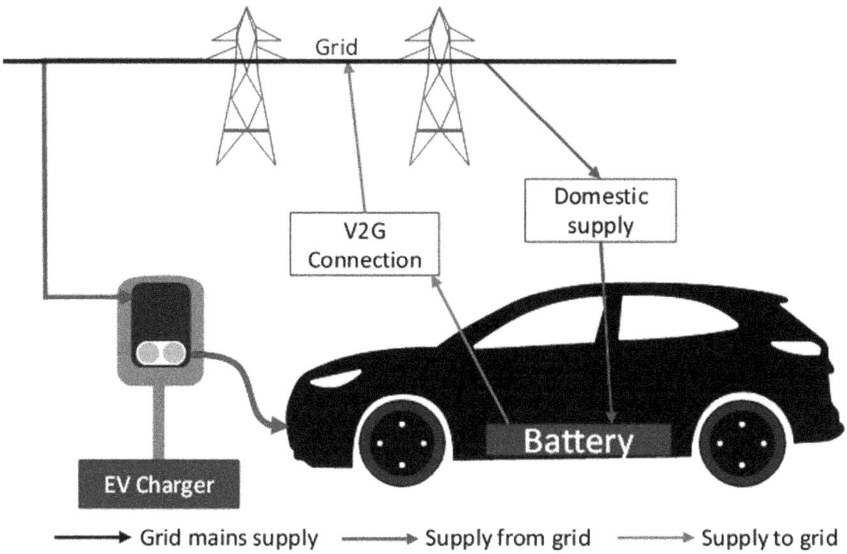

FIGURE 13.1 The concept of V2G system.

FIGURE 13.2 Word cloud of vehicle to grid (V2G) through bibliometric analysis.

Advanced control algorithms have capabilities of seamless integration of EVs into the grid for smoother operation of the power supply [12–13]. This integration is like intermittent renewable energy sources [14]. The integration technology will surely enable the EV and the grid to ensure power capacity with a bidirectional flow of power at a faster response time. Economic incentives and regulatory frameworks are necessary to incentivize EV owners to participate in grid services [14].

13.2 STATE-OF-THE-ART ANALYSIS

Grid infrastructure requirements are crucial for successful V2G integration and efficient provision of grid services. Charging and discharging of EVs have a direct impact on grid stability, power quality, and distribution system. The accommodation of bidirectional power flow requires infrastructure upgrades and modifications such as distribution transformers, smart meters, and communication systems [15]. The integration of renewable energy sources and energy storage systems with the V2G technology enhances grid reliability and supports the integration of intermittent renewables [16] along with optimizing the energy flow and enhancement of system flexibility [17]. Table 13.1 presents the details of the bibliometric data on V2G.

TABLE 13.1 Summary of Bibliometric Data on V2G.

Parameter	Timespan	Sources (Journals, Books, etc.)	Documents	Annual Growth Rate %	Average citations/ doc	Authors
Results	2003:2023	152	550	19.58	48.87	1555

A total of 550 documents from 152 sources revealed in the WoS database in the bibliometric analysis from 2003 to 2023. The research is growing at an annual growth rate of about 20%. A high citation rate of 50 citations per document reveals the future of the research in the selected domain of V2G technologies. Figure 13.3 demonstrates the annual scientific production of research articles in V2G technology (2003–2023).

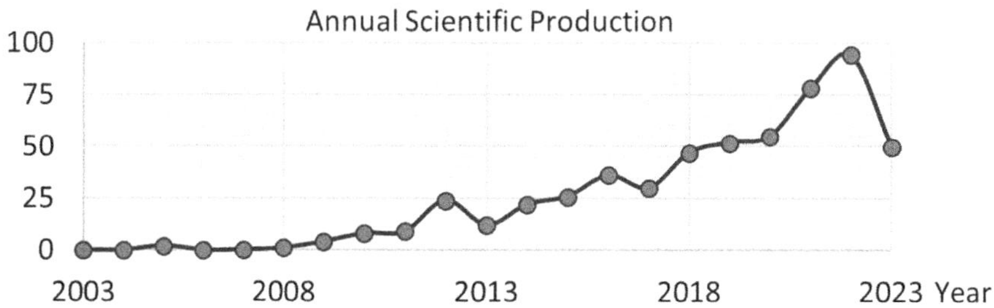

FIGURE 13.3 Annual scientific production of articles in V2G technology (2003–2023).

Electric vehicles, optimization, management, integration, impact, system, energy, power, systems, model, V2G, renewable energy, design, demand, strategy, operation, storage, cost, algorithm, and capacity are top 20 words that are occurred in the bibliometric analysis from 2003 to 2023 from the WoS database. Figures 13.4 and 13.5 demonstrate the cumulative frequency and tree diagram of major keywords over time for the most trending words, respectively.

The exchange of information and control signals between EVs, charging infrastructure, and the grid is facilitated by communication protocols. Developing standardized communication protocols, such as OpenADR, OCPP, and ISO/IEC 15118, shall enable interoperability and seamless integration between different stakeholders [18]. The communication requirements for efficient scheduling and coordination of V2G transactions also include real-time data exchange, demand forecasting, and billing systems [19]. Data privacy mechanisms must be integrated with secure communication protocols to protect EV owners' information and ensure grid security [20]. Table 13.2 presents the comparative analysis of communication protocols for V2G.

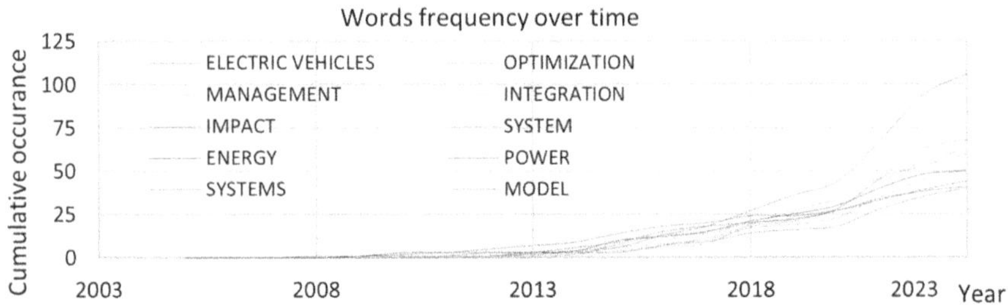

FIGURE 13.4 Cumulative frequency of major keywords over time.

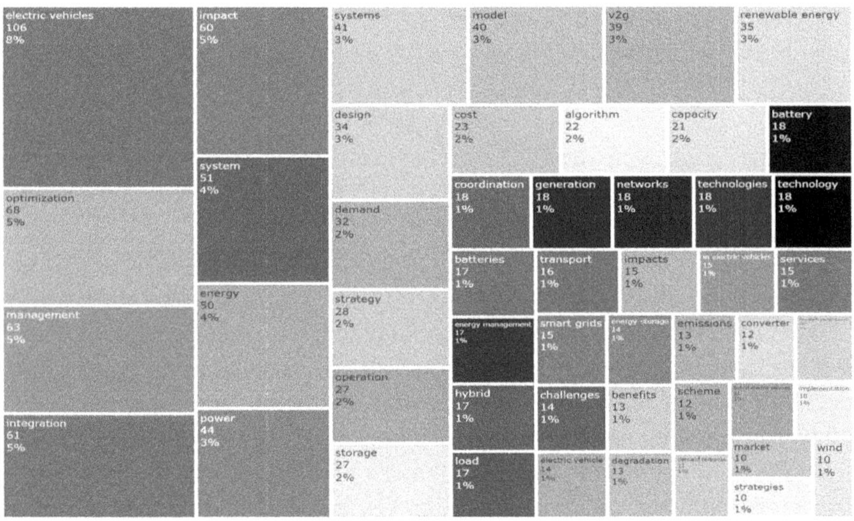

FIGURE 13.5 Tree diagram of the trending words.

The development of advanced energy management strategies, such as optimal power flow algorithms, distributed control systems, and intelligent algorithms, plays a crucial role in optimizing the utilization of EVs for grid services and in maximizing the economic and environmental benefits of V2G integration [21]. System performance evaluation provides insights into the effectiveness and benefits of V2G integration and grid services.

TABLE 13.2 Comparative Analysis of Communication Protocols for V2G.

Parameter	Open Automated Demand Response (OpenADR) Protocol	Open-Source Communication (OCPP) Protocol	ISO/IEC 15118 Protocol
Developed by	Demand Response Research Center	Developed by ElaadNL	ISO and IEC
Developed in	2009	2009	2010
Published in	2009	2010	2010
Purpose	Standardize, automate, and simplify demand response	Security and smart charging of networked EV chargers	Integrate EVs into smart grid
Type	Paid	Open source	Paid
Application	Energy management	Monitor charging station	Cryptographic mechanisms
Approach	Manage information and signals	Share data between EV chargers and charging station management	Secure confidentiality, integrity, and authenticity of all communications
Methodology	Manage peak load demands by turning off electrical devices during high demand	Manage complex multivendor charging stations	Enable EV to identify charging facility and get authorized access for charging
Security level	Secure information exchange	Secure firmware, logging, event notification, and profiles	Security through cryptography in plug and charging
Security type	Two-way information exchange	Client-side certificates and TLS authentication	Digital certificates and public-key infrastructure
Type of information exchange	No information exchange between EV and charging station	No information exchange between EV and charging station	Dynamic exchange of info between EV and charging station
Charging rates	No information on charging rates and tariffs	Provide information regarding rate and tariff	(Re)negotiate charging rates
Scheduling	No	No	Yes
Information about demand and availability of electricity	Does not provide such information	Does not provide such information	Provides information about state of grid, EV energy demand, and mobility needs

The impact of V2G on grid stability, power quality, and system reliability is evaluated through simulation studies, field trials, and real-world deployments [22]. Factors such as market price signals, incentive mechanisms, and revenue sharing are used to determine the economic benefits of V2G participation for EV owners, grid operators, and electricity markets [23]. V2G integration provides environmental benefits such as emission reductions and increased renewable energy penetration [24].

13.3 IMPLEMENTATION OF V2G TECHNOLOGY

The implementation of grid services in V2G integration poses various challenges such as grid compatibility, market barriers, regulatory frameworks, charging infrastructure availability, and customer acceptance that need to be addressed [25]. Solutions such as grid reinforcement, demand response programs, market design modifications, and policy incentives are introduced to overcome these challenges [26]. The development of interoperability standards, cybersecurity protocols, and data management frameworks ensures the security and reliability of V2G operations [27]. Figure 13.6 represents the outline of the power demand and supply network for EVs.

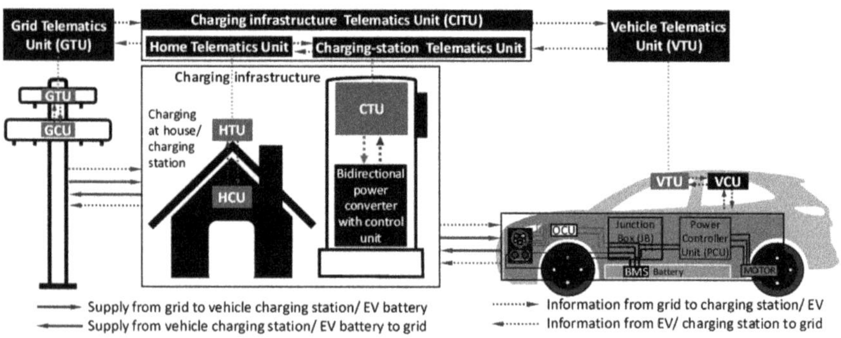

FIGURE 13.6 Outline of power demand and supply network for EVs.

Integration of grid services played a vital role in maximizing the benefits of V2G integration and adoption of renewable energy sources. Recent developments in grid services for V2G integration had majorly focused on services provided by EVs, grid infrastructure requirements, V2G communication protocols, energy management, and system performance evaluation. For efficient V2G integration grid infrastructure upgrades are necessary along with the grid services provided by the EVs. Interoperability and seamless integration have been facilitated by standardized communication protocols. Modern energy management strategies enhance utilization of EVs for grid services. Researchers can improve grid stability, promote the integration of renewable energy sources, and maximize the use of EVs for grid services by comprehending and advancing grid services in V2G integration.

Power flow control mechanisms enable efficient and reliable bidirectional energy exchange between EVs and the grid. Power flow control strategies, such as voltage regulation, power factor control, and reactive power support, have capabilities to ensure grid stability and quality. Integration of advanced power electronics devices like bidirectional converters and DC–DC converters can enable seamless energy exchange and power flow control [28]. The control algorithms and communication protocols are suitable for coordination between power flow control among multiple EVs [29].

Optimization algorithms like mathematical programming, evolutionary algorithms, and game theory have capabilities to minimize energy costs, optimize grid services, and satisfy user constraints [30]. Multiobjective optimization has possibilities to ensure grid stability with optimum EV preference and economic incentives [31]. Adding real-time data of grid conditions, electricity prices, and EV availability open the paths for dynamic optimization and decision-making in V2E [32]. There is a strong need for development of interoperability standards, cybersecurity protocols, and data management frameworks for secure and reliable V2G operations [33].

13.3.1 ENERGY MANAGEMENT IN V2G TECHNOLOGY

Energy management in V2G is possible with application of several different communication protocols. Figure 13.7 presents various communication protocols at different levels.

An efficient energy management is necessary for effective V2G integration. Researchers have developed intelligent systems to optimize the charging and discharging schedules of EVs based on grid conditions, energy prices, and user preferences. The efficiency of the V2G system depends on the coordination between EVs as energy storage systems, renewable generation, and the grid [34]. Figure 13.8 depicts concept of energy management with V2G connectivity.

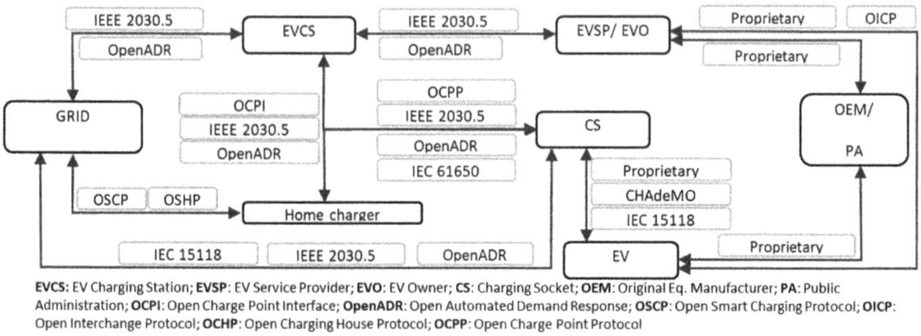

FIGURE 13.7 Vehicle-grid integration ecosystem and communication protocols.

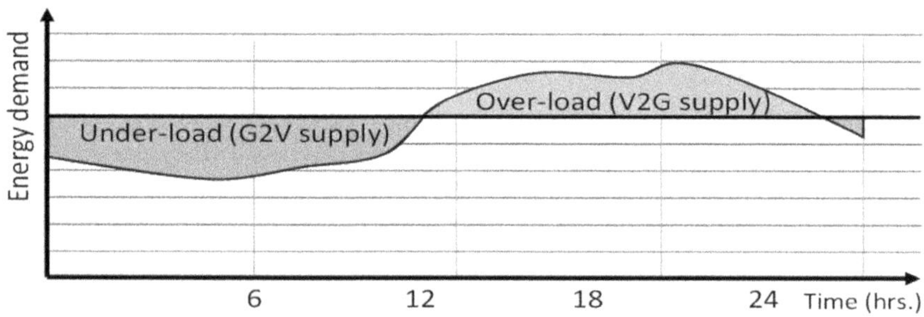

FIGURE 13.8 Energy management with V2G connectivity.

Energy management optimizes the utilization of EV batteries, catering the grid and transportation needs. Hierarchical control architectures have capabilities to coordinate charging and discharging of operations with focus on central control, aggregate level control, and individual vehicle-level control [35]. The integration of V2G with smart grid infrastructure and renewable energy sources can enhance flexibility and renewable energy integration [17]. Integration of V2G with demand response programs and energy market mechanisms enables efficient energy trading and develops more optimized charging–discharging strategies with focus on state of charge (SoC), grid demand, electricity prices, and EV user preferences [36]. Coordinated scheduling of charging and discharging operations balancing the grid load optimizing the energy flow and minimizing the cost has been worked in the past [37].

13.3.2 REGULATIONS FOR V2G TECHNOLOGY

The policy frameworks and market mechanisms of incentivization for EV owners need to be developed to promote V2G programs. Regulatory frameworks address issues related to tariff structures, grid access, liability, and standards for communication protocols and interoperability. Aggregators acting as intermediaries between EV owners and grid operators need to be promoted to ensure the larger participation of EVs EV owners in V2G programs. Supportive policies like mandates, incentives, and subsidies are useful in the promotion of V2G deployment and adoption [22]. Researchers need to develop grid regulations and technical requirements to address grid interconnection standards, power quality requirements, and operational constraints [23]. Interoperability and international standards [18], legal barriers [38] in existing energy and transportation laws, regulations, and licensing requirements also need to be addressed [39].

13.4 CHALLENGES AND FUTURE RESEARCH DIRECTIONS FOR V2G TECHNOLOGY

For the deployment to be successful, several technical issues with V2G integration must be resolved. Bidirectional power flow control, battery degradation management, communication protocols, and cybersecurity are just a few of the subjects being researched [11]. New control algorithms and procedures allow safe and effective power exchange between EVs and the grid while lessening the influence on battery life [40]. To ensure seamless integration and compatibility, studies also examine how V2G systems interact with various grid infrastructures, EV models, and charging standards [41]. For the implementation and operation of V2G systems to be successful, a few technological issues with V2G integration must be resolved. The technical difficulties of V2G integration are thoroughly examined in this literature review, with focus on power electronics, communication protocols, grid integration, battery management, and system optimization. To advance V2G technology, ensure grid stability, and maximize advantages of V2G integration, it is essential to comprehend these difficulties.

Power electronics are a crucial component of V2G integration because they enable efficient energy conversion between grid and EV batteries as well as bidirectional power flow. To enable the seamless integration and control of V2G systems, research is concentrated on developing advanced power electronics systems. Research is also looking into how to combine power electronics with energy storage systems, renewable energy sources, and grid management systems to enhance the performance and dependability of V2G systems [17].

Communication protocols are essential for smooth and secure communication between V2G components, such as EVs, charging infrastructure, aggregators, and grid operators. The purpose of the project is to develop real-time data sharing, authentication, and encryption communication protocols for V2G systems [18]. Studies investigate V2G integration methods and communication interfaces outlined in ISO/IEC 15118 and IEC 61850 standards [38]. Research is also looking toward the integration of wireless communication technologies like cellular networks and WiFi to provide dependable and efficient communication in V2G systems [38].

Voltage management, power quality, and grid stability during V2G operation are being researched. The ability to ride through faults, voltage profile analysis, and load flow analysis are some of the methods examined in studies for figuring out how the grid influences the grid [42]. To facilitate the seamless integration of V2G systems into the grid, research also examines the development of grid management systems, such as demand response programs, distribution automation, and smart grid technologies [23]. Battery management is essential for V2G integration to maintain the longevity, security, and functionality of EV batteries. To maximize battery performance and minimize deterioration, research is focused on creating battery management systems (BMS). Studies have examined estimate approaches, control tactics, and battery modeling techniques to precisely forecast and control battery behavior during V2G operation [43]. To increase the functionality and adaptability of V2G systems, research is also looking at fast-charging and high-energy-density batteries [44].

To fully utilize the advantages and effectiveness of V2G integration, system optimization is essential. For effective energy management, load balancing, and grid services in V2G systems, research is focused on creating optimization algorithms, control techniques, and scheduling methodologies. Studies investigate multiobjective optimization methods that account for factors such as energy prices, grid stability, consumer preferences for EVs, and environmental effects. Research is also looking into the integration of real-time data, such as grid conditions, power prices, and EV availability, to assist dynamic optimization and decision-making in V2G systems [32]. Figure 13.9 shows challenges in implementation of V2G technology in the present scenario.

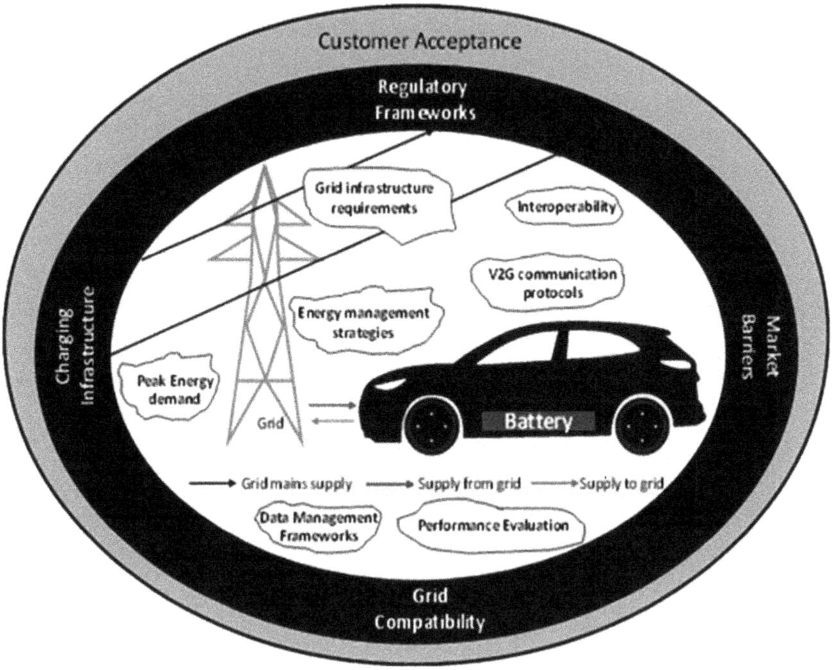

FIGURE 13.9 Challenges in implementation of V2G technology.

13.5 CONCLUSION

There are technological challenges that must be overcome for V2G integration. Identifying and resolving problems including battery deterioration, grid compatibility, communication reliability, and power electronics efficiency are the main goals of research. Better power electronics architectures, dependable communication protocols, grid management strategies, battery health monitoring, and optimization algorithms are some of the solutions that are the subject of studies to address these issues. Development of testing and validation procedures, interoperability standards, and technological advancements ensure the dependability and performance of V2G systems.

Power electronics, communication protocols, grid integration, battery management, and system optimization are among the technological issues that V2G integration faces. The development of effective power electronics systems, dependable communication protocols, grid-compatible integration techniques, sophisticated BMS, and ideal system optimization algorithms are main recent advances in these fields. Interdisciplinary research, partnerships between business and academia, and improvements in power electronics, communication technology, grid management systems, battery technology, and optimization algorithms are all necessary to meet these problems. Researchers, policymakers, and stakeholders may support the effective deployment and operation of V2G systems, guarantee grid stability, and maximize the benefits of V2G integration by comprehending and tackling the technological hurdles in V2G integration.

KEYWORDS

- **bibliometric analysis**
- **electric vehicle**
- **smart grid**
- **telematics**
- **vehicle to grid**

REFERENCES

1. Global EV Outlook. (2022). *IEA, Paris.* https://www.iea.org/reports/global-ev-outlook-2022, License: CC BY 4.0.
2. Zimm, C. (2021). Improving the understanding of electric vehicle technology and policy diffusion across countries. *Transport Policy*, *105*, 54–66.
3. Alanazi, F. (2023). Electric vehicles: Benefits, challenges, and potential solutions for widespread adaptation. *Applied Sciences*, *13*(10), 6061.
4. Liu, W., Placke, T., & Chau, K.T. (2022). Overview of batteries and battery management for electric vehicles. *Energy Reports*, *8*, 4058–4084.

5. Anderson, J. (2019). Batteries are a scalable energy storage technology. *Solar Today Magazine, American Solar Energy Society*, March 2.
6. The future of battery technology. *SP Global.* November 25, 2022.
7. Wu, J., Liu, J., Wang, C., & Lee, J. (2019). A review of vehicle-to-grid projects for frequency regulation in power systems. *IEEE Transactions on Smart Grid, 11*(3), 2671–2682.
8. Knapen, L., Bienert, M., & Claessens, B. (2021). A review on vehicle-to-grid applications in a smart grid environment. *Energies, 14*(7), 2071.
9. Morstyn, T., Kumar, S., Timms, D., & Steinberger-Wilckens, R. (2020). Vehicle-to-grid (V2G) integration: Key developments, challenges and opportunities. *Energy Policy, 146*, 111799.
10. Bessa, R., Matos, M. A., & Soares, J. (2018). Vehicle-to-grid systems for improved grid integration of renewable energies: A review. *Applied Energy, 206*, 354–366.
11. Kazemi, A., Parhizgar, M. M.., & Savkin, A. V. (2020). A review on vehicle-to-grid applications in power systems considering plug-in electric vehicle aggregators. *Renewable and Sustainable Energy Reviews, 122*, 109716.
12. Hawkes, A., Brett, D. J., Brandon, N. P., & Kesicki, F. (2017). Vehicle-to-grid integration of electric drive vehicles using loss-minimisation scheduling. *Applied Energy, 190*, 67–77.
13. Anvari-Moghaddam, A., Siano, P., Soroudi, A., & Mohammadi-Ivatloo, B. (2018). Optimal energy management of a V2G-based DC fast charging station considering renewable energy resources. *Applied Energy, 212*, 202–213.
14. Ma, Z., Lv, Z., Yang, D., & Chen, D. (2022). Traffic management in connected vehicle systems: A comprehensive survey. *IEEE Transactions on Intelligent Transportation Systems, 23*(3), 1211–1230.
15. Barbato, A., De Dominicis, L., & Todeschini, C. (2022). V2G-based solutions for the integration of electric vehicles into distribution grids: A review. *Sustainable Energy, Grids and Networks, 29*, 100586.
16. Huang, X., Huang, L., Huang, J., Xu, W., & Zeng, Y. (2021). Energy internet: A review. *Journal of Cleaner Production, 298*, 126869.
17. Mashhour, E., Mohsenian-Rad, H., & Wozniak, T. (2021). Vehicle-to-grid systems: A comprehensive review. *IEEE Transactions on Industrial Informatics, 17*(2), 1346–1360.
18. Fan, S., Zhang, J., & Yang, Z. (2020). Vehicle-to-grid communication protocols: A review. *Electric Power Systems Research, 189*, 106711.
19. Zhang, Y., Li, L., & Qiu, M. (2022). Cybersecurity in intelligent transportation systems: Challenges, solutions, and future directions. *IEEE Transactions on Intelligent Transportation Systems, 23*(5), 2176–2194.
20. Lopez, G., Zhu, X., Huang, L., & Almasoud, A. (2021). Legal framework for vehicle-to-grid integration: A review. *IEEE Transactions on Intelligent Transportation Systems, 22*(4), 2417–2433.
21. Saboori, H., Zadehbagheri, M., Safari, A., & Kazemi, A. (2022). A comprehensive review of the vehicle-to-grid optimization strategies. *Journal of Cleaner Production, 318*, 128447.
22. Bessa, R., Cardoso, G., & Soares, F. (2020). Vehicle-to-grid services: Potential, benefits, and challenges. *Wiley Interdisciplinary Reviews: Energy and Environment, 9*(3), 365.

23. Hafeez, A., Ullah, A., & Zhang, Y. (2021). A comprehensive review on vehicle-to-grid: Technological, regulatory, and business challenges. *Renewable and Sustainable Energy Reviews*, *137*, 110685.
24. Wu, L., Gan, C., Yang, X., Li, M., & Huang, Q. (2022). Interoperability of wireless charging systems for electric vehicles: A review. *Journal of Cleaner Production*, *335*, 130702.
25. Sun, S., Wang, Z., & Lu, Y. (2022). A review of policy and regulatory frameworks for autonomous vehicles. *Transportation Research Part A: Policy and Practice*, *155*, 255–273.
26. Gomes, J. P., Leite, P., & Ferreira, L. A. F. M. (2021). Vehicle-to-grid: From stabilizing the grid to creating a prosumer energy market. *Renewable and Sustainable Energy Reviews*, *138*, 110539.
27. Gong, M., Xiao, Y., Guo, J., Liu, L., & Zhang, X. (2020). A comprehensive review of vehicle-to-grid (V2G) technology: Configurations, requirements, and effects on power systems. *Journal of Cleaner Production*, *255*, 120191.
28. Bessa, R., Cardoso, G., & Soares, F. (2019). Vehicle-to-grid services: Overview, prospects, and challenges. *Electric Power Systems Research*, *170*, 97–108.
29. Clement-Nyns, K., Haesen, E., & Driesen, J. (2020). The impact of vehicle-to-grid on the distribution grid. *Electric Power Systems Research*, *81*(1), 185–192.
30. Li, G., Xu, Z., Liu, J., & Yang, J. (2020). Optimal energy management strategy for vehicle-to-grid systems. *Applied Energy*, *262*, 114594.
31. Ma, Z., Hu, Z., & Yu, Z. (2021). Energy management of vehicle-to-grid systems with deep reinforcement learning. *Applied Energy*, *281*, 116014.
32. Zhu, J., Chen, N., Yang, H., & Chen, X. (2022). A review of data-driven modeling and optimization for vehicle-to-grid integration. *Applied Energy*, *309*, 117827.
33. Fan, X., Lu, J., & Li, L. (2021). Review on vehicle-to-grid control strategies for aggregated electric vehicles. *Energy Reports*, *7*, 4797–4813.
34. Liu, W., Xu, Y., He, Z., Li, Y., & Xiong, H. (2021). Electric vehicle charging station site selection: A comprehensive review. *Journal of Cleaner Production*, *280*, 124349.
35. Acha, S., Brignone, M., & Vicuña, L. G. (2022). A review of vehicle-to-grid system architecture: Application, design, and control. *Sustainable Energy, Grids and Networks*, *29*, 100637.
36. Wu, Q., Lin, Y., & Wu, F. (2021). Integrated energy management of vehicle-to-grid systems: A comprehensive review. *Applied Energy*, *282*, 116234.
37. Zhang, H., Luo, X., & Gooi, H. B. (2021). Vehicle-to-grid: A comprehensive review. *IEEE Transactions on Industrial Informatics*, *17*(1), 315–330.
38. Gong, M., Zhang, X., Guo, J., Li, C., & Yang, Y. (2021). A comprehensive review on regulatory frameworks of vehicle-to-grid (V2G) technology. *Applied Energy*, *287*, 116528.
39. Jin, H., Jiang, H., Wang, Y., & Zhang, C. (2022). A comprehensive review on vehicle-to-grid (V2G) regulation framework in China. *Applied Energy*, *311*, 117852.
40. Ou, C., Cao, J., & Mei, S. (2018). A survey of vehicle-to-grid implementation: Enabling technologies, optimization methodologies, and control strategies. *Journal of Modern Power Systems and Clean Energy*, *6*(4), 539.

41. Cheng, Y., Zhang, J., & Jiang, C. (2019). A review of vehicle-to-grid projects from the perspective of global energy internet. *Renewable and Sustainable Energy Reviews*, *113*, 109253.
42. Clement-Nyns, K., Haesen, E., & Driesen, J. (2020). The impact of vehicle-to-grid on the distribution grid. *Electric Power Systems Research*, *81*(1), 185–192.
43. Zhang, H., Luo, X., & Gooi, H. B. (2022). Vehicle-to-grid: A comprehensive review. *IEEE Transactions on Industrial Informatics*, *17*(1), 315–330.
44. Huang, L., Zhang, X., Li, Z., Luo, S., & Chen, J. (2021). Review of smart charging and discharging of electric vehicles: Key technologies, challenges and future trends. *Energy Reports*, *7*, 2030–2045.

CHAPTER 14

SMART HELMET WITH ENHANCED SAFETY FEATURES INCLUDING ALCOHOL DETECTION AND RASH DRIVING

SMITA GUPTA, JUNAID MD. KHAN, GAUTAM KUMAR, and YAKA BULO

Department of Electronics and Communication, NIT Arunachal Pradesh, Papum Pare, Arunachal Pradesh, India

ABSTRACT

This chapter presents a smart helmet used by bike riders for safety purpose. It is integrated with an embedded system whose main objective is to detect the presence of alcohol particles and prevents rash driving. It utilizes majorly an alcohol sensor, pressure sensor, a 6-axis motion-tracking device MPU 6050, and a 16×2 LCD display among others. LCD displays the current status of the ride. Basically, it prevents the ignition of engine if the helmet is either not worn properly or alcohol consumption is detected. Other than this with the help of MPU 6050, the rash driving above a preset threshold is detected and a warning is fed back to the rider, making use of Bluetooth module for maintaining connectivity between helmet and the bike within short range. By doing so it additionally ensures the presence of helmet within the short range.

Cyber-Physical Systems: Applications, Challenges, and Research Directions. Rajanikanth Aluvalu, Samarjeet Borah, & Lalit Garg (Eds.)

DOI: 10.1201/9781779643421-14

14.1 INTRODUCTION

We are witnessing the burden of roads increasing day by day; more traffic has started moving and therefore safe driving has become a more important aspect to avoid accidents. As per the report of National Crime Records Bureau (NCRB), unfortunate bike riding cases have claimed highest number of lives, that is, around 70,000 people in India during 2021. This chapter presents important aspects of safety measures for bike riders.

There are four major objectives resolved in this chapter: (a) to enforce wearing helmet for every ride, (b) to prevent the ignition of engine if the helmet is fastened loosely, (c) to prevent the ignition of engine if the rider is drunk, and (d) to warn the rider if the driving is rash or impulsive.

Several works have already been done to achieve at least one among many objectives of safety of bike riders. In Ref. [1], the authors have designed a smart helmet that works on radio frequency (RF), transmitter (Tx), and receiver (Rx) between bike and helmet to detect the presence of helmet; they also have incorporated the mechanism of accident detection into it. This chapter has an improvement over Ref. [2], which was not ideal for near field communication. Also, it is an advantage over Ref. [3], as this chapter does not incorporate the acceleration aspect in detecting rash driving. In Ref. [4], the major shortcoming is to not display the status of the ride as it does not use any display or audio device for the same. This problem has been solved in our research. In Ref. [5], there is no connectivity to motorcycle via a circuit; instead, it uses only one circuit integrated in helmet to achieve its objectives including the presence of helmet, alcohol, and accident detection. In Ref. [6], they have proposed the smart helmet consisting of two modules: one at bike and the other at helmet. The unique feature about Ref. [6] is that it uses a rear camera to detect the approaching vehicles and communication between bike and helmet is done through RF module. Smart Helmet based on Internet of Things (IoTs) [7] incorporated IoT in it and focused on the material of which the helmet is to be manufactured as well as it sends SMS to family members if accident is detected.

Smart Helmet for Accident Prevention and Detection [8] proposes a system that prevents ignition of the bike if helmet is not detected or if the driver is drunk. Along with this it detects the accident and sends the location to family using GSM and uses RF module to communicate between bike and helmet. In Ref. [9], the authors have proposed the system where after detection of sudden change in accelerometer readings it waits for 20 seconds for the rider to confirm if it is not a major accident and then it reports the accident to the contacts. Ref. [9] only focuses on two issues: one is helmet detection and the other is accident detection but it does not facilitate alcohol detection. Some commercially available helmets are there, but the costs are too high to be afforded by low- to medium-income countries. This chapter presents a smart helmet used by bike riders for safety purpose. It is integrated with an embedded system whose main objective is to detect the presence of alcohol particles and prevents rash driving. We have described the parameters of rash driving as impulsive riding and the angle of inclination of the bike with respect to road. The remaining section of the chapter comprises components used, further it elaborates the proposed helmet circuitry and working principle along with flow charts and algorithm used.

14.2 COMPONENTS USED

14.2.1 BLUETOOTH

We have utilized the Bluetooth module HC-05 in the helmet as well as in motorcycle circuitry as a mean of connecting link. All the detection, command is passed by utilizing this wireless short-range communication medium.

14.2.2 MPU 6050

It is a 6-axis motion-tracking device equipped with an accelerometer and a gyroscope. It is integrated with bike's Arduino board circuitry and acts as the detector of rash driving, that is detecting sudden and impulsive acceleration and more inclination of the bike.

14.2.3 ALCOHOL SENSOR

It detects weather the rider has drunk alcohol. We have used a gas sensor-MQ3 suitable for our purpose. It has been integrated with the helmet circuitry in the area around the mouth. When alcohol is detected by the sensor, the Bluetooth circuit sends signal to motorcycle and it in turn prevents the ignition of the engine. Hence, it makes sure that the driver is not drunk.

14.2.4 FORCE-SENSITIVE RESISTORS

It is integrated in helmet's circuitry to check if the helmet is fastened properly. Force sensitivity is measured in terms of the resulting voltage by this sensor and is measured to detect if the applied pressure is good enough to go. The characteristic of FSR sensors is that it tends to show a nonlinear relationship within the force range of 0–4 N [10–12].

14.2.5 ARDUINO BOARD

Since there are two circuitries, one at bike and the other attached to helmet, we have used two boards, one Arduino UNO and the other is Arduino Mega, respectively. Arduino has the versatility to be compatible with our project.

14.2.6 16×2 LCD DISPLAY, BUZZER, AND LI-ION BATTERY

The LCD display is equipped here for the purpose to show the status of the ride. For example, it displays whether helmet is fastened properly and if the alcohol is detected. Buzzer is turned on in the case of rash driving to alert the rider and also in case if helmet is not fastened properly and alcohol particles are detected near mouth. Battery is used to power up the Arduino board.

14.2.7 STEP-UP MODULE AND CHANNEL RELAY

Step-up module is used for voltage boosting to meet the threshold calibration. The channel relay is playing a major role here. Relays are switch that operate short and open circuits electromechanically or electronically. It is used to connect battery to engine when bike receives "OK" command from helmet.

14.3 PROPOSED CIRCUIT AND WORKING PRINCIPLES

14.3.1 CIRCUIT INTEGRATED IN HELMETS

The helmet circuit shown in Figure 8.1 incorporates Alcohol Sensor, Force Sensor Resistor (FSR), SD Card Module, Bluetooth Module HC-05, a speaker, Arduino Board, and a battery.

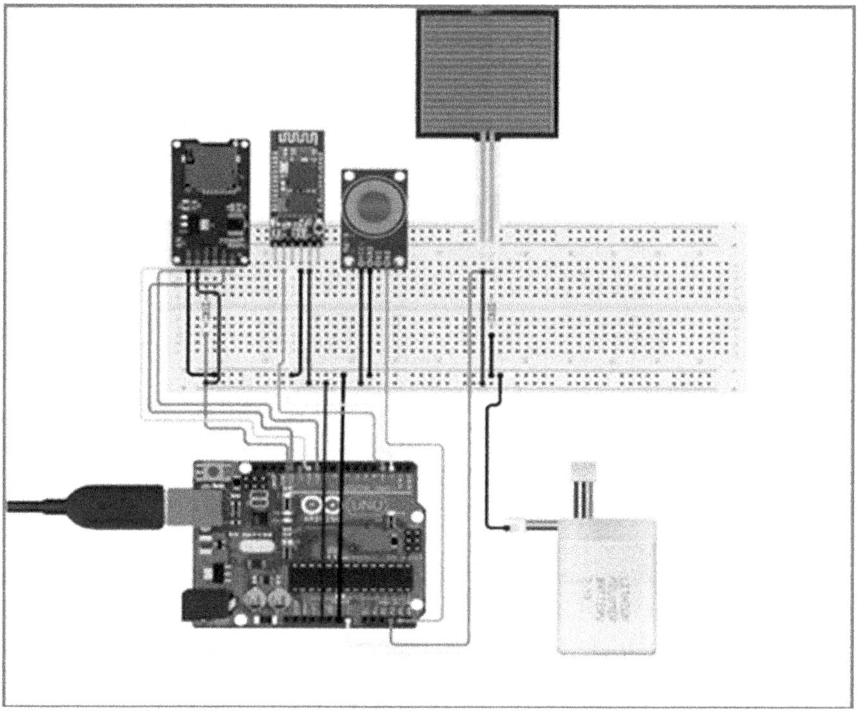

FIGURE 14.1 Helmet circuitry.

14.3.2 CIRCUIT INTEGRATED IN MOTORCYCLE

Figure 8.2 shows the assembly of component at motorcycle. It combines a Bluetooth module HC-05, Relay (engine switch), Buzzer, 16×2 LCD display, MPU 6050, and an Arduino Mega Board.

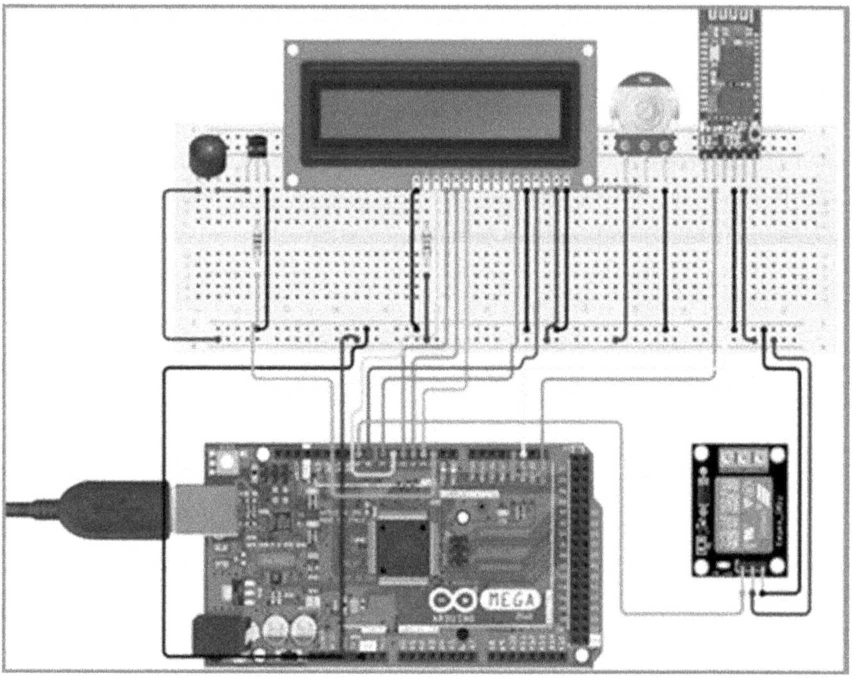

FIGURE 14.2 Circuit integrated at motorcycles.

14.3.3 WORKING PRINCIPLE

As soon as the person wears helmet and switches it on, the helmet checks for pressure on the head. If the helmet is not worn tightly, it issues a voice warning. If the helmet is fastened properly, then it checks for alcohol particles within the potential vicinity of the rider's mouth. If any alcohol is detected, again a voice warning is issued. If no alcohol is detected, the helmet sends an "OK" command to the bike. The bike on receiving the "OK" command connects the battery to the engine, as a result of which ignition can be achieved. When the engine is connected, the bike continuously checks for rash driving conditions. If a high acceleration or an angle of bike with the horizontal ground of more than 35° is detected, the bike squeezes the buzzer and sends a command to the helmet, which in turn issues a voice warning to the rider. Meanwhile, the helmet continuously checks for helmet pressure and alcohol content.

14.4 FLOW CHARTS

14.4.1 FLOW CHART FOR IGNITION CONTROL

The algorithm for the ignition control of the bike is shown in Figure 8.3, where it shows that the bike's circuit waits for the data from helmet and sees if the helmet is worn. If it receives the positive signal then next check is for the alcohol. If alcohol sensor does not find any alcohol particle around the mouth of driver, then the relay switch allows the battery connection to engine and the bike gets started. In case of unfulfillment of any of the two prerequisite conditions, the engine does not start as in those cases the battery is not connected to engine. Therefore, engine ignition could happen only if both the conditions are satisfied, ensuring safety.

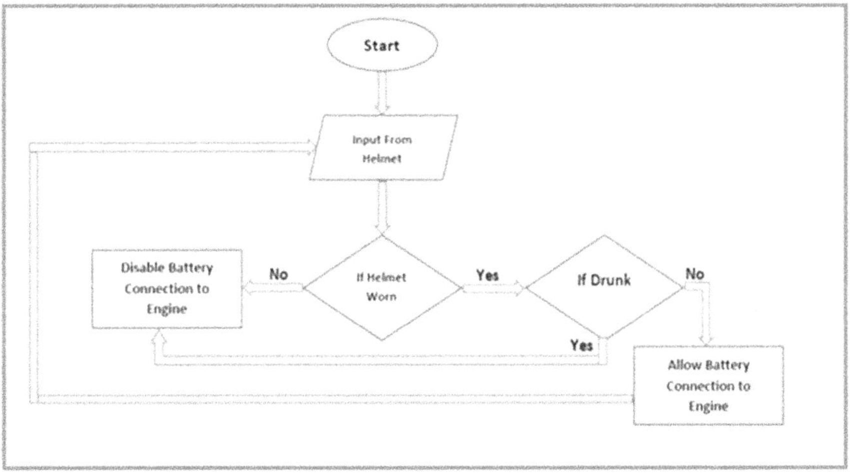

FIGURE 14.3 Ignition control flow chart.

14.4.2 FLOW CHART FOR DRIVING CONTROL

The MPU-6050 is monitored continuously by the microcontroller. If the speed and angle of inclination of the bike are found not to be in range, it is understood that this is rash driving and in turn warns the driver by operating the buzzer. This algorithm is shown as a flowchart in Figure 8.4.

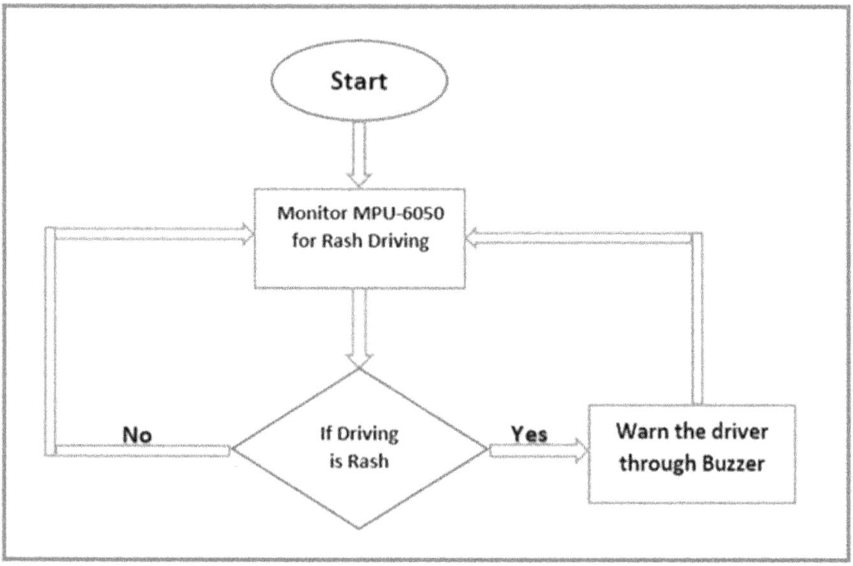

FIGURE 14.4 Driving control flowchart.

14.5 CONCLUSION

Our project has realized the objective of life safety on roads that proactively plans to prevent any mishap while riding a bike. We have seen in most of the cases of road accident that if the rider would have fastened the helmet or if he or she had not drunk or at least did not drive rashly the accident could have been not the case and the harm could be lessened. Therefore, these are the foundational elements of our project, it alarms the rider about the careless and impulsive riding. It prevents ignition of the engine by disconnecting the battery if the bike rider has not properly worn helmet or has consumed alcohol. In the future we can add the feature of sending SMS to family if the speed limit exceeds a predefined value along with traffic-monitoring feature integrated on helmet. By doing so, an additional sense of responsibility comes to a driver.

KEYWORDS

- **alcohol sensor**
- **force sensitive sensor**
- **MPU 6050**
- **smart helmet**
- **16×2 LCD**

REFERENCES

1. Nataraja, N., Mamatha, K. S., & Keshavamurthy, S. (2018). Smart helmet. In *3rd IEEE International Conference on Recent Trends in Electronics, Information and Communication Technology* (pp. 2338–2341).
2. Paulchamy, B., Sundhararajan, C., Xavier, R., Ramkumar, A., & Vigneshwar, D. (2018). Design of smart helmet and bike management system. *Asian Journal of Applied Science and Technology*, *2*(2), 207–211.
3. Saravana Kumar, K., Anjana, B. S., Thomas, L., & Rahul, K.V. (2016). Smart helmet. *International Journal of Science, Engineering and Technology Research*, *5*(3).
4. Lakshmi Devi, P., Bindushree, R., Deekshitha, N. M., Jeevan, M., & Likhith, M. P. (2016). Helmet using GSM and GPS technology for accident detection and reporting system. *International Journal on Recent and Innovation Trends in Computing and Communication*, *4*(5), 18–21.
5. Sobhana, S., Sowmeeya, S. R., Srinathji, M., & Tamilselvan, S. (2021). Smart helmet. *IOP Conference Series Material Science and Engineering*, 1–10.
6. Sattar, M. K., Fayyaz, S., Waseem, M., Saddique, M. S., Usama, M., & IIyas, H. B. (2021). Smart modular helmet with an innovative information relaying system. *Engineering Proceedings*, *12*(1), 1–6.
7. Dwivedi, A., Singh, A; Narwat, K., & Singh, L. (2022). Smart helmet based on IOTs. *AIP Conference Proceedings*, *2413*(1), 020007.
8. Jebanazer, J., Dhanasekaran, A., Chinnaraj, Y., & Ravanan, Y. (2021). Smart helmet for accident prevention and detection. *International Research Journal of Engineering and Technology*, *8*(4), 1164–1170.
9. Navya Sri, K., Pavan Reddy, M., & Lavan Kumar, G. (2016). Smart helmet with accident detection. *International Journal of Science and Research*, *7*(10), 1537–1540
10. Sadun, S., Jalani, J., & Sukor, J. A. (2016). Force sensing resistor (FSR): A brief overview and the low-cost sensor for active compliance control. *Proceedings of the SPIE*, *10011*, 2016.

11. Bhoi, A. K., Sherpa, K. S., & Mallick, P. K. (2014). A comparative analysis of neuropathic and healthy EMG signal using PSD. *International Conference on Communication and Signal Processing*, 1375–1379.
12. Samanta, S. R., Mallick, P. K., Pattnaik, P. K., Mohanty, J. R., & Polkowski, Z. (2022). *Cognitive computing for risk management.* Springer.

CHAPTER 15

DESIGN OF DEPTH CONTROLLER USING OPTIMAL FRACTIONAL ORDER PID BASED ON EVOLUTIONARY ALGORITHMS

LINKAN PRIYADARSINI, SHUBHASRI KUNDU, and
MANOJ KUMAR MAHARANA

School of Electrical Engineering, KIIT University, Bhubaneswar, Odisha, India

ABSTRACT

In recent years, autonomous underwater vehicles (AUVs) have gained a lot of popularity due to their versatile application. It has the advantage of monitoring the underwater environment, surveying the geological and biological parameters present under the sea, and so on. But due to the complexity and time variance it becomes difficult to model the AUV and design a proper controller for controlling its trajectory. In this chapter, the depth control problem of AUV has been considered using dual loop in which inner loop represents the pitch control and outer loop represents the depth control. A FOPID (fractional order proportional-integral-derivative) controller has been used to control the inner loop and a PD (proportional-derivative) controller has been used to control the outer loop. The controller has been designed taking into consideration the nonlinearity and effect of the integrator. The parameters of FOPID have been optimized using Ant Colony Optimization (ACO) and Firefly Algorithm (FA). The simulated results show the robustness of the system, performance improvement, and less error.

Cyber-Physical Systems: Applications, Challenges, and Research Directions. Rajanikanth Aluvalu, Samarjeet Borah, & Lalit Garg (Eds.)

DOI: 10.1201/9781779643421-15

15.1 INTRODUCTION

Nowadays, scientists have found a lot of interest in AUVs due to their large-scale underwater sea applications. The use of AUVs has made it possible for researchers to explore the vast number of unused resources like petroleum, minerals, natural gas, and so forth, that are present in the sea floor bed. The difficulty of AUVs lies in its hydrodynamic forces which make it highly nonlinear, thereby making it difficult to derive its mathematical model [1]. With the development of control engineering there has been a lot of techniques developed for controlling the dynamics of AUVs such as linear controller, sliding controller [5, 6], controller that adapt themselves to the system [7], and so forth. But the performance of these controllers degrades when the AUVs are subjected to a wide range of operating conditions. Hence, the combination of classical and intelligent control methods can be used to design a suitable controller to overcome the environmental disturbances and noise.

The mathematical modeling of AUV itself gives a major challenge for designing a controller to control it. Since the dynamics of the vehicle is highly nonlinear, it becomes quite difficult to find all the hydrodynamic parameters with reasonable accuracy [3]. Keeping these challenges in mind a fractional order PID controller has been developed as its performance does not take into account the definiteness of the system model [8]. In order to control the depth subsystem a two-loop controller is required. For example, a dual-loop controller with an inner loop representing the pitch control and outer loop representing the depth control [3].

The arrangement of the chapter is as given below:

The modeling of AUV regarding its kinematics and dynamics has been described in Section 15.2. A dual-loop PID controller has been designed in Section 15.3. Section 15.4 gives a brief description about the fractional order PID design using Ant Colony Optimization and Firefly algorithm. The conclusion with simulation graph and comparative analysis among Ant Colony Optimization and Firefly Algorithm has been presented at last.

15.2 MODELING OF AUV

The study of both kinematics and the dynamics is required to formulate the mathematical model of an AUV (Figure 15.1). The geometrical aspects give a detailed analysis of the kinematics while for analyzing the force that causes the motion is being described by the dynamics of the vehicles [2].

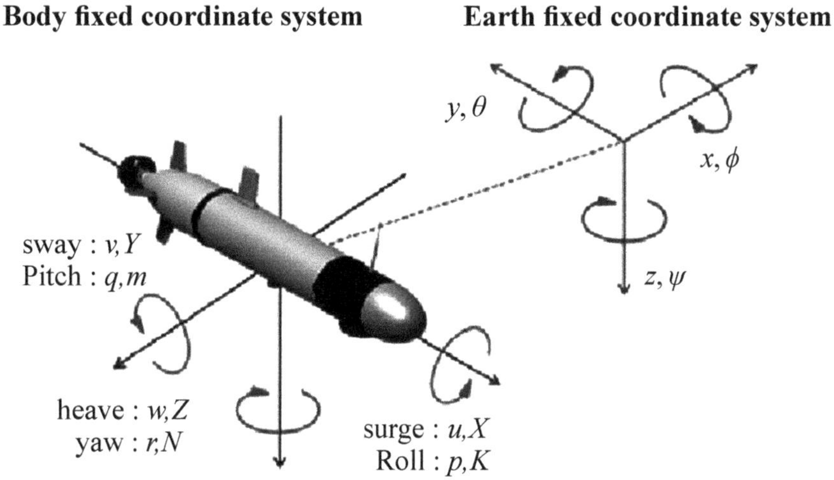

FIGURE 15.1 Prototype of AUV.

To decide the location and direction of the AUV it is required to find the differential equation of 6-DOF motion of the vehicle [3] in Table 15.1. The location as well as translational movement of AUV along the x, y, and z axes [1] is characterized by flow, amplitude, and Heave [3] respectively, while the inclination and circular movement has been outlined in the form of Roll, Pitch, and Yaw [1].

TABLE 15.1 Parameters of AUV.

Variable Quantity	Movement	Forces and Moments	Speed	Position and Euler Angles [1]
I.	Surge	X	u	x
II.	Sway	Y	υ	y
III.	Heave	Z	w	z
IV.	Roll	K	P	φ
V.	Pitch	M	q	θ
VI.	Yaw	N	R	ψ

15.2.1 VEHICLE KINEMATICS

In order to analyze the path of an AUV in 6 DOF a two-coordinate system is required [14]. Since the moving reference frame remains attached to the body of an AUV, it is also called as fixed body reference frame whose path is described with the help of an inertial frame. The frame fixed to the body is considered to define the linear and angular speed while the location and inclination of the vehicle is described with the help of an inertial structure [1]. The path of the vehicle having 6 DOF can be described with the help of the following vectors [1, 12].

$$\eta = \left\|\begin{bmatrix}\eta_1 \\ \eta_2\end{bmatrix}\right\| \qquad \eta_1 = \begin{bmatrix}x \\ y \\ z\end{bmatrix} \text{direction of location} \qquad \eta_2 = \begin{bmatrix}\phi \\ \theta \\ \psi\end{bmatrix}$$

Euler angular vector

$$v = \begin{bmatrix}v_1 \\ v_2\end{bmatrix} \quad v_1 = \begin{bmatrix}u \\ v \\ w\end{bmatrix} \text{uniform speed vector [31]} \quad v_2 = \begin{bmatrix}p \\ q \\ r\end{bmatrix} \text{rotational speed vector}$$

$$\tau = \begin{bmatrix}\tau_1 \\ \tau_2\end{bmatrix} \quad \tau_1 = \begin{bmatrix}X \\ Y \\ Z\end{bmatrix} \text{forces direction} \quad \tau_2 = \begin{bmatrix}K \\ M \\ N\end{bmatrix} \text{moments vector}$$

15.2.2 VEHICLE DYNAMICS

The vehicle dynamics consists of location and rotational motion [4, 12]. The equation for the translatory movement of the vehicle is as follows:

$$m\,(v_0 + \omega * v_0 + \dot{\omega} * r_g + \omega * (r_g)) = f_0 \tag{15.1}$$

and the rotational movement of the vehicle is as follows:

$$I_0\dot{\omega} + \omega * (I_0\omega) + mr_g * (v_0 + \dot{\omega} * v_0) = m_0 \tag{15.2}$$

With the help of Newton–Euler's equation the 6 DOF equation of an AUV can be written as [2, 14]:

$$m\,(\dot{u} - vr + wq - x_g(q^2 + r^2) + y_g(pq - \dot{r}) + z_g(pr + \dot{q})) = X \tag{15.3}$$

$$m\,(\dot{v} - wp + ur - y_g(r^2 + p^2) + z_g\,(qr - \dot{p}) + x_g(qp + \dot{r})) = Y \tag{15.4}$$

$$m\,(\dot{\omega} - uq + vp - z_g(p^2 + q^2) + x_g\,(rp - \dot{q}) + y_g(rq + \dot{p})) = Z \tag{15.5}$$

$$\begin{aligned} I_x\dot{p} + (I_z - I_y)qr - (\dot{r} + pq)I_{XZ} + (r^2 - q^2)I_{yz} + (pr - \dot{q})I_{xy} \\ + m[y_g(\dot{w} - uq + vq) - z_g(\dot{v} - wp + ur)] = K \end{aligned} \tag{15.6}$$

$$\begin{aligned} I_y\dot{q} + (I_x - I_z)rp - (\dot{p} + qr)I_{Xy} + (p^2 - r^2)I_{xz} + (qp - \dot{r})I_{yz} \\ + m[z_g(\dot{u} - vr + wq) - x_g(\dot{w} - uq + vp)] = M \end{aligned} \tag{15.7}$$

$$\begin{aligned} I_z\dot{r} + (I_y - x)pq - (q + rp)I_{yZ} + (q - p^2)I_{xy} + (rq - \dot{p})I_{xz} \\ + m[x_g(\dot{v} - wp + ur) - y_g(\dot{u} - vr + wq)] = N \end{aligned} \tag{15.8}$$

The kinematics for AUV pitch and depth subsystem by linearizing the kinematics and dynamics of AUV and neglecting the terms not required for modeling the subsystem is given by

$$\begin{aligned} \dot{x} &= \cos(\theta u) + \sin\,(\theta w) \\ \dot{z} &= -\sin(\theta u) + \sin(\theta w) \\ \dot{\theta} &= q \end{aligned} \tag{15.9}$$

With the help of Maclaurin series of trigonometric terms, the linearized equation is obtained after neglecting the higher order terms; we get

$$\dot{x} = u + \theta w \tag{15.10}$$

$$\dot{z} = u_1\theta + w \tag{15.11}$$

$$\dot{\theta} = q \tag{15.12}$$

Similarly, the dynamic equations are given by

$$m(\dot{u} + wq - x_g q^2 + z_g\dot{q}) = X \tag{15.13}$$

$$m(\dot{w} - uq - z_g q^2 - x_g\dot{q}) = Z \tag{15.14}$$

$$I_y\dot{q} + m[z_g(\dot{u} + wq) - x_g(\dot{w} - uq)] = M \tag{15.15}$$

Linearizing the above equation, we get

$$m(\dot{u} + z_g\dot{q}) = X \tag{15.16}$$

$$m(\dot{w} - x_g\dot{q} - u_1 q) = Z \tag{15.17}$$

$$I_y\dot{q} + m[z_g\dot{u} - x_g(w - u_1\dot{q}) = M \tag{15.18}$$

With the assumption that the speed of the heave is negligible for which it can be ignored then the matrix form of the kinematics and dynamics of the vehicle can be written as

$$\begin{bmatrix} I_y - M_q & 0 & 0 \\ 0 & 1 & 0 \\ 0 & 0 & 1 \end{bmatrix} \begin{bmatrix} \dot{q} \\ \dot{z} \\ \dot{\theta} \end{bmatrix} + \begin{bmatrix} -M_q & 0 & M_\theta \theta \\ 0 & 0 & u_1 \\ -1 & 0 & 0 \end{bmatrix} \begin{bmatrix} q \\ z \\ \theta \end{bmatrix} = \begin{bmatrix} M_{fs} \\ 0 \\ 0 \end{bmatrix} \tag{15.19}$$

The pitch transfer function obtained from the above matrix is given by

$$G_\theta(s) = \frac{\theta(s)}{f_s(s)} = \frac{\dfrac{M_{fs}}{I_y - M_q}}{s^2 \dfrac{M_{fs}}{I_y - M_q} s - \dfrac{M_\theta}{I_y - M_q}} \tag{15.20}$$

Considering the input pitch angle, the depth can be represented by the following transfer function [11]:

$$G_z(s) = \frac{z(s)}{\theta(s)} = \frac{-u_1}{s} \tag{15.21}$$

The transfer function for the AUV can be obtained considering the following data in Table 15.2.

TABLE 15.2 Parameter Values.

Parameter	Value
Fin lift (M_{fs})	-1575.9 kg.m²/s
I_y	469 kg/m²
Added mass (M_q)	-458 kg.m²
Hydrostatic (M_θ)	13719.6 kg.m²/s²

$$G_\theta(s) = \frac{1.818}{0.00077s^2 + 0.507s + 1} \tag{15.22}$$

Assuming a linear velocity of 11 m/s the transfer function of depth is given by

$$G_z(s) = \frac{-11}{s} \tag{15.23}$$

15.3 DESIGN AND DEVELOPMENT OF DUAL CONTROL LOOP

A dual PID controller has been designed for controlling the depth of AUV in Figure 15.2. The inner loop represents the pitch control loop whose transfer function is of second order and has a faster response as compared to the outer loop [9, 10]. For controlling the depth of an AUV the pitch needs to be controlled first; that is the pitch needs to be controlled at a certain angle. The output of the pitch control loop acts as an input to the depth control loop. The inner loop is controlled by FOPID optimized using ACO as well as FA whereas the outer loop is controlled by the PD controller [3].

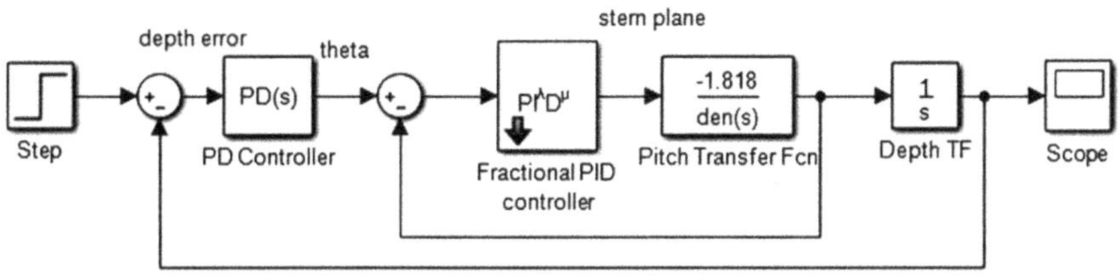

FIGURE 15.2 Dual-loop control.

15.4 STRUCTURE OF FRACTIONAL ORDER PID (FOPID)

The fractional order PID controller can be described as [8]

$$G_{PID}(s) = k_p + \frac{k_i}{s^{\lambda}} + k_d s^{\mu} \tag{15.24}$$

where λ and μ are fractional numbers. The FOPID behaves as a conventional PID controller when both λ and μ become equal to 1 in Figure 15.3.

15.4.1 ANT COLONY OPTIMIZATION (ACO) [13]

The Ant Colony Optimization is a swarm intelligence and metaheuristics technique. This method was proposed by Marco Dorigo in the 1990s. Ant Colony Optimization technique has been inspired by the foraging behavior of the ants. The ants in a colony communicate with each other with the help of sound, touch, and pheromone. Pheromone is a type of chemical used by the ants to deposit on the path in which they travel so that the other ants in the colony can also follow the same path. This behavior of the ants has been used to develop the algorithm. The artificial "ants" are used as the simulating agents to find the optimum as well as suitable track for the control objective moving through a boundary area having all kinds of feasible solutions. The Ant Colony Optimization (ACO) Algorithm starts with the generation of initial random population. Then the iteration is carried out until all the nodes are dead. The objective function chosen for the ACO Algorithm is IAE. For example, Integral of Absolute Error is given by

$$IAE = \int_0^\infty |e(t)|\, dt \tag{15.25}$$

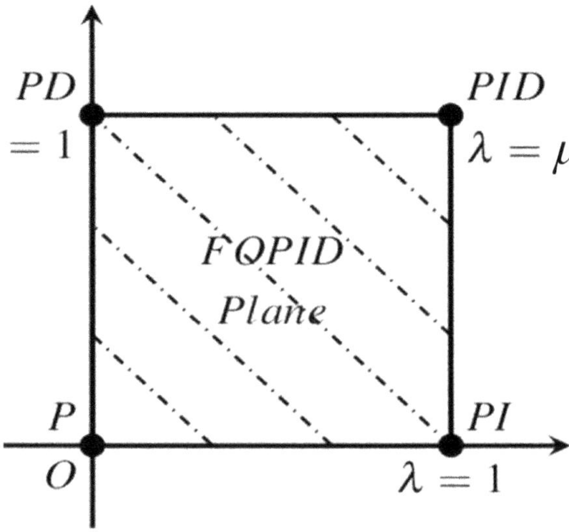

FIGURE 15.3 XY plane.

15.4.2 *FIREFLY ALGORITHM*

Firefly algorithm was initially proposed by Krishnanad in 2005 and was developed by Dr Xin She Yang [12]. It is a type of meta-heuristic optimization technique which is inspired by nature. This optimization technique is adapted as per the flashing pattern and behaviors of the fireflies. This algorithm is based upon three idealized rules:

1. The fireflies that flash the brightest light are attracted toward other flies that exhibit the same brightness and form a group irrespective of their gender.
2. The alluring of the fireflies is proportionate to its illumination. The lesser the distance between each firefly in the group the more

brightness the group will be. In the absence of brighter firefly, they will proceed arbitrarily.

3. The alluring or attractiveness of the fireflies depends upon the objective function chosen.

The attractiveness of the firefly is given by

$$\beta = \beta_0 e^{-\gamma r^2} \tag{15.26}$$

where r is the span between firefly i and firefly j at x_i and x_j positions, respectively. The initial solution is given by

$$X_j = rand(Ub - Lb) + Lb \tag{15.27}$$

The firefly i is attracted toward the brighter firefly j as per the following equation:

$$X_i^{r+1} = X_i^r + \beta e^{-\gamma r_{ij}^2} + \alpha_t (rand - \frac{1}{2}) \tag{15.28}$$

The Firefly Algorithm starts with the Objective Function and generation of initial random population. Then the parameters that are to be optimized are defined. The objective function chosen for the Firefly Algorithm is IAE. For example, Integral of Absolute Error is given by

$$IAE = \int_0^\infty |e(t)|\, dt \tag{15.29}$$

15.5 SIMULATED OUTPUT

This chapter accounts for the study of controller design for the depth subsystem using FOPID based upon ACO and FA technique in the MATLAB/SIMULINK environment. For controlling the depth of an AUV the parameters involved are speed of heave $\omega(t)$, pitch rate $q(t)$, pitch angle $\theta(t)$, and depth $z(t)$. The parameters are tuned based upon minimizing the objective function IAE. The FAFOPID gives improved performance result in comparison to ACO FOPID controller.

Table 15.3 contains specific values of control parameters for ACO and FA. The individual algorithms are implemented for 30 runs. With each iteration the objective function should be reduced to achieve the desired convergence value. The firefly FOPID gives a satisfied converged value before the 30th iteration with a fitness value of 1.1 as compared to ACOFOPID in Figure 15.4.

TABLE 15.3 Control Parameters for Metaheuristics Algorithms.

Algorithms	Details of Control Parameters
ACO FOPID:	Iteration number = 30; population size = 100; no. of ants = 5; Pheromone decay = 0.01
FA FOPID:	Iteration number = 30; population size = 100; $\beta = 0.3$; $\alpha = 0.4$; $\gamma = 0.6$

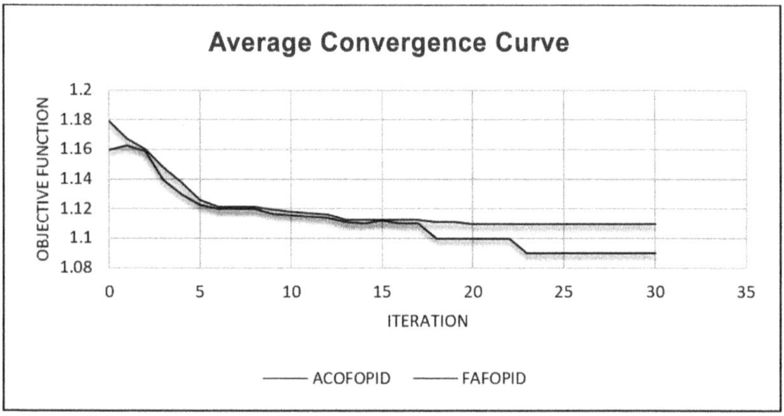

FIGURE 15.4 Convergence curve.

The optimization process for pitch response of AUV can be seen in Figure 15.5. The AUV dives vertically for a short time period and then decreases suddenly within the time period between 0 and 2 s. After 2 s the AUV again dives vertically in the forward direction till 4 s and after 4 s it dives in the downward direction. This process continues and finally it dives in the horizontal direction after 10 s. This has been depicted in Figure 15.5 showing the optimization process for Pitch Angular Velocity.

Figure 15.6 gives the waveform of pitch error optimized using integral absolute error (IAE). With the use of input step, the value of the desired pitch angle changes and the error in pitch reduces and finally reaches zero. The parameters optimized using IAE are presented in Table 15.4.

TABLE 15.4 Optimized Parameters using ACO and FA for Pitch Control.

Controller	K_p	K_i	K_d	λ	μ
ACO FOPID	0.3695	0.0367	0.3367	0.386	0.98
FA FOPID	0.1250	0.0167	0.1057	10.1764	0.837

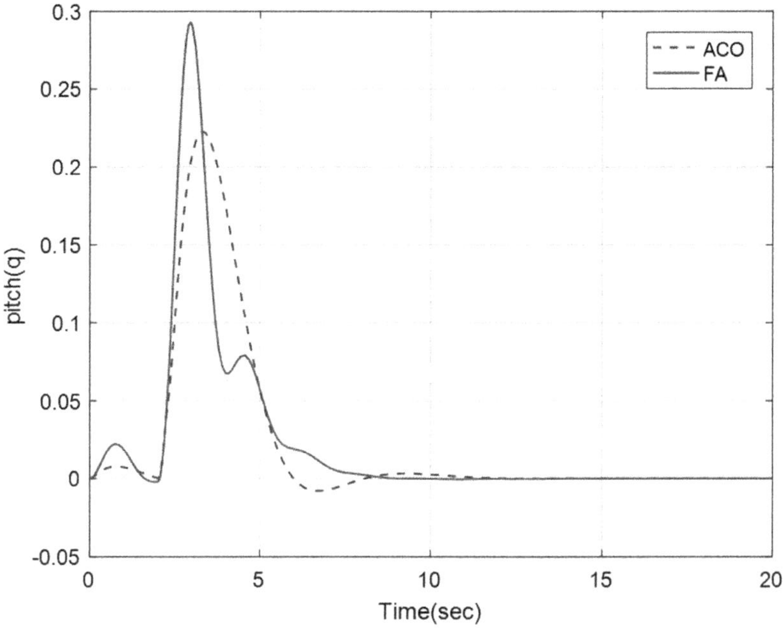

FIGURE 15.5 Pitch angular velocity.

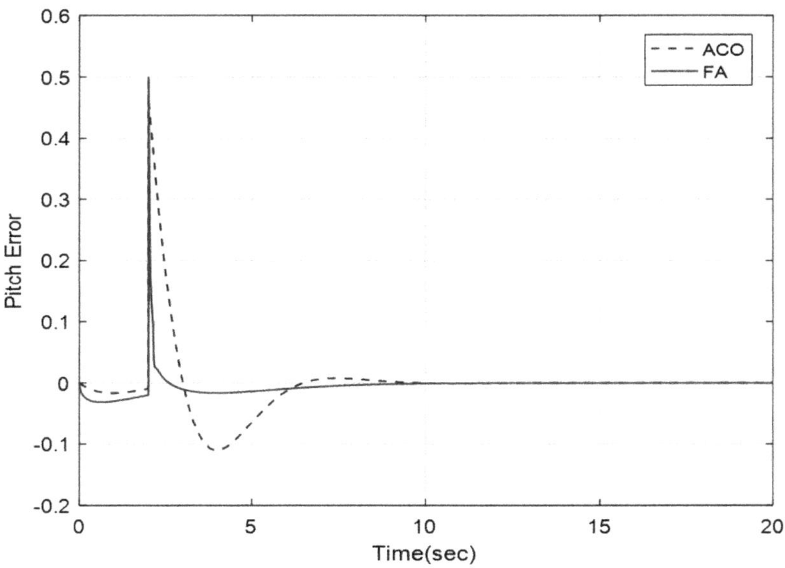

FIGURE 15.6 Pitch error.

Table 15.5 shows the transient response performance characteristics of AUV with FOPID optimized using Ant Colony Optimization (ACO) and Firefly Algorithm (FA). The system when optimized using the FA technique gives less error as compared to the ACO technique. The response shows that the pitch response for AUV with FOPID optimized using Firefly Algorithm (FA) is more stable and the actual depth gets settled to the commanded depth [14–19].

TABLE 15.5 Transient Response Characteristics of Pitch Angle.

Optimization Techniques	Time to Rise (Tr)	Time to Settle (Ts)	Overshoot	IAE
ACOFOPID	0.389	4.75	31	4.457
FAFOPID	0.235	1.83	13.9	1.472

Figure 15.7 shows the linear depth control of AUV. In order to control the depth of an AUV there is deflection in the horizontal plane fins due to which there is change in the moment of pitch. This change in the pitch moment causes change in the pitch angle thereby changing the depth of an AUV. This change characteristic has been depicted in Figure 15.7 where the AUV dives in vertical direction 9 s and finally gets settled to the commanded depth horizontally after 10 s.

Figure 15.8 gives the waveform of depth error optimized using integral absolute error (IAE). With the application of commanded input of 5 m the AUV dives in the upward direction from 0 to 3 s. After 3 s the error in depth reduces and finally reaches zero; the parameters optimized using IAE are presented in Table 15.6.

TABLE 15.6 Optimized Parameters using ACO and FA for Depth Control.

Controller	K_p	K_d
ACO FOPID	0.18077	1.56199
FA FOPID	1.092	0.0663

Table 15.7 shows the transient response performance characteristics for depth control of AUV with FOPID optimized using Ant Colony Optimization (ACO) and Firefly Algorithm (FA). The system when optimized using the FA technique gives less error as compared to the ACO technique. The response shows that the pitch response for AUV with FOPID optimized

using Firefly Algorithm (FA) is more stable and the actual depth gets settled to the commanded depth as depicted in Figure 15.7.

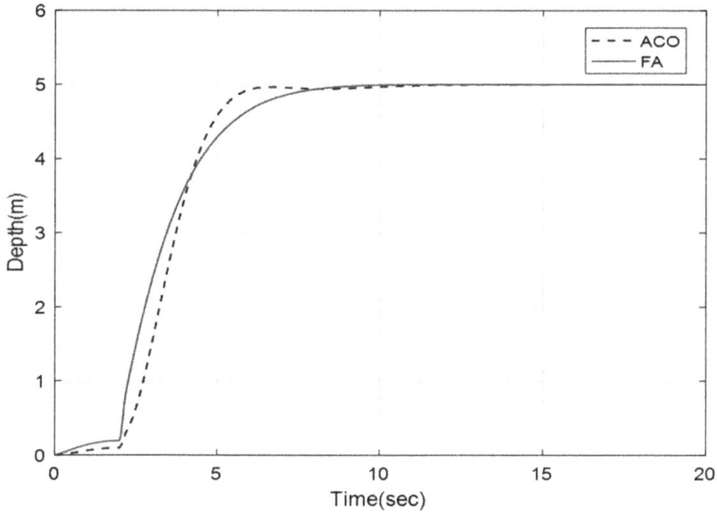

FIGURE 15.7 Depth response.

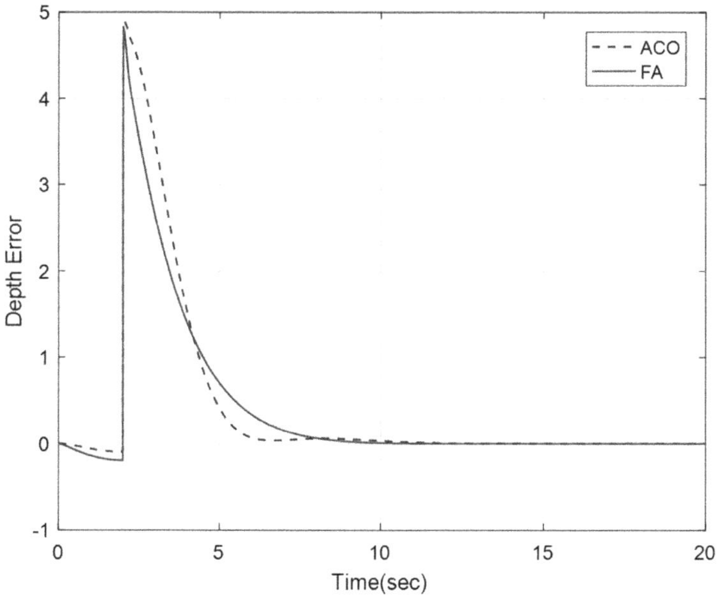

FIGURE 15.8 Depth error.

TABLE 15.7 Transient Response Characteristics for Depth Response.

Optimization Technique	Time to Rise (Tr)	Time to Settle (Ts)	Overshoot	IAE
ACOFOPID	3	5.85	9.63	8.145
FAFOPID	0.575	5.43	7.56	7.658

15.6 CONCLUSION

Here the ACO and FA design for optimizing the parameters of fractional order PID controller was suggested. IAE was selected as the objective function to be reduced. The simulation result shows that Firefly Algorithm Fractional Order PID (FA-FOPID) gives a better result in terms of faster concurrence and stronger strength in the presence of nonlinearity and integrator in comparison with Ant Colony Optimization Algorithm Fractional Order PID (ACO-FOPID) controller.

KEYWORDS

- **autonomous underwater vehicles**
- **depth control**
- **firefly algorithm**
- **FOPID**
- **pitch control**

REFERENCES

1. Fossen, T. I. (1994). *Guidance and control of ocean vehicles*. Wiley: New York.
2. Nag, A., Patel, S. S., & Akbar, S. A. (2013). Fuzzy logic based depth control of an autonomous underwater vehicle. In *2013 International Mutli-Conference on Automation, Computing, Communication, Control and Compressed Sensing (iMac4s)* (pp. 117–123). IEEE.
3. Vahid, S., & Javanmard, K. (2016). Modeling and control of autonomous underwater vehicle (AUV) in heading and depth attitude via PPD controller with state feedback. *International Journal of Coastal and Offshore Engineering, 4*(2016), 11–18.

4. Yang, C. (2008). *Modular modeling and control for autonomous underwater vehicle (AUV)*. ScholarBank@NUS Repository.
5. Soylu, S., Buckham, B. J., & Podhorodeski, R. P. (2008). A chattering-free sliding-mode controller for underwater vehicles with fault-tolerant infinity-norm thrust allocation. *Ocean Engineering, 35*(16), 1647–1659.
6. Zeinali, M., & Notash, L. (2010). Adaptive sliding mode control with uncertainty estimator for robot manipulators. *Mechanism and Machine Theory*, *45*(1), 80–90.
7. Mohan, S., & Kim, J. (2012). Indirect adaptive control of an autonomous underwater vehicle-manipulator system for underwater manipulation tasks. *Ocean Engineering*, *54*, 233–243.
8. Tepljakov, A., Alagoz, B. B., Yeroglu, C., Gonzalez, E., Hosseinnia, H., & Petlenkov, E. (2018). FOPID controllers and their industrial applications: A survey of recent results. *IFAC-PapersOnLine, 51*(4), 25–30. AU: Please provide text citations for references 9 and 10.
9. Medagoda, L., & Williams, S. B. (2012, May 21–24). Model predictive control of an autonomous underwater vehicle in an *in situ* estimated water current profile. In *OCEANS 2012* (pp. 1–8), Yeosu, South Korea. IEEE.
10. Subudhi, B., Mukherjee, K., & Ghosh, S. (2013). A static output feedback control design for path following of autonomous underwater vehicle in vertical plane. *Ocean Engineering*, *63*, 72–76.
11. Das, B., Subudhi, B., & Pati, B. B. (2016). Cooperative formation control of autonomous underwater vehicles: An overview. *International Journal of Automation and Computing, 13*(3), 199–225.
12. Kommula, B. N., & Kota, V. R. (2020). Direct instantaneous torque control of brushless DC motor using firefly algorithm based fractional order PID controller. *Journal of King Saud University-Engineering Sciences, 32*(2), 133–140.
13. Herlambang, T., Rahmalia, D., & Yulianto, T. (2019 November 24). Particle swarm optimization (PSO) and ant colony optimization (ACO) for optimizing PID parameters on autonomous underwater vehicle (AUV) control system. *Journal of Physics: Conference Series*, In *The 2nd International Conference of Combinatorics, Graph Theory, and Network Topology,* East Java, Indonesia (Vol. 1211). IOP Publishing.
14. Samanta, S. R., Mallick, P. K., Pattnaik, P. K., Mohanty, J. R., & Polkowski, Z. (Eds.). (2022). *Cognitive computing for risk management*. Springer.
15. Polkowski, Z., Mishra, S. K., Mishra, B. K., Borah, S., & Mohanty, A. (2023). Impact of internet of things in social and agricultural domains in rural sector: A case study, *International Journal of Cloud Computing* (pp. 90–105). *Inderscience Publishers*. https://doi.org/10.1504/IJCC.2023.129776
16. Prasad, S. S., Mishra, J. P., Mishra, S. K., & Borah, S. (2020). Challenges toward blockchain and renewable energy linked to IoT: A survey. *Advances in energy technology, Proceedings of ICAET 2020* (pp. 49–55). Springer Nature. https://doi.org/10.1007/978-981-15-8700-9
17. Nath, M. P., Priyadarshini, S. B. B., Mishra, D., & Borah, S. (2020). A comprehensive study of contemporary IoT technologies and varied machine learning (ML) schemes. *Soft computing techniques and applications* (pp. 623–634). Springer Nature. https://doi.org/10.1007/978-981-15-7394-1_1

18. Mishra, J. P., Borah, S., & Mishra, S. K. (2021). Performance of scalable data linked to internet of things: A case study. In S. Borah, R. Pradhan, N. Dey, & P. Gupta (Eds.), *Soft computing techniques and applications*, Advances in intelligent systems and computing (Vol. 1248). Singapore: Springer. https://doi.org/10.1007/978-981-15-7394-1_54
19. Verma, U., Sohani, M., Borah, S., Nagwanshi, K. K., & Pathak, S. (2022). Role of fog-assisted internet of things-enabled system for managing the impact of COVID-19. In S. Borah, S. K. Mishra, B. K. Mishra, V. E. Balas, & Z. Polkowski (Eds.), *Advances in data science and managemen*t, Lecture Notes on Data Engineering and Communications Technologies (Vol. 86). Singapore: Springer. https://doi.org/10.1007/978-981-16-5685-9_39

CHAPTER 16

MACHINE LEARNING IN LIBRARY AUTOMATION: ENHANCING SERVICES AND USER EXPERIENCE

RAJANI MESHRAM[1], MANOHAR KEDAR[2], ANIL BHANUDAS PAWAR[3], K. R. SENTHILKUMAR[4], RANJANA JADHAV[5], and YOGESH NAGARGOJE[6]

[1]*Jayakar Knowledge Resource Centre, Savitribai Phule Pune University, Pune, India*

[2]*SNJB's Late Sau. K. B. Jain College of Engineering, Neminagar, Chandwad, Nashik, India*

[3]*Art's Science and Commerce College, Kolhar TAL, Rahata, Ahmednagar, India*

[4]*Sri Krishna Arts and Science College, Coimbatore, India*

[5]*Bharati Vidyapeeth (deemed to be) University, Institute of Management and Entrepreneurship Development, Erandwane, Pune, India*

[6]*CSMSS CHH. Shahu College of Engineering, Kanchanwadi, Paithan Road, Chhatrapati Sambhaji Nagar (Aurangabad), India*

ABSTRACT

Library automation systems have been created with the intention of optimizing and managing various tasks and procedures within a library. Considering the escalating demand for services, the utilization of machine learning techniques has been implemented to offer superior user solutions. Within this chapter, there is a comprehensive discussion on

Cyber-Physical Systems: Applications, Challenges, and Research Directions. Rajanikanth Aluvalu, Samarjeet Borah, & Lalit Garg (Eds.)

DOI: 10.1201/9781779643421-16

the implementation of machine learning techniques in library automation software. The focal points where machine learning may play a significant role in library automation have been emphasized. Lastly, a generalized machine learning-based library automation model has been deliberated, which can be amalgamated with the software solution of a library system.

16.1 INTRODUCTION

Library automation refers to the employment of technology and software to optimize various library processes and services, thereby increasing the efficiency and accessibility of these services to library staff and patrons. In the past, libraries relied on manual methods for cataloguing, circulation, and resource management. However, with the advent of computers and digital technologies, integrated library systems (ILS) [1, 2] were adopted to automate these tasks.

An ILS is a centralized software that combines multiple functions, such as acquisitions, circulation, and cataloguing, into a single platform. This system enables libraries to manage their collections, track loans, handle patron records, and provide an online public access catalogue (OPAC) for users to search and locate library resources with ease. The integration of physical devices with effective software systems provides a smart interactive solution to users.

Cyber-physical systems (CPS) [3, 4] refer to the integration of computational elements (cyber) with physical processes (physical) to create intelligent, interconnected systems. These systems sense and respond to the physical world through sensors and actuators, making them adaptive and efficient. By integrating CPS principles, libraries can create an interconnected system where digital elements (software, databases) interact with physical elements (books, devices) to enhance user experiences and operational efficiency. Libraries can collect data on user preferences, borrowing patterns, and resource utilization, using data analytics and machine learning to make informed decisions about resource acquisition, allocation, and user services.

16.1.1 MACHINE LEARNING

Machine learning, a constituent part of artificial intelligence, assumes a pivotal role in the progression of library automation. Its modus operandi involves the indoctrination of algorithms with copious amounts of data in order to glean patterns and engender predictions or decisions sans explicit programming for each task [5]. The foundational tenets of machine learning are delineated thus.

1. **Learning from Data**

At the center of machine learning lies the fundamental principle of acquiring knowledge from data. Rather than being explicitly programmed with predetermined rules or instructions, machine learning models undergo training on extensive datasets in order to identify and comprehend patterns, correlations, and tendencies within the data.

2. **Generalization**

The goal of machine learning is to generalize from the training data and apply what it has learned to new, unseen data. This ability to make accurate predictions on unseen data is crucial for the practical application of machine learning models.

3. **Feature Extraction and Representation**

Feature extraction involves selecting relevant information from raw data to create a meaningful representation that the machine learning algorithm can work with [6]. The quality of features significantly impacts the performance of the model.

4. **Model Evaluation and Validation**

The performance of machine learning models is assessed through evaluation and validation techniques. This ensures that the model's predictions are accurate and reliable when applied to real-world scenarios.

5. **Bias and Fairness**

Machine learning models can inherit biases present in the training data, leading to unfair or discriminatory outcomes. Addressing bias and ensuring fairness is a critical ethical consideration in machine learning development.

6. **Model Interpretability**

As machine learning is increasingly used in critical applications, model interpretability becomes important for understanding how decisions are made. Interpretability helps build trust and enables humans to comprehend and potentially challenge the model's decisions.

16.1.2 CATEGORY OF MACHINE LEARNING ALGORITHMS

Machine learning algorithms can be categorized into categories [7]. These are listed below:

1. Supervised Learning

In supervised learning, the machine learning algorithm is trained using labeled data, where the input data is paired with corresponding correct outputs. The model learns to map inputs to outputs based on the provided examples:

1. Linear Regression: Used for predicting continuous numeric values, such as predicting house prices based on features like area, number of rooms, and so forth.
2. Logistic Regression: Applied to binary classification problems, like determining whether an email is spam or not.
3. Support Vector Machines (SVM): Used for classification tasks, especially when dealing with nonlinear decision boundaries, such as image classification.
4. Decision Trees and Random Forests: Suitable for both classification and regression tasks, used in applications like credit scoring and medical diagnosis.
5. K-Nearest Neighbors (KNN): Used for classification and regression, KNN predicts the target value based on the k-nearest data points.
6. Naive Bayes: Effective for text classification tasks like sentiment analysis and spam detection.

2. Unsupervised Learning

In unsupervised learning, the algorithm is trained on unlabeled data and must identify patterns or structures within the data on its own. This type of learning is used for clustering, dimensionality reduction, and other exploratory tasks:

a. K-Means Clustering: Used to partition data into k clusters based on similarities. Applications include customer segmentation and image compression.

b. Hierarchical Clustering: Arranges data into a tree-like structure of clusters, useful for taxonomic studies and document clustering.
c. Principal Component Analysis (PCA): Reduces the dimensionality of data while preserving its most important features, commonly used for data visualization and feature extraction.
d t-Distributed Stochastic Neighbor Embedding (t-SNE): Another dimensionality reduction technique often used for visualizing high-dimensional data.
e. Autoencoders: A type of neural network used for unsupervised feature learning and anomaly detection.

3. Semi-supervised Learning

Semi-supervised learning combines elements of supervised and unsupervised learning, where the algorithm is trained on a combination of labeled and unlabeled data. This approach is useful when obtaining labeled data is expensive or time-consuming.

4. Reinforcement Learning

In reinforcement learning, an agent learns to make decisions in an environment to achieve specific goals [8, 9]. The agent is capable of receiving constructive criticism through incentives or penalties, which enhances its competence and boosts its effectiveness over time. In the realm of game playing, reinforcement learning techniques have been efficaciously utilized to play games such as chess, video games, and so forth.

In Robotics, the application of reinforcement learning allows robots to acquire new skills such as walking or object manipulation. Reinforcement learning is useful in Recommendation Systems, as it enables the optimization of recommendation algorithms that suggest items for users that maximize their engagement. Moreover, reinforcement learning has a critical role to play in training autonomous vehicles to make appropriate decisions while driving.

Machine learning is a dynamic field with various algorithms and approaches, and it continues to evolve with ongoing research and advancements in AI. Its applications span across numerous domains, including image and speech recognition, natural language processing, recommendation systems, autonomous vehicles, and more.

16.2 UNDERSTANDING MACHINE LEARNING IN LIBRARY AUTOMATION

In the context of library automation, machine learning brings several key benefits.

16.2.1 PERSONALIZED RECOMMENDATIONS

Machine learning algorithms analyze users' borrowing histories, search behavior, and preferences to provide personalized reading recommendations. This helps patrons discover new materials tailored to their interests.

16.2.2 RESOURCE MANAGEMENT AND PREDICTIVE ANALYTICS

Machine learning models can analyze historical library data, such as circulation records and resource usage, to predict future demand for certain materials. Libraries can then optimize collection development and resource allocation based on these predictions.

16.2.3 NATURAL LANGUAGE PROCESSING (NLP) FOR CATALOGUING

NLP is utilized to extract and understand information from textual descriptions, enabling more efficient and accurate cataloguing and metadata generation for library resources.

16.2.4 VIRTUAL ASSISTANTS AND CHATBOTS

Machine learning-driven chatbots can handle routine patron inquiries, provide instant answers, and guide users to relevant resources, enhancing the library's support services.

16.2.5 USER BEHAVIOR ANALYSIS

Machine learning algorithms can analyze user behavior on library websites and OPACs, providing insights into how patrons interact with the library's online services, leading to better interface design and improved user experiences.

16.2.6 FRAUD DETECTION AND SECURITY

Machine learning can help detect and prevent fraudulent activities, such as identifying potential copyright violations or unauthorized access to electronic resources.

Overall, the implementation of machine learning techniques can enhance a library's comprehension of their users, streamline their operational procedures, and provide tailored and effective services. This approach reinforces a user-centric methodology and supports libraries in remaining technologically advanced, thereby rendering resources and information more accessible to their patrons. As the field of machine learning continues to mature, its involvement in library automation is presumed to broaden, resulting in additional enhancements and advancements in library services.

16.3 APPLICATIONS OF MACHINE LEARNING IN LIBRARIES

16.3.1 PERSONALIZED RECOMMENDATIONS

Personalized recommendations in library automation, which are enabled by the implementation of machine learning (ML) technology, have the potential to revolutionize the manner in which libraries cater to their patrons [10]. Through the analysis of user data and preferences, ML algorithms can provide customized book recommendations, resource suggestions, and an array of other personalized services. The incorporation of personalized recommendations in library automation, which are powered by ML technology, can result in heightened user engagement, increased resource utilization, and enhanced user satisfaction. By providing pertinent and captivating suggestions, libraries can enrich the user experience and foster stronger relationships with their patrons, thereby augmenting the value and impact of their service offerings.

16.3.2 PREDICTIVE ANALYTICS FOR RESOURCE MANAGEMENT

Predictive analytics for library resource management involves using historical library data and machine learning algorithms to make predictions about future library trends and resource demands [11, 12]. By analyzing past patterns and user behavior, predictive analytics can help libraries optimize collection development, allocate resources efficiently, and enhance overall library services. Predictive analytics is an iterative process that improves over time with more data and feedback. Libraries can regularly evaluate the accuracy and effectiveness of predictions to fine-tune their models and ensure ongoing improvement.

By leveraging predictive analytics for library resource management, libraries can proactively respond to user needs, optimize resource allocation, and deliver more personalized and efficient services. However, it is essential to handle data responsibly and consider privacy and ethical implications when implementing predictive analytics in library operations. Some more area of implementation may be:

a. Forecasting book demand and optimizing collection development.
b. Data-driven decisions for resource allocation and budgeting.

16.3.3 AUTOMATED CATALOGUING AND METADATA GENERATION

Automated cataloguing in libraries using machine learning (ML) is a transformative approach that streamlines the process of adding new resources to the library's collection and generating metadata for these resources. Traditionally, cataloguing and metadata generation have been manual and time-consuming tasks, but ML can automate and accelerate these processes, leading to more efficient library operations. Automated cataloguing and metadata generation through ML not only save time and effort for library staff but also improve the accuracy and consistency of metadata. By utilizing machine learning algorithms, libraries can create comprehensive and standardized metadata for their resources, leading to better resource discoverability and an enhanced user experience. However, it is essential to regularly update and improve ML models, validate their performance, and ensure the appropriate handling of user data and privacy during the automated cataloguing process.

Machine learning techniques can also be used in:

a. Natural Language Processing (NLP) for cataloguing and classification.
b. Improving accuracy and efficiency in data entry.

16.3.4 ENHANCING USER SERVICES

Enhancing library user services through machine learning (ML) can significantly improve the overall user experience and satisfaction. ML can automate and optimize various aspects of library services, making them more personalized, efficient, and user centric. ML algorithms can enhance content discovery by suggesting related resources, enabling users to explore and discover materials beyond their initial search. By leveraging ML to enhance library user services, libraries can create a more engaging, efficient, and user-friendly environment. However, it is crucial to prioritize data privacy and ensure transparency in ML-driven processes to build trust and confidence among library users. Regular evaluation and user feedback are essential for continuously improving and fine-tuning ML-based services to meet the evolving needs of library patrons. The following functionalities also will be useful for an automated library system.

a. Chatbots and virtual assistants for patron support.
b. 24/7 availability and instant responses to common queries.

Further, there may be tailoring services based on user preferences and behavior.

16.4 BENEFITS OF MACHINE LEARNING IN LIBRARY AUTOMATION

16.4.1 IMPROVED USER ENGAGEMENT AND SATISFACTION

Machine learning-based recommendation systems can significantly enhance library automation by providing personalized and relevant content to library patrons [13]. These systems leverage user data, such as borrowing history, preferences, and interactions with library resources, to deliver tailored recommendations and improve user experiences. Here is how machine learning-based recommendation systems can be applied in library automation.

1. Personalized Book Recommendations

Machine learning algorithms can analyze a user's borrowing history, ratings, and reading preferences to suggest books that align with their interests. By understanding user preferences, the recommendation system can offer more accurate and personalized book suggestions.

2. Article and Research Paper Recommendations

For academic and research libraries, recommendation systems can use machine learning to analyze users' search queries, academic interests, and past reading behavior to suggest relevant research papers, articles, and scholarly resources.

3. Digital Resource Recommendations

Libraries increasingly offer digital resources, such as e-books, audiobooks, and online journals. A machine learning-based recommendation system can analyze user interactions with these resources and provide personalized suggestions to users based on their digital reading habits.

4. Similarity-Based Recommendations

Machine learning algorithms can identify similar items (books, articles, etc.) based on content features, topic modeling, or collaborative filtering techniques. This enables the system to suggest related items to users, encouraging them to explore resources beyond their initial interests.

5. Context-Aware Recommendations

Recommendation systems can take into account contextual factors such as the user's location, time of day, and current events to provide relevant recommendations. For example, the system can suggest travel guides when a user is near a vacation destination or academic resources during exam periods.

6. Hybrid Recommendation Approaches

Combining different recommendation techniques, such as collaborative filtering and content-based filtering, in a hybrid approach can improve the accuracy and diversity of recommendations.

7. Reinforcement Learning for Continuous Improvement

Reinforcement learning-oriented recommendation systems possess the ability to perpetually acquire knowledge from user interactions and feedback in order to enhance and perfect their recommendations over the course of time. Explicit feedback on suggested items, ratings, and reviews are all viable means through which user feedback may be collected.

8. Adaptive Interfaces

The utilization of machine learning algorithms can enable the personalization of the online interface of the library in accordance with the preferences and behavior of its users. The system can present tailored content, such as book covers or featured resources, with the purpose of augmenting the user's browsing experience.

By means of implementing recommendation systems based on machine learning, libraries have the ability to produce a more captivating and individualized encounter for their clients. Nevertheless, it is of utmost importance to accord priority to data privacy and security, guaranteeing the responsible handling and consent of user data. Consistent evaluation and feedback from users are crucial for the enhancement and optimization of recommendation algorithms, thereby bestowing upon users more precise and valuable recommendations as the automation of libraries continues to progress.

16.4.2 DATA-DRIVEN DECISION-MAKING FOR RESOURCE MANAGEMENT

Data-oriented decision-making constitutes a potent methodology for the management of library resources. This approach empowers libraries with the capacity to make judicious and empirically supported decisions that optimize their collections, services, and operations [14]. Through the exploitation of data from numerous sources, libraries can obtain invaluable insights that foster more streamlined and efficacious resource management. The following are ways in which data-driven decision-making can be implemented in the context of library resource management:

1. Collection Development and Acquisition

The utilization of data analytics provides libraries with the ability to evaluate usage patterns and the level of popularity pertaining to various resources within their collections. Through the analysis of borrowing statistics, user preferences, and circulation trends, libraries can detect materials that are in high demand, as well as popular genres and subject areas that require further attention. This information can serve as a guide for the selection and procurement of new resources, aimed at meeting the ever-evolving needs of patrons.

2. Weeding and Deselection

Utilization of data-driven analysis can be of great assistance in identifying underutilized or outdated materials within the library collection. By carefully scrutinizing circulation data, publication dates, and user feedback, libraries can make informed decisions based on the collected data, which can guide the process of weeding or deselecting items, thereby creating additional space on the shelves for more pertinent and highly sought-after resources.

3. Resource Sharing and Interlibrary Loan (ILL)

Analyzing the demand for resources that are not available within the library's collection can help libraries prioritize resource sharing and interlibrary loan services. Data on ILL requests and unmet user needs can inform decisions on expanding partnerships with other libraries or consortia to enhance resource accessibility.

4. Budget Allocation and Spending

Data-driven analysis of resource usage and collection performance can help libraries allocate their budgets more efficiently. By understanding which resources generate the most value and usage, libraries can prioritize spending on materials that are in high demand and align with user interests.

5. Usage and Access Monitoring

Data analytics can provide real-time insights into how patrons interact with digital resources and online platforms. By tracking usage patterns, libraries can assess the effectiveness of their digital collections, databases, and online services, leading to improvements and better user experiences.

6. Resource Duplication and Format Decisions

Analyzing data on resource duplication and format preferences can help libraries make informed decisions about the most suitable formats and quantities of resources to acquire. For instance, data might reveal that certain resources are highly sought after in both print and digital formats, prompting the library to invest in multiple formats to cater to diverse user preferences.

7. Space and Facility Management

Data on library space usage, foot traffic, and seating patterns can aid in optimizing the layout and arrangement of library spaces. Libraries can use this data to create more comfortable and user-friendly environments, ensuring that resources and services are easily accessible to patrons.

By adopting a data-driven approach to resource management, libraries can align their collections and services with user needs, optimize resource allocation, and continuously improve the overall library experience. It is crucial for libraries to invest in data collection, storage, and analysis tools and to develop the necessary expertise to interpret and act on the insights derived from data analytics. Regular assessment and evaluation will enable libraries to adapt and thrive in an ever-changing information landscape.

16.4.3 ENHANCED USER EXPERIENCE THROUGH PERSONALIZED SERVICES

Enhancing library user experience through personalized services is a powerful way to engage and meet the diverse needs of library patrons. By tailoring services to individual preferences and interests, libraries can create a more user-centric and enjoyable experience. Here are some strategies to achieve personalized services and improve library user experience.

1. Personalized Book Recommendations

Implement a recommendation system that suggests books, articles, or other resources based on a user's borrowing history, ratings, and interests. Provide personalized reading lists and recommendations to help users discover content that aligns with their preferences.

2. Customized Content Displays

Tailor the library's website or online catalog interface to display personalized content to users. Show recommended materials, new releases in preferred genres, or resources related to the user's recent searches to facilitate easier and quicker access to relevant information.

3. Individualized Communication

Use patron data to send personalized communication, such as email newsletters, alerts for upcoming events or workshops, and notifications about borrowed items nearing due dates.

4. Personalized Research Assistance

Offer personalized research assistance by analyzing user queries and providing tailored information and resource suggestions to support their research needs.

5. Preferred Format Options

Allow users to set their preferred format options for borrowing materials, such as print, digital, or audiobooks, to accommodate individual preferences.

6. User-Driven Service Development

Engage with users through surveys, feedback forms, and focus groups to understand their needs and preferences. Use this input to inform service development and improvements.

7. Tailored Events and Workshops

Offer events, workshops, and programs that cater to specific user interests and demographics. Customized events can range from book clubs focused on preferred genres to technology workshops based on individual skill levels.

8. User Profiles and Accounts

Enable users to create personalized profiles with preferred genres, topics of interest, and reading history. This information can guide the delivery of personalized recommendations and services.

9. Loyalty Programs and Rewards

Implement loyalty programs or rewards for frequent users, offering incentives such as extended borrowing periods, priority access to popular resources, or special library privileges.

10. Personalized Support and Assistance

Utilize chatbots or virtual assistants to provide personalized support and answer user inquiries. These systems can learn from user interactions and offer more accurate and relevant responses over time.

By embracing personalized services, libraries can create a more welcoming and inclusive environment that fosters stronger connections with their patrons. It is essential to prioritize user data privacy and obtain explicit consent for data usage to ensure that personalized services are delivered responsibly and ethically. Regular evaluation and feedback from users will enable libraries to continuously improve and refine their personalized offerings to enhance the overall library user experience.

16.5 CHALLENGES AND LIMITATIONS

While machine learning (ML) offers numerous benefits and opportunities for library automation, it also comes with several challenges and limitations that need to be addressed for successful implementation. Some of the key challenges and limitations of using ML in library automation are given in the following sections.

16.5.1 DATA QUALITY AND AVAILABILITY

ML algorithms heavily rely on high-quality and diverse data for effective learning. Libraries may face challenges in accessing and curating large and relevant datasets, especially for specialized or niche subjects. Incomplete or biased data can lead to inaccurate or unfair recommendations and decisions.

16.5.2 DATA PRIVACY AND ETHICS

Using ML in library automation involves handling user data, which raises privacy and ethical concerns. Libraries must ensure compliance with data protection regulations and establish robust security measures to protect user information.

16.5.3 BIAS AND FAIRNESS

ML algorithms can inadvertently inherit biases present in the training data, leading to biased recommendations or decision-making. Ensuring fairness and mitigating bias in ML models is a critical challenge that requires careful data selection, preprocessing, and model evaluation.

16.5.4 INTERPRETABILITY

Many ML algorithms, especially complex deep learning models, lack interpretability, making it challenging to understand the reasoning behind their decisions. In critical applications, interpretability is crucial to gain user trust and ensure accountability.

16.5.5 OVERFITTING AND GENERALIZATION

ML models can sometimes be overfit to the training data, meaning they perform well on the training data but fail to generalize to unseen data. Ensuring that ML models generalize well to new data is essential for their practical application in library services.

16.5.6 MODEL COMPLEXITY AND RESOURCE REQUIREMENTS

Some advanced ML models, such as deep learning models, can be computationally expensive and require significant computational resources and expertise for training and deployment. This can be a limitation for smaller libraries with limited resources.

16.5.7 CONTINUOUS LEARNING AND UPDATES

Library user preferences and needs may evolve over time, requiring ML models to adapt and update continuously. Implementing mechanisms for continuous learning and model updates is a challenge that libraries need to address.

16.5.8 USER ACCEPTANCE AND ADOPTION

Introducing ML-based systems may face resistance or hesitation from users who may be sceptical about the accuracy, privacy, or usefulness of the recommendations. Libraries need to communicate the benefits and value of ML-driven services to gain user acceptance.

16.5.9 INTEGRATION WITH EXISTING SYSTEMS

Integrating ML solutions into existing library automation systems can be complex and require careful planning and coordination to ensure compatibility and seamless operation.

16.5.10 EXPERTISE AND TRAINING

Implementing ML in library automation requires specialized expertise in data science and ML. Libraries may need to invest in training staff or collaborate with external experts to effectively deploy ML solutions.

Addressing these challenges and limitations requires a thoughtful and well-planned approach to ensure the responsible and successful use of ML in library automation. Libraries should prioritize data quality, privacy, fairness, and user-centric design in their ML-driven initiatives to deliver valuable and user-friendly services to their patrons.

16.6 A MACHINE LEARNING BASED MODEL FOR LIBRARY AUTOMATION

A generalized machine learning based model has been presented in this section. A simplified diagram is presented in Figure 16.1, illustrating how machine learning can be integrated into library automation to enhance various aspects of library services.

a. The Library Catalogue serves as the central repository of all library resources, including books, articles, multimedia materials, and digital content.

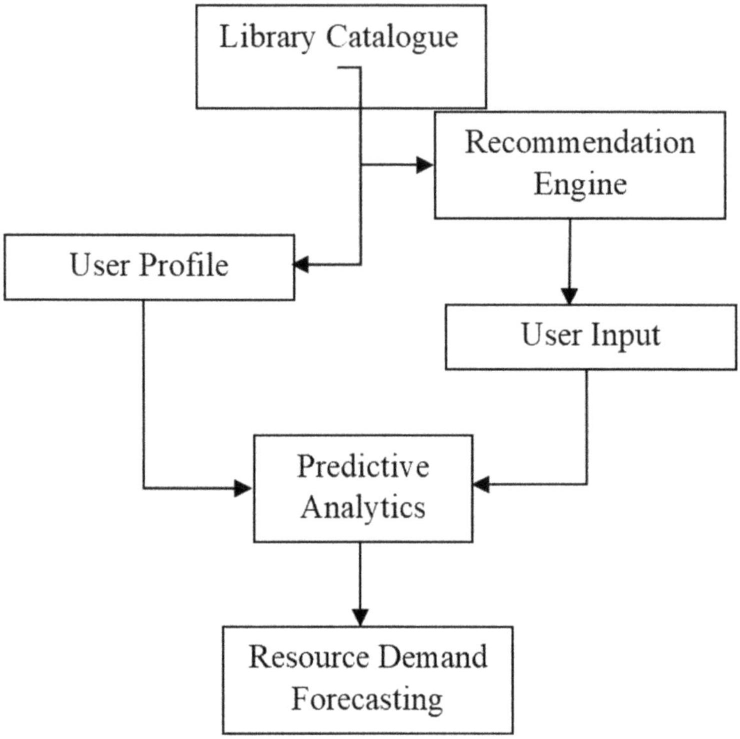

FIGURE 16.1 A generalized machine learning based library automation model.

b. User Profiles store information about individual users, including their borrowing history, preferences, and interests.
c. The Recommendation Engine uses machine learning algorithms to analyze user profiles and provide personalized recommendations to users based on their preferences and behavior.
d. User Input represents the interactions and queries from library users, such as search queries, borrowing requests, and feedback.
e. The Predictive Analytics component employs ML algorithms to analyze user behavior, historical borrowing data, and resource usage patterns to forecast future resource demands and optimize collection development.
f. The Resource Demand Forecasting module uses ML-based predictive analytics to estimate the demand for specific resources, enabling the library to plan resource allocation and collection development more efficiently.

By integrating machine learning into library automation, libraries can offer personalized services, improve resource management, and enhance the overall user experience for their patrons.

16.7 CONCLUSION

This chapter provides a detailed discussion on implementation and usefulness of machine learning techniques in library automation systems. Automation has been adopted by majority of the large-sized libraries by aligning the physical devices with software solutions; together which can be referred as a cyber physical system. Various services provided by such kinds of systems may be budget allocation and spending, usage and access monitoring, resource duplication and format decisions, space and facility management, resource sharing and interlibrary loan, and so forth. Machine learning techniques can be implemented in various phases of such kind of a system. However, several issues are also there pertaining to this implementation such as data quality and availability, data privacy and ethics, interpretability, and so forth, which need to be taken care of.

KEYWORDS

- **cyber-physical systems**
- **ILS**
- **library automation**
- **machine learning**
- **services**
- **user experience**

REFERENCES

1. Chow, A. S., & Bucknall, T. (2012). 6-Emerging technology trends in libraries. *Library technology and user services, chandos information professional series* (pp. 105–130). Chandos Publishing. https://doi.org/10.1016/B978-1-84334-638-8.50006-0.

2. Sikhosana, R. B., Rajkoomar, M., & Naresh, S. R. (2022). Adoption and maintenance of the next generation integrated library systems (ILS) in academic libraries. In *Innovative technologies for enhancing knowledge access in academic libraries*. New York: IGI Global. https://doi.org/10.4018/978-1-6684-3364-5.ch013
3. Sedigh, S., & Hurson, A. (2012). Chapter one - Introduction and preface. *Advances in Computers*, *87*, 1–6. Elsevier. https://doi.org/10.1016/B978-0-12-396528-8.00001-8
4. Sharma, N. V., Yadav, N. S., & Sharma, S. (2022), Machine learning and security in cyber physical systems. *Cyber-Physical Systems*, 171–187. Academic Press. https://doi.org/10.1016/B978-0-12-824557-6.00015-7
5. Brown, S. (2021). Machine learning, explained. https://mitsloan.mit.edu/ideas-made-to-matter/machine-learning-explained (accessed August 12, 2023)
6. Ippolito, P. P. (2019). Feature extraction techniques: Towards data science. https://towardsdatascience.com/feature-extraction-techniques-d619b56e31be (accessed August 14, 2023)
7. Machine Learning Algorithms. (2021). https://www.javatpoint.com/machine-learning-algorithms (accessed August 14, 2023)
8. Barto, A. G. (1997). Chapter 2 - Reinforcement learning. *Neural Systems For Control*, 7–30. Academic Press. https://doi.org/10.1016/B978-012526430-3/50003-9
9. Belyadi, H., & Haghighat, A. (2021). Chapter 3 - Machine learning workflows and types. *Machine Learning Guide for Oil and Gas Using Python*, 97–123. Gulf Professional Publishing. https://doi.org/10.1016/B978-0-12-821929-4.00001-9
10. Liu, M. (2022). Personalized recommendation system design for library resources through deep belief networks. In Y. Tang (Eds.), *Mobile information systems*. Hindawi. https://doi.org/10.1155/2022/7870724
11. Massis, B. E. (2012). Using predictive analytics in the library. *New Library World*, *113*(9/10), 491–494. https://doi.org/10.1108/03074801211273957
12. Iqbal, N., Jamil, F., Ahmad, S., & Kim, D. (2020). Toward effective planning and management using predictive analytics based on rental book data of academic libraries. *IEEE Access*, *8*, 81978–81996. https://doi.org/10.1109/ACCESS.2020.2990765
13. Masrek, M. N., Razali, M. H., Ramli, I., & Andromeda, T. (2018). User engagement and satisfaction: The case of web digital library. *International Journal of Engineering and Technology*, *7*(4.15). https://doi.org/10.14419/ijet.v7i4.15.21364
14. Massis, B. (2016). Data-driven decision-making in the library. *New Library World*, *117*(1/2), 131–134. https://doi.org/10.1108/NLW-10-2015-0081

CHAPTER 17

MULTI-OBJECTIVE META-HEURISTIC ALGORITHMS FOR BLDC MOTOR DESIGN PARAMETER OPTIMIZATION

SURYA NARAYAN TRIPATHY, SHUBHASRI KUNDU, and ARJYADHARA PRADHAN

School of Electrical Engineering, KIIT, Bhubaneswar, India

ABSTRACT

Nowadays permanent magnet brushless DC motors are highly sought in electrical vehicles, robotics, and other industrial applications. They provide higher efficiency, better torque, and a wider range of speed control than conventional motors. Based on the needs of the applications, performance improvement is always desirable yet challenging owing to non-linear behavior and intricate design. In the current work, the design dimension of a BLDC motor has been optimized using meta-heuristic algorithm. Based on fifteen motor dimensions as design factors, a single objective optimization technique as well as multi-objective optimization have been used to find the motor's optimal size, electromagnetic torque, power density, and efficiency. The results have been supported by comparison with the reference publication.

17.1 INTRODUCTION

At present permanent magnet brushless DC motor (PMBLDC) is widely used in a variety of applications due to its improved performance, high

Cyber-Physical Systems: Applications, Challenges, and Research Directions. Rajanikanth Aluvalu, Samarjeet Borah, & Lalit Garg (Eds.)

DOI: 10.1201/9781779643421-17

power density, speed dynamic response, low maintenance requirements, and wide speed control [l]. Many efforts have been made over the past 40 years to identify strategies to maximize the performance of electrical machines. Improvement can be achieved in two ways, first to maximize power density, and efficiency, and reduce torque ripple as well as the size and weight of the motor, among other things, power electronics are used in control [5]. Second, improved design and new materials have been created, among other things, to manage thinner lamination and improve performance and affordability [7].

In general, the performance of the machine is determined by its dimensions. Machines are often made with the least amount of production and material expenses while still retaining an acceptable operational cost [4]. There are several applications, such as aircraft, robotics, and medical, where volume and weight are more important factors than price. The complexity of the machine makes it difficult and time consuming to choose the appropriate design. Many design criteria are conflicting, including power density and the smallest possible magnet volume [8]. Because of the complex relationships between many geometrical factors and the material nonlinearities, optimal machine design is inherently a multi-objective problem. Recent surveys have been used to determine the most effective design parameters for BLDC motors by applying advanced optimal techniques [9].

Using three design variables—slot opening, magnet length, and air gap length, Kamal et al. [11] employed PSO and GA in 2017 to reduce cogging torque. With GA and PSO, the cogging torque was decreased from 0.0510 to 0.04 Nm and 0.0510 to 0.025 Nm, respectively. Noori et al. [10] with a new approach to optimization with modified version of the social spider algorithm was used to design a slotless permanent magnet brushless DC motor while concurrently optimizing cost, volume, and efficiency across the ten motor design parameters. The motor design that was produced has been compared to the GA, SSA, CBA (Chaotic Bat Algorithm), and BA (Bees Algorithms), demonstrating that the proposed design has fewer deviations and can provide outcomes with fewer iterations.

Chakkarapani [6] uses three algorithms, PESA (Pareto Envelope-based Selection Algorithm), PAES (Pareto Archived Evolutionary Strategy), and NSGA-II with three objective functions—output torque, total loss, and volume and six design variables to build a BLDC motor. In 2021, Barba et al. [3] used an evolutionary algorithm by taking into account five design variables and applying Estra's single- and multi-objective optimization techniques, five objective functions have been optimized. According to Zaaraoui and Mansouri [2], two swarm optimization techniques—improved PSO multi-objective optimization and the speed-constrained multi-objective PSO—are used to maximize performance and decrease the weight of the in-wheel outer rotor surface mounted permanent magnet motor while simultaneously satisfying seven mechanical and magnetic constraints. For the design of BLDC with 3D structure in the recent year 2022, Jo et al. [1] used the Response Surface Methodology (RSM). Cogging torque reduction and back emf reduction were chosen as the objective functions for enhancing BLDC performance.

From the review, it was observed that numerous evolutionary algorithms have been used to optimize the cogging torque, efficiency, cost, volume, weight, power density, and so forth of BLDC motor. A maximum of nine machine dimensions with three or four objective functions have been chosen as objective function for the design of BLDC motors. Finding the ideal design variable values to maximize performance was challenging when there were more design considerations to take into account, but it became much easier with additional design analysis. In the present work, fifteen design parameters of the BLDC motor are used as design variables to optimize power density, efficiency, electromagnetic torque, and power losses by using both single objective and multi-objective optimization using GA.

17.2 FORMULATION OF OBJECTIVE FUNCTION

An objective function that simultaneously takes power losses, volume, electromagnetic torque, and motor efficiency into consideration has been established here in addition to the mechanical and electrical constraints. The construction of a brushless DC motor is fairly similar to ac synchronous motor, with the exception that a hall optical sensor is used to detect the magnetic poles, that is, the location of the rotor [15]. The design of permanent magnets takes into account a number of internal parameters that should fall within the range.

Various design objectives are considered for the optimization as minimization of joule loss, and volume, maximization of electromagnet torque, power density, and efficiency. The design variables are considered, such as the bore diameter, axial length, magnet thickness, winding thickness, air gap, yoke thickness, current density, form factor, pole arc per pole pitch, magnetic flux density, and interpolar leakage coefficient.

The basic design of an electric motor starts with the equations for size, power demand, and torque for a given load [4]. The sizing equation is given by:

$$S = 11k_w B * AC * \left(\frac{D}{100}\right)^2 \left(\frac{L}{100}\right)\eta \tag{17.1}$$

The total output power of the machine:

$$Q = 1.11P^2 B_{av} ACK_w D^2 Ln_s \tag{17.2}$$

The total volume of the machine:

$$V_T = \frac{\pi}{4} L\left[D + 2\left(h_s + h_{bis}\right)\right]^2 \tag{17.3}$$

So, the power density can be calculated as it is total apparent output power per volume as:

$$PD = \frac{1.11P^2 B_{av} ACK_w D^2 Ln_s}{\frac{\pi}{4} L\left[D + 2\left(h_s + h_{bis}\right)\right]^2} \tag{17.4}$$

To calculate the efficiency of the machine, first, it is necessary to find the losses in the machine. The losses in machine can be divided into three types: mechanical, magnetic, and electrical.

The major power loss occurs in the resistance of the winding which can be calculated as:

$$P_R = \rho K_f K_c K_{et} A_w LJ_{cu}^{\ 2} \tag{17.5}$$

The winding cross section can be found as:

$$A_w = \pi t_w \left(2R + 2M_a + t_w\right) \tag{17.6}$$

By the application of maximum flux density, the hysteresis and eddy current loss can be calculated as:

$$P_H = K_h B_m^{1.6} f \tag{17.7}$$

$$P_E = K_e B_m^2 f^2 \tag{17.8}$$

The mechanical power losses in the machine which are due to friction and bearing (P_F), windage losses (P_W) (which depends on the rotor parameters) as 1.5% of the total output power.

In the magnetic core, the losses occur in the form of eddy current loss and hysteresis loss depending on the magnetic flux density which can be expressed as [13]:

$$B_m = \frac{\pi\beta K_l B_r l_m}{2Pl_y \ln(\frac{R+M_a+t_w}{R-t_m})} \tag{17.9}$$

Now the total losses in the machine can be calculated as:

$$P_{TL} = P_R + P_H + P_E + P_w + P_F \tag{17.10}$$

After putting the values, it becomes as:

$$P_{TL} = \rho K_f K_c K_{et} A_w L J_{cu}{}^2 + K_h B_m^{1.6} f + K_e B_m^2 f^2 + 0.015Q \tag{17.11}$$

The efficiency of the machine can be calculated as:

$$\eta = \frac{\text{Outputpower}}{\text{Inputpower}} = \frac{\text{OutputPower}}{\text{Outputpower} + \text{PowerLosses}}$$

$$\eta = \frac{1.11P^2 B_{av} ACK_w D^2 Ln_s}{1.11P^2 B_{av} ACK_w D^2 Ln_s + P_{TL}} \tag{17.12}$$

The developed electromagnetic torque can be calculated by considering the magnetic losses as [13]:

$$T_{em} = \frac{\pi K_f K_c K_l K_\beta B_r t_m L t_w (2R + 2M_a + t_w) J_{cu}}{\ln(\frac{R+M_a+t_w}{R-t_m})} \tag{17.13}$$

Now the main objective function can be designed as follows:

$$F = w_1 * PD + w_2 * P_{TL} + w_3 * T_{em} + w_4 \eta \tag{17.14}$$

Where w1, w2, w3, and w4 are the weighting factors, PD is the power density, P_{TL} is the total power loss, T_{em} is the electromagnetic torque, and η is the efficiency.

17.2.1 IMPLEMENTATION OF OPTIMIZATION TECHNIQUES

The ultimate objective of manufacturing machine is to achieve high quality performance while adhering to the strictest financial, volumetric, dimensional, and temporal constraints. The optimum balance between productivity, cost, and sustainability can be achieved by employing the right set of conditions. The best values for the parameter may be found using an optimization method. The application of optimization techniques in electrical machine design has expanded over the past three decades to meet fiercely competitive market demand. Multi-objective optimization is more difficult in comparison with single objective problem because there is no single solution rather there is a set of acceptable trade-off optimal solutions, known as the pare to front. It has the advantage of providing a designer with a wide range of alternatives from an overall standpoint.

In this design process, well-known optimization techniques including Genetic Algorithm and Particle Swarm Optimization have been used to increase the power density, efficiency, torque, loss, and volume of the BLDC motor while taking 15 design variables and 100 iterations into account. One of the most important factors in accelerating the optimization process is the parameter selection, which is based on human judgment and expertise [15–18].

17.3 GENETIC ALGORITHM

The genetic algorithm randomly generates the starting binary population for each of the fifteen design variables [8]. In this instance, the length of each population has been set to 10, which will affect the accurate value of the design variable. The flowchart for implementing GA in present research problem is shown in Figure 17.1. The population's fitness values have been evaluated using Equation 17.14. Strong species from the population were chosen for further processing based on the best fitness values [18]. Through cross-over and mutation, these chosen populations are permitted to reproduce and give rise to new population.

With the advancement of the search process, the population becomes more fit. Cross-over and mutation are the two key agents that increase fitness.

17.3.1 PSO

Despite certain similarities and variations between GA and PSO, PSO has some advantages, including increased computational efficiency due to the use of formal hypothesis testing and statistical analysis [8]. Multi-objective PSO demonstrated success in achieving good convergence when applied to the solution of real-world issues. The solutions that PSO creates are referred to as particles. To begin the PSO approach, the positions and velocities of fifteen design variables must first be assigned randomly. The flowchart for the PSO strategy is shown in Figure 17.2. The flight paths of each particle are altered in a quasi-stochastic manner, enabling the swarm to search in the vicinity of the desired boundaries [18]. The pattern of each particle is moved by combining global and local searches. In Figure 17.2, the velocity and position updating are two crucial PSO phases that can be summed up as follows:

$$v_j^{t+1} = v_j^t + \alpha c_1 \left(g_b^* - x_j^t\right) + \beta c_2 \left(P_b^* - x_j^t\right) \tag{17.15}$$

Where x_j^t and v_j^t are the position and velocity of jth particle at time t.

Then the position is updated as:

$$x_j^{t+1} = x_j^t + v_j^{t+1} \tag{17.16}$$

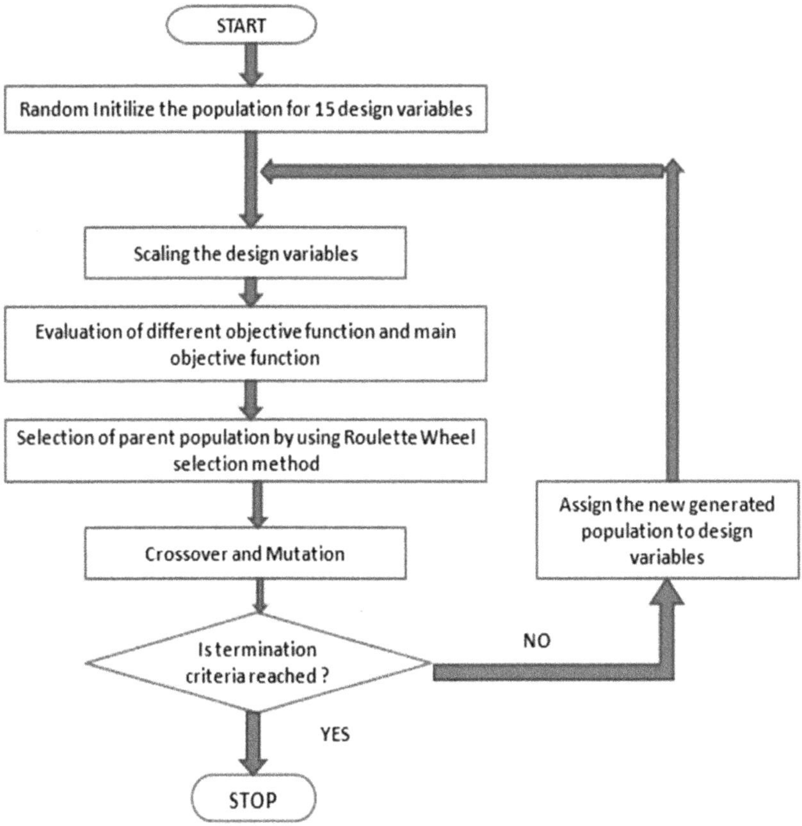

FIGURE 17.1 Flowchart of genetic algorithm.

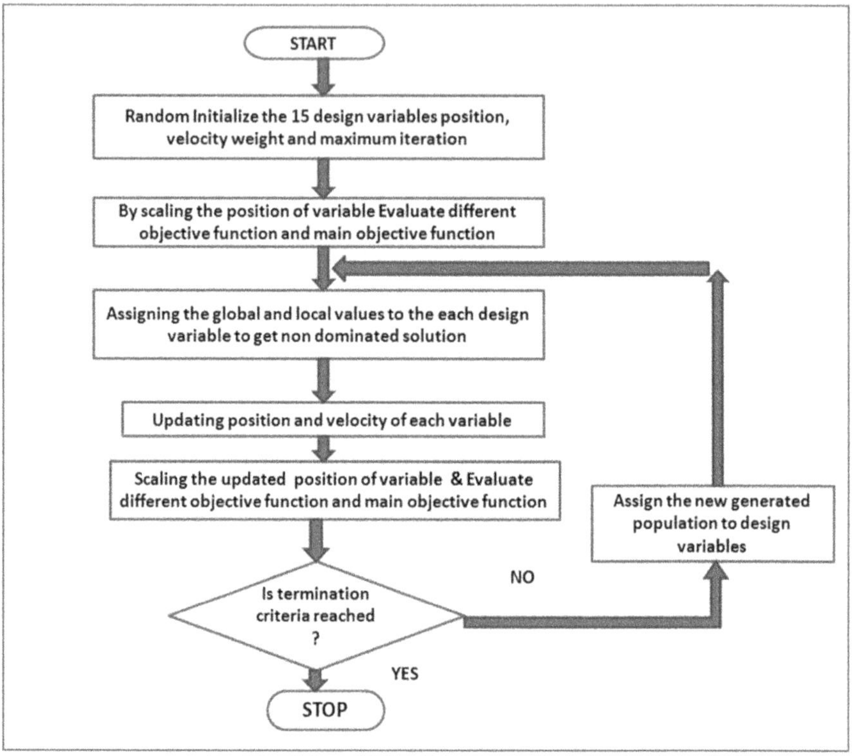

FIGURE 17.2 Flowchart of PSO.

Until they reach saturation or the maximum number of iterations, the particles continue to seek out and refine their global optimal position.

17.3.2 SIMULATED OUTPUT

The goal of this study is to simultaneously minimize joule losses and volume while maximizing power density, electromagnetic torque, and efficiency. To balance all the objectives and manage the relative importance of each one, four weighting considerations are taken into account. In this study, fifteen different design parameters should be optimized to get the best values of the objective function.

The parameters of motor as well as the upper and lower bounds of the parameter are displayed in Table 17.1.

TABLE 17.1 Design Variables and their Range.

Sl. No.	Parameter	Notation	Minimum Value	Maximum Value
1.	Number of pole pair	P	1	10
2.	Cross-sectional area of the winding (mm^2)	Ac	0.1	2
3.	Pole arc/pole pitch	B	0.5	1
4.	Magnet thickness (m)	Lm	$1*10^{-3}$	$50*10^{-3}$
5.	Stator/rotor yoke thickness	Ly	0.002	0.01
6.	Winding thickness (m)	Tw	0.001	0.0055
7.	Mechanical air gap	Mg	0.001	0.004
8.	Rotor radius (m)	R	0.005	0.1
9.	Axial length (m)	L	0.06	0.1
10.	Pole pitch (mm)	τ_p	40	80
11.	Average flux density (T)	B_{av}	0.4	0.9
12.	Specific electric loading	Ac	8000	30,000
13.	Stator back iron height (mm)	h_{bis}	5	15
14.	Slot height (mm)	h_s	10	25
15.	Current density	Jcu	3,000,000	6,000,000

17.3.3 DESIGN PARAMETER'S IMPACT ON OBJECTIVE FUNCTION

The following graphics illustrate how the objective functions can vary when a motor's geometrical parameters are changed while the rest of the parameters are kept constant. Each figure depicts the impact of changing each geometrical parameter of the motor on the objective function.

According to Figure 17.3, a large number of poles increase the cost of producing motor and lower magnetic losses due to the stator and rotor's low density, which increases power density. It should be emphasized that this problem has little effect on torque generation and efficiency but no effect on volume. It was determined by taking into account the specified objective function and figuring that using more poles might improve the design of the motor. However, this problem results in a rise in the magnetic flux leakage and a fall in the generating torque and efficiency.

The cost and volume of the motor are both impacted differently by the presence of significant winding space. However, it can generally result in an enhancement in torque production. It is important to remember that for a motor to deliver the necessary torque, the winding thickness must not go below an allowed limit. The variation of objective functions with winding thickness is shown in Figure 17.4.

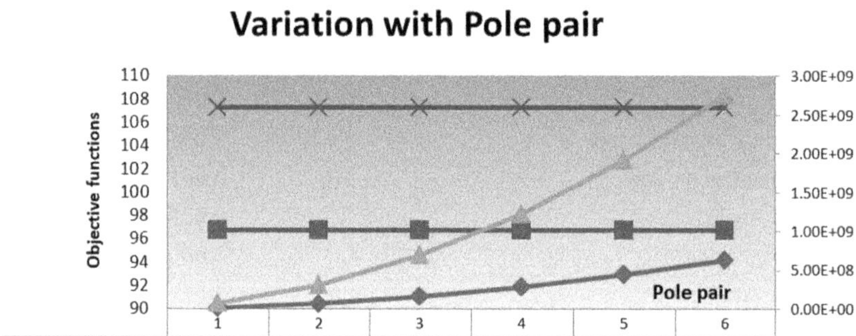

	1	2	3	4	5	6
Torque	10.53	10.5358	10.5317	10.526	10.5178	10.5098
Efficiency	96.76	96.78	96.78	96.79	96.79	96.79
Power Density	5.79E+07	2.36E+08	5.30E+08	9.43E+08	1.47E+09	2.12E+09
Fitness	1.77E+07	7.01E+07	1.59E+08	2.83E+08	4.42E+08	6.36E+08

FIGURE 17.3 Variation of objective function with pole pair.

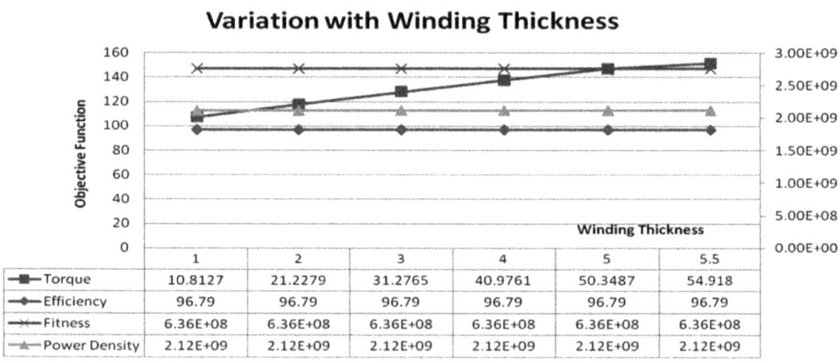

	1	2	3	4	5	5.5
Torque	10.8127	21.2279	31.2765	40.9761	50.3487	54.918
Efficiency	96.79	96.79	96.79	96.79	96.79	96.79
Fitness	6.36E+08	6.36E+08	6.36E+08	6.36E+08	6.36E+08	6.36E+08
Power Density	2.12E+09	2.12E+09	2.12E+09	2.12E+09	2.12E+09	2.12E+09

FIGURE 17.4 Variation of objective function with winding thickness.

According to the preceding Figure 17.5, decreasing the magnetic thickness (l_m) improves the volume and losses of the BLDC motor but decreases torque generation and has no impact on the motor's power density.

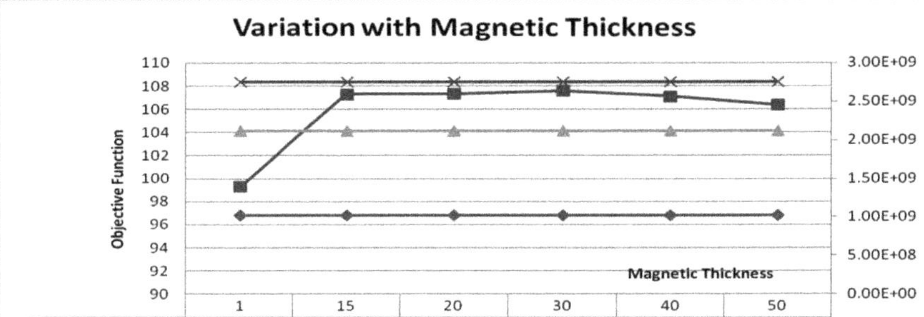

	1	15	20	30	40	50
Torque	2.5378	10.5098	10.5472	10.8127	10.3218	9.5738
Efficiency	96.79	96.79	96.79	96.79	96.79	96.79
Fitness	6.36E+08	6.36E+08	6.36E+08	6.36E+08	6.36E+08	6.36E+08
Power Density	2.12E+09	2.12E+09	2.12E+09	2.12E+09	2.12E+09	2.12E+09

FIGURE 17.5 Variation of objective function with magnetic thickness.

The cost and volume of the motor are both impacted differently by the presence of significant winding space. However, it can generally result in an enhancement in torque production. It is important to remember that for a motor to deliver the necessary torque, the winding thickness must not go below an allowed limit.

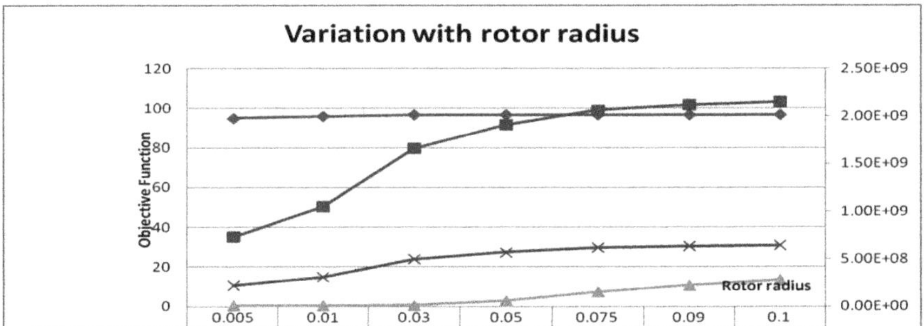

	0.005	0.01	0.03	0.05	0.075	0.09	0.1
Efficiency	95.05	96.07	96.68	96.76	96.78	96.79	96.79
Torque	0.5	0.5	0.7	2.8285	7.2611	10.8127	13.551
Fitness	2.21E+08	3.14E+08	4.98E+08	5.72E+08	6.19E+08	6.36E+08	6.45E+08
Power Density	7.36E+08	1.05E+09	1.66E+09	1.91E+09	2.06E+09	2.12E+09	2.15E+09

FIGURE 17.6 Variation of objective function with rotor radius.

As seen in the accompanying Figure 17.6, one of the most important characteristics of BLDC motors is rotor radius. Volume and losses will both decrease when Rr is reduced, but it should be remembered that a short rotor radius has a detrimental impact on creating the required output torque. On the other hand, a large rotor radius has an inverse relationship with the speed of the motor and increases torque, efficiency, and power density.

The chart indicates that the value of pole arc per pole pitch (β) has no impact on volume, power density, or efficiency. However, it should be noted that a large value of β can result in a reduction in torque generation because of its effect on increasing the magnetic leakage flux.

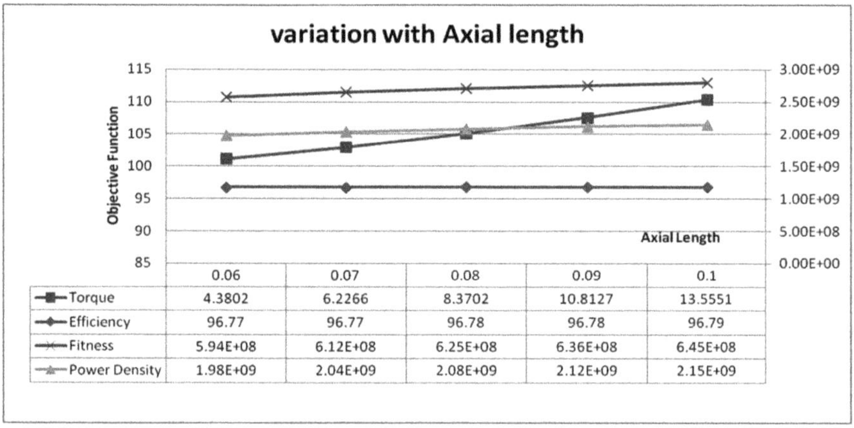

	0.06	0.07	0.08	0.09	0.1
Torque	4.3802	6.2266	8.3702	10.8127	13.5551
Efficiency	96.77	96.77	96.78	96.78	96.79
Fitness	5.94E+08	6.12E+08	6.25E+08	6.36E+08	6.45E+08
Power Density	1.98E+09	2.04E+09	2.08E+09	2.12E+09	2.15E+09

FIGURE 17.7 Variation of objective function with axial length.

Another crucial factor in motor design is axial length, as shown in Figure 17.7, which has a significant impact on volume, torque output, and performance but less so on efficiency. Although a small value of L is advantageous for obtaining less volume, this tiny amount might unfortunately have an unfavorable impact on the output torque and power density, which is undesirable. The performance and torque of the motor will rise with an increase in axial length.

17.3.4 RESULTS OF SINGLE OBJECTIVE OPTIMIZATION ALGORITHMS

To obtain the optimal parameter values of design variables, the objective functions have been optimized using single objective optimization for the BLDC motor using GA and PSO. The table contains the parameter values, which have been verified using the reference data. The majority of the geometrical optimization parameters, such as β, la, ly, lw, Rr, and h_{bis}, have been found to have their optimal values, which enhance losses, efficiency, torque, and power density while also reducing volume and cost.

The values of the design factors to obtain the best values of the objective functions, such as power density, efficiency, joule power loss, and torque individually using SOGA, are shown in Tables 17.2 and 17.3.

Similarly, the outcomes of optimal design variables by the use of Single objective PSO is presented in Table 17.3.

17.3.5 RESULTS OBTAINED BY CONSIDERING MULTI-OBJECTIVE FUNCTIONS

To get the optimum values of all objectives simultaneously multi-objective optimization approach using GA has implemented. The values of the design factors to obtain the best values of the objective functions such as power density, efficiency, joule power loss, and torque simultaneously using MOGA are shown in Table 17.4.

According to the results of the multi-objective optimization, GA performs better than the reference and single objective GA. The values of the objective functions from SOGA and MOGA algorithms are superior to the reference results.

17.4 CONCLUSION

In this study, the optimal design of a PMBLDC motor has been examined. Key motor specifications are originally defined in this study as functions of the motor's geometrical parameters. In order to simultaneously increase the power density, efficiency, and torque of the motor while reducing losses, and volume, the objective function has been established. In order to determine the optimal way to construct a BLDC motor using SOGA and MOGA optimization techniques, four objective functions have been taken into account independently and simultaneously. This research topic has, therefore, become more important due to the simultaneous optimization of design characteristics based on the requirements of many applications. By adding more objective functions and design variables, the design can be further enhanced to make the actual design of the motor simpler, but the

TABLE 17.2 Optimization Parameter by SOGA.

Sl. No.	Parameter	Notation	Single Objective GA			
			Power Density	**Efficiency**	**Torque**	**Power Loss**
1.	Number of pole pair	P	5.97	5.9853	5.99	1.0147
2.	Cross-sectional area of the winding (mm^2)	Ac	1.355	1.6768	0.9265	1.0454
3.	Pole arc/pole pitch	B	0.6432	0.6657	0.6549	0.9355
4.	Magnet thickness (m)	lm	0.0114	0.0116	0.0066	0.0051
5.	Stator/rotor yoke thickness	ly	0.0035	0.0067	0.0041	0.0036
6.	Winding thickness (m)	Tw	0.0016	0.0055	0.0054	0.0052
7.	Mechanical air gap	Mg	0.0036	0.0017	0.0020	0.0031
8.	Rotor radius (m)	R	0.0939	0.05	0.0849	0.0708
9.	Axial length (m)	L	0.0951	0.0688	0.0730	0.0880
10.	Pole pitch (mm)	τ_p	62.5611	77.3803	49.0714	53.0596
11.	Average flux density (T)	B_{av}	0.7700	0.7402	0.5437	0.5764
12.	Specific electric loading	Ac	15,376	26,989	8688	12,753
13.	Stator back iron height (mm)	h_{bis}	0.0115	8.8	0.0135	11
14.	Slot height (mm)	h_s	0.1189	22	0.199	24.3
15.	Current density	Jcu	5,642,200	4,026,400	3,853,400	4,539,600

TABLE 17.3 Optimization Parameter by SOPSO.

Sl. No.	Parameter	Notation	Single Objective GA			
			Power Density	**Efficiency**	**Torque**	**Power Loss**
1.	Number of pole pair	P	6	6	5.4790	1
2.	Cross-sectional area of the winding (mm^2)	Ac	1.5560	2	0.3721	2
3.	Pole arc/pole pitch	B	1	0.6324	0.6310	1
4.	Magnet thickness (m)	Lm	0.0119	0.0150	0.0150	0.001
5.	Stator/rotor yoke thickness	ly	0.0020	0.0020	0.0020	0.001
6.	Winding thickness (m)	lw	0.0055	0.001	0.0042	0.001
7.	Mechanical air gap	Lg	0.0014	0.0040	0.0040	0.001
8.	Rotor radius (m)	Rr	0.100	0.1	0.0690	0.005
9.	Axial length (m)	L	0.100	0.1	0.0987	0.1
10.	Pole pitch (mm)	τ_p	80	80	50.2847	40
11.	Average flux density (T)	B_{av}	0.9	0.9	0.5987	0.4
12.	Specific electric loading	AC	11,092	30,000	13,371	8000
13.	Stator back iron height (mm)	h_{bis}	12.5	15	0.0150	9.4
14.	Slot height (mm)	h_s	25	10	0.1346	10
15.	Current density	Jcu	6,000,000	6,000,000	6,000,000	3,000,000

complexity would rise. Another way to get the best outcomes faster and more accurately is by using optimization algorithms.

TABLE 17.4 Optimization Parameter by MOGA.

Sl. No.	Parameter	Notation	From Reference	MOGA
1.	Number of pole pair	P		6
2.	Cross-sectional area of the winding (mm^2)	Ac	1.9551	0.1
3.	Pole arc/pole pitch		0.6912	0.5
4.	Magnet thickness (m)	Lm	0.0120	0.015
5.	Stator/rotor yoke thickness	Ly	0.0081	0.001
6.	Winding thickness (m)	Lw	0.0033	0.001
7.	Mechanical air gap	Lg	0.001	0.0040
8.	Rotor radius (m)	Rr	0.0592	0.0900
9.	Axial length (m)	L	0.0730	0.100
10.	Pole pitch (mm)	τ_p		80
11.	Average flux density (T)	B_{av}		0.900
12.	Specific electric loading	AC		30,000
13.	Stator back iron height (mm)	h_{bis}		0.0050
14.	Slot height (mm)	h_s		0.0100
15.	Current density	Jcu	5,819,800	6,000,000
16.	Electromagnetic torque		10.53	10.58
17.	Efficiency		96.61	96.79
18.	Power density		$1.95*10^{-9}$	$2.52*10^{-9}$
19.	Volume		0.0033	0.0032

KEYWORDS

- **BLDC motor**
- **efficiency**
- **multi-objective optimization**
- **power density**
- **torque**

REFERENCES

1. Jo, S. T., Shin, S. H., Lee, Y. G., Lee, J. H., & Choi, J. Y. (2022). Optimal design of a BLDC motor considering three-dimensional structures using the response surface methodology. *Energies, 15*(2), 461.
2. Zaaraoui, L., & Mansouri, A. (2021). Optimization and finite element analysis of an in-wheel permanent magnet motor. *Malaysian Journal of Fundamental and Applied Sciences, 17*(1), 104–108.
3. Barba, P. D., Mognaschi, M. E., Petkovska, L., & Cvetkovski, G. V. (2022). Optimal shape design of a class of permanent magnet motors in a multiple-objectives context. *COMPEL - The International Journal for Computation and Mathematics in Electrical and Electronic Engineering.*
4. Nekoubin, A., Soltani, J., & Dowlatshahi, M. (2020). Design optimization and prototype of a multi-phase fractional slot concentrated windings surface mounted on permanent magnet machine, *Majlesi Journal of Electrical Engineering, 14*(4), 75–84.
5. Abd-Rabou, A. S., Marei M. I., El-Sattar A. A., & Basha, M. A. (2019). Multiobjective design optimization of axial flux permanent magnet brushless DC micromotor using response surface methodology and multi-verse optimization algorithm. In *Jordan International Joint Conference on Electrical Engineering and Information Technology (JEEIT)* (pp. 13–18), *Amman, Jordan.* IEEE.
6. Chakkarapani, K., Thangavelu, T., & Dharmalingam, K. (2020). Thermal analysis of brushless DC motor using multiobjective optimization. In *International transaction electrical energy system.* John Wiley & Sons Ltd.
7. Golilarz, N. A., Gao, H., Addeh, A., & Pirasteh, S. (2020). Orca Optimization Algorithm: A New Meta-Heuristic Tool for Complex Optimization Problems. In *2020 17th International Computer Conference on Wavelet Active Media Technology and Information Processing (ICCWAMTIP)* (pp. 198–204). IEEE.
8. Baeku, S. (2018). Optimal design of surface-permanent-magnet-type BLDC motor for radiator fan to reduce cogging torque using genetic algorithm. *International Journal of Engineering and Technology, 7*(4), 39.
9. Qiu, H., & Duan, H. (2015). Multi-objective pigeon-inspired optimization for brushless direct current motor parameter design. *Science China Technological Sciences, 58*(11), 1915–1923.
10. Noori, A., & Noruzi, N. (2017). A new method for optimal design of slotless permanent magnet BLDC motor with surface mounted magnets using modified social spider algorithm. *Computational Research Progress in Applied Science and Engineering, 3*(3), 109–114.
11. Kamal, C., & Thyagarajan, T. (2017). Cogging torque minimization in brushless DC motor using PSO and GA based optimization, *Trends in Industrial Measurement and Automation (TIMA)*, 1–5
12. Tripathy, S. N., Kundu, S., & Pradhan, A. (2022 January). Multi-objective optimization technique based design of squirrel cage induction motor. In *2nd International Conference on Power Electronics and IoT Applications in Renewable Energy and its Control (PARC)* (pp. 1–6), *Mathura, India.* IEEE.

13. Samanta, S. R., Mallick, P. K., Pattnaik, P. K., Mohanty, J. R., & Polkowski, Z. (Eds.). (2022). *Cognitive computing for risk management*. Springer.
14. Polkowski, Z., Mishra, S. K., Mishra, B. K., Borah, S., & Mohanty, A. (2023). Impact of internet of things in social and agricultural domains in rural sector: A case study. *International Journal of Cloud Computing*, *12*(1), 90–105. https://doi.org/10.1504/IJCC.2023.129776
15. Prasad, S. S., Mishra, J. P., Mishra, S. K., & Borah, S. (2020). Challenges toward blockchain and renewable energy linked to IoT: A survey. *Advances in energy technology* (pp. 49–55). Singapore: Springer Nature.
16. Nath, M. P., Priyadarshini, S. B. B., Mishra, D., & Borah, S. (2020). A comprehensive study of contemporary IOT technologies and varied machine learning (ML) schemes. In S. Borah, R. Pradhan, N. Dey, & P. Gupta (Eds.), *Soft computing techniques and applications*: *Advances in intelligent systems and computing* (Vol. 1248, pp. 623–634). Singapore: Springer Nature. https://doi.org/ 10.1007/978-981-15-7394-1_56
17. Mishra, J. P., Borah, S., & Mishra, S. K. (2021). Performance of scalable data linked to internet of things: A case study. In S. Borah, R. Pradhan, N. Dey, & P. Gupta (Eds.), *Soft computing techniques and applications*: *Advances in intelligent systems and computing* (Vol. 1248). Singapore: Springer. https://doi.org/10.1007/978-981-15-7394-1_54
18. Verma, U., Sohani, M., Borah, S., Nagwanshi, K. K., & Pathak, S. (2022). Role of fog-assisted internet of things-enabled system for managing the impact of COVID-19. In S. Borah, S. K. Mishra, B. K. Mishra, V. E. Balas, & Z. Polkowski (Eds.), *Advances in data science and management: Lecture notes on data engineering and communications technologies* (Vol. 86, pp. 397–406). Singapore: Springer. https://doi.org/10.1007/978-981-16-5685-9_39

CHAPTER 18

NOWCASTING OF PRECIPITATION USING CONVOLUTIONAL NEURAL COMMUNITY AND INTERNET OF THINGS

KISHOR KUMAR REDDY C.[1], MARLIA M. HANAFIAH[2,3], ANISHA P. R.[1], and SRINATH DOSS[4]

[1]*Department of Computer Science and Engineering, Stanley College of Engineering and Technology for Women, Hyderabad, India*

[2]*Department of Earth Sciences and Environment, Faculty of Science and Technology, Universiti Kebangsaan Malaysia (UKM), Bangi, Selangor, Malaysia*

[3]*Centre for Tropical Climate Change System, Institute of Climate Change, Universiti Kebangsaan Malaysia (UKM), Bangi, Selangor, Malaysia*

[4]*Botho University, Gaborone, Botswana*

ABSTRACT

Rainfall nowcasting is vital for the public to make the right decision at the right time. With traditional methods, accuracy is limited in numerical weather prediction (NWP) because the resolution of a solid thermal climate model must be smaller than the minimum scale that the model can capture. It frequently requires a supercomputer. In addition, the optical buoyancy approach for nowcasting precipitation has proven itself. On

Cyber-Physical Systems: Applications, Challenges, and Research Directions. Rajanikanth Aluvalu, Samarjeet Borah, & Lalit Garg (Eds.)

DOI: 10.1201/9781779643421-18

the other hand, it is complex to conclude the release parameters because the traceability and induction procedures are separate. In contrast, current machines that gain knowledge about software rely primarily on carefully selected entire datasets, ignoring the information that actual datasets repeatedly hold misplaced datasets that require more attention. In this chapter, authors used a real Hubei dataset missing some radar echo recordings and suggested a suitable mechanism to address the situation. Here, authors present a convolutional neural community for nowcasting of radar-based precipitation. RainNet has been trained to forecast continuous rainfall intensity with a 5-min lead time using world-class climate radar vehicles, equipped with world-class climate radar vehicles for many years. From the investigational effects, authors conclude that our technique can predict extreme climate activity with high accuracy despite its lack of statistics and that it is superior to the traditional optical slaying method. In addition, our version can be used for different types of radar recordings with species-specific feature extraction, making the technology more flexible and suitable for extreme conditions.

18.1 INTRODUCTION

Nowcasting can predict precipitation and warn of any possibility of sudden climate adaptations that may cause additional risks to humans and tissues. Most traditional nowcasting techniques gather the retrieval of weather radar. Still, forecasting rainfall remains a dispute, mostly because of rapid adjustments in meteorological procedures and the time needed for numerical simulations. The occurrence of extreme weather sports, which include storms, hurricanes, lightning, and high winds, can harm individual life and the social, financial, and environmental property of groups and nations. According to Golnaraghi et al., amid 1970 and 2012, floods, mass movements, and storms accounted for 85% of faults associated with severe climate, water and weather, are accountable for more than one million death and caused trade and industry damage estimated at approximately $2 billion worldwide. In addition, floods are anticipated to be additional intricate and have an effect on added public as the temperature rises due to the climate change.

Floods affect not only people in situations where extreme weather events can occur but also decision makers in civil protection, electricity, agriculture, and aviation companies who may be accountable for security and control. The ability to receive accurate warnings about severe weather at slightest 1 h in progress should help plan and mobilize responsible businesses and can help reduce losses, mitigate damage, and save lives. Precipitation nowcasting equipment has been used to predict extreme climate activity. In summary, this provides an in-depth description of today's climate to make predictions over a small range (from meters to a few kilometers) and for a short period (from 0 to 6 h). They are localized in large and excessive storms, with a little spatial location and a rapid period [1].

One type of precipitation nowcast relies on weather radars due to its increased spatial (about 1 km) and temporal (about 5 min) accuracy. Several methods related to nowcasting can be identified in the literature, from evaluating meteorological variables along with evaluating images from weather radars or satellites, evaluating some meteorological variables such as wind and precipitation, and indicators of lack of stability from radiosondes to prediction techniques that include patterns of Euler stability, Lagrangian endurance models (e.g., optical walking with gliding), and mathematical image extrapolation models from weather radars or satellites and computer intelligence models. Teleportation systems can only use these technologies or combine several. The most common prediction technique in teleportation is the inductive version; it is either based primarily on cell monitoring or area monitoring [2, 3].

The purpose of the study is to predict rainfall using old monthly statistics based mainly on AI methods such as lead vector tools and artificial neural networks. Extraction strategies/algorithms generate output by categories of statistics according to SVM and ANN usage methods. Similar statistics are collected for the correct and exact dates to predict precipitation more effectively and with better numbers. Correct and accurate forecasts help to develop more suitable approaches for agriculture and water assets, and they are well-versed about flooding to enforce preventive measures. Precipitation is the most important component of all the climatic events of human existence.

This chapter aims to provide an end to hardware abandonment to gain insights into the lifestyle cycle, from data pre-processing to fashion imposition and evaluation. The data pre-processing steps consist of missing value input, feature transformation, feature fast coding, feature scaling, and feature selection. We applied the models along with logistic regression, decision trees, nearest neighbors, existing bases, and groups. As the reason for evaluation, we used accuracy, precision, recall, F-Score, and area under the curve as rating scales. For our experiments, we mark our classifier with Australian weather records collected from various weather stations in Australia.

The study is significant in terms of its crucial contribution to agriculture, water resource management, flood forecasting, and management to comfort people by keeping them informed about weather and precipitation predictions. It is also important for rural industries to use it to maintain the integrity of their crops and ensure seasonal end product and vegetable production through updated rainfall forecasts. The investigation may also be of great importance to flood control establishments, as a more accurate prediction of heavy rains will keep authorities attentive to an upcoming event that will minimize the devastation through preventive measures. Precipitation forecasts will help admirably in dealing with the growing problem of managing beneficial water resources; because water is a scarce resource and must be stored for the advantage of the people themselves. Also, it will benefit people to accomplish and plan their social activities.

The primary contributions of the chapter are as follows:

1. Since some of the recordings may be missing, to reserve as numerous examples as potential, we observed that the samples of the radar recordings were ignored and the significance was tested, which showed a very good effect. Likewise, it can resolve any existing dataset containing missing data, which is why it can address this lost data vulnerability.
2. We advocate a rainfall nowcasting method appropriate for individual or cumulative radar reflection facts.
3. The experimental outcomes demonstrated have a low RMSE for forecasting rainfall and outperformed the standard method based entirely on visual walk downstream.

This chapter is planned as follows: Section 18.2 provides the "Related Work," an in-depth outline of present techniques applied to previous work and the standards elaborated. Section 18.3 provides the "Proposed Methodology," containing the proposed technique and statistics along with the system design. A unique framework for the proposed approach is presented in the same chapter. Section 18.4 indicates the "Results and Discussion" and Section 18.5 describes the "Conclusion."

18.2 RELATED WORK

In Ref. [1], effective DNN (deep neural network) training requires a huge number of attributes and computational power. As a baseline, more extended periods are required to train complex patterns that need huge units of information, which can hardly limit standard development studies and the use of all known evidence. This chapter examines this experiment in the context of rainfall forecasting, a period utilized to define reliable short-term rainfall predictions and various dangerous climates. Convolutional Neural Networking (CNN) is a powerful class of methods best suited to this task; the excessive decision to insert climate images, combined with the complexity of the model required to organize these facts, makes training CNNs solve this endeavor. A parallel copy of the data is run to overcome this difficulty. The CNN is replicated across a few computing nodes, and modulation groups are spread across more than one node. Using more than one GPU, we show that the training time for engineering a telecast version can be condensed from 59 h to just over 1 h. That enables faster iterations to improve CNN frameworks and ease future developments in telecasting [8].

In Ref. [2], precipitation forecast is vital because heavy rain can cause many errors. The forecast allows individuals to take preventive measurements. Moreover, the forecast must be accurate. There are forecast patterns for short-term precipitation forecasting and long-term precipitation forecasting. Forecasts can give us the exact result for short periods in most cases. The mainstream paper is to build a long-range version of the rain forecast. The expectation of heavy rainfall may be a major flaw in the Earth technology industry because it is carefully connected to the economy and hominoid life. It is a cause of natural tragedies like floods and deficiencies that occur on this scale every year with the help of people worldwide. The accuracy of calculating rainfall makes interesting sense for international locations such as India, whose financial system is largely based on agriculture [7].

The dynamic nature of the environment and the mathematical techniques used do not provide realistic accuracy of precipitation readings. Precipitation forecasting using automated survey strategies can also use regression. This endeavor aims to provide non-experts with easy access to strategies and techniques used in rainfall forecasting and a comparative study of several machine learning strategies. Regression assessment provides the requirement of one variable on one or more other variables valuable for approximating and predicting the mean or regular responsibility of the former in combinations of acceptable or constant values from the latter. For instance, someone's profit depends mainly on their experience, immersion in the trait is an unbiased variable, and revenue depends on the variable. Simple linear regression regulates the association between an unpaired fixed variable and a single independent variable.

Current weather forecasts rely heavily on massive digital simulation systems that nationwide climate agencies worldwide automate. The chain of processes accompanying the generation of these numerical weather predictions can be divided into several closely interrelated steps (origin 1, left column). An incredible variety of meteorological observations from around the world have been compiled to determine the Earth's original landform (atmosphere, soil, and ocean). Along with classic weather stations and radiosondes, aircraft measurements and remote sensing products (including radar and satellite observations) have become a necessary part of the global monitoring network. Although tens of millions of different direct and indirect measurements are obtained daily, these observations are still insufficient to describe the entire kingdom of the ecosystem and the various machine components of the Earth with which the ecosystem exchanges electricity or mass [3–6].

18.3 PROPOSED METHODOLOGY

The proposed work is divided into two categories: first, we classify the vulnerabilities that may be narrowly relevant to rainfall fate traits. Second, we outline the release results to achieve the predicted precipitation. For the assessment, we generated a new procedural dataset that includes radar echo maps, general precipitation measurements via geometry stations, and altimetry measurements from positioning stations. The illustrations are then provided in several machine learning methods, namely, a neural network (NN), convolutional neural network (CNN), and a gradient boosting decision tree (GBDT) to predict rainfall. Given that some radar facts are ignored, obtaining as numerous reconnaissance models as feasible, and we have already cached the models, the primary type of cutting radar records now does not exceed the three during the creation of the information set. The set of facts generated in this way has little effect on the expected development. In addition, we use traditional light wave-based techniques from the Rain Motion Library as recording standards. Finally, the usual scoring metrics are applied, which include root-suggestion square errors (RMSE) and mean absolute percentage errors (MAPE).

The proposed mechanism includes four main steps: tracing, feature extraction, transcription, and precision-weighted transcription results. In the first step, we harvest the pricing problem by calculating the offset of the best matching element between the maps in the respective cases. In feature extraction, we extract four types of capabilities: radar mirror image, cloud trajectory description, cloud trajectory diagram, and ground station altitude. Then, we train three poses to look at devices or apply specific skills. The dominant model is the convolutional neural network (CNN), and the supporting patterns are the NN and the GBDT. It can trace the cloud pattern parallel to the price. Lastly, the three ways were regulated, weighed, and totaled to produce the expected final sedimentation.

18.3.1 DATA COLLECTION

With the help of an Arduino board and the KNIME Analytics Platform, we have created an IoT system. Except for the Arduino scripts, the majority of the project was completed without any coding. This office weather station is a complete IoT system since it moved from gathering weather information from a sensor board to storing and processing that information to make forecasts for the next day. Our IoT system's design consists of a sensor board for data collection, a repository system for data storage, a backup repository system, a web service for data transfer, and a SARIMA model for making predictions for the following day. Using the Arduino sensor board "MKR IoT Carrier" and the sampled signal, we measured the temperature data. Hygrometers, or humidity sensors, gauge how much water vapor is present in the air or other gases. In order to monitor and forecast the weather, humidity sensors are commonly utilized in commercial and residential HVAC systems. IoT pressure sensors are used to analyze the pressure level. Pressure variations can be detected via pressure sensors [9].

Sensors are linked by gateways, which allow them to transmit the data they have gathered to a cloud server. Your computer or mobile device will then get the information, giving you immediate access to all monitored activity.

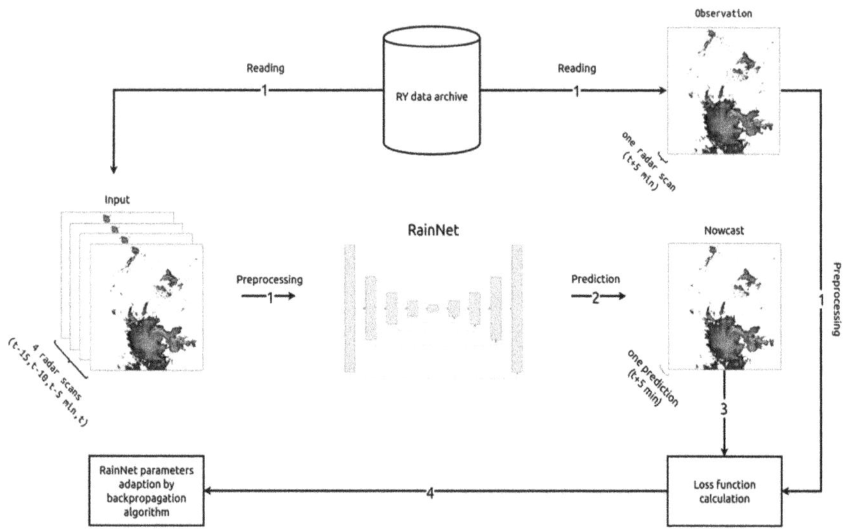

FIGURE 18.1 Proposed architecture to nowcast precipitation.

Since learning with gadgets is an information-dependent area based on large and distinct samples, it is important to build a dataset that includes several actual international scenes. First, we scan hourly precipitation records to keep rainfall records (>0 mm/h). In the second zone, we choose plots within four contiguous hours from the time of rainfall (forty radar images if no longer available). Finally, we cropped the location around the ground station (301 × 301 km) in each image in conjunction with the station location.

Specified the lost data overhead, we discard unfinished samples from less than 40 frames; the diversity of the final samples is now insufficient to train the model, which requires sufficient input to achieve accurate performance. To get enough variety of samples provided, we keep all samples with at least 37 frames. Assuming that precipitation should be predicted according to the plots within 1–2 h and the example is wide-ranging, the 10–20 maps can be fixed in sequence. In addition, the incomplete sample selection process remains unchanged. If there is a framework missing from the collection within 0–2 h, the first image can be selected in practice within 2–3 h. It will not distress the designated tires if the loss occurs in 2–4 h. In the state of the art, we frequently substitute lost archives with timely information.

Therefore, our set of statistics corresponds to the actual international set. They are also rare cases due to the diversity of unfinished examples is much smaller than that of whole samples. Because the representations used to study the system are robust, prediction accuracy is not unnecessarily compromised. Therefore, our file is easy to use and cheaper. In total, we received 2502 patterns, 1099 patterns with 40 radar echoes, 413 patterns with 39 charts, 538 patterns with 38 charts, and 452 patterns with 37 charts. Now that precipitation depth was no longer readily available, we provided information on high-quality precipitation and obtained the ratio of different depths.

18.3.2 FEATURE EXTRACTION

Since the contemplative images are radar echo maps with a superior spatio-temporal correlation and the management of rainfall evolution, it is critical to classify the image that can recollect the rainfall type. Here, we extract three types of capabilities: radar reflectivity, general cloud report, cloud trajectory presentation, and ground station elevation. For comparison, we chose specific skills as entry points into fashion. Radar reflection image: here, we have extracted three different functions of the radar reflection image. Close-up reflections on the track. Understanding the chronological course of every aspect, we get the immortality coordinates for the clouds to go to the target station in destiny. Then we clip the spot 41 × 41 above the coordinates [10].

A convolutional neural network (CNN) for radar-based rainfall forecasting. RainNet was introduced to forecast continued rainfall power with a 5-min lead time utilizing multiple years of adequately sustained weather radar vehicles supplied by the German Weather Service (DWD). RainNet greatly exceeds standard standards for all 60-min lead times for common validation metrics, absolute error (MAE), and vital achievement index (CSI) at power thresholds of 0.125, 1, and 5 mm h^{-1}. Nevertheless, rain has been confirmed to be ahead in predicting exceeding higher intensity thresholds [11].

RainNet's finite capacity to forecast substantial precipitation intensities is an unpleasant aspect that we feature to the lofty point of spatial flattening performed by the copy. With an uncertainty of five minutes, an assessment of the power spectral consistency demonstrated a significant lack of spectral power at scales 16 km in length and less. RainNet has learned a high-quality smoothing level to give the live stream a 5-min time slot. For this reason, the loss of spectral power on small rankings is also helpful because it contemplates the boundaries of predictability ingrained at the spatial scale. After the 5-min timeout, however, increasing smoothing is a minor artifact—an analog to scalar propagation—not always owned by RainNet itself but through its iterative program. In the context of initial cautionary, smoothing is particularly harmful because the mentioned functions of excessive precipitation tend to migrate with longer periods. Consequently, we advocate several options for addressing this issue in potential studies, such as changing the loss process for model training, modeling for more extended lead samples, and indicating threshold intersection in a binary hash challenge. Also, we keep more input data that can support sufficiently perceiving problems with the dynamics of upcoming precipitation [12].

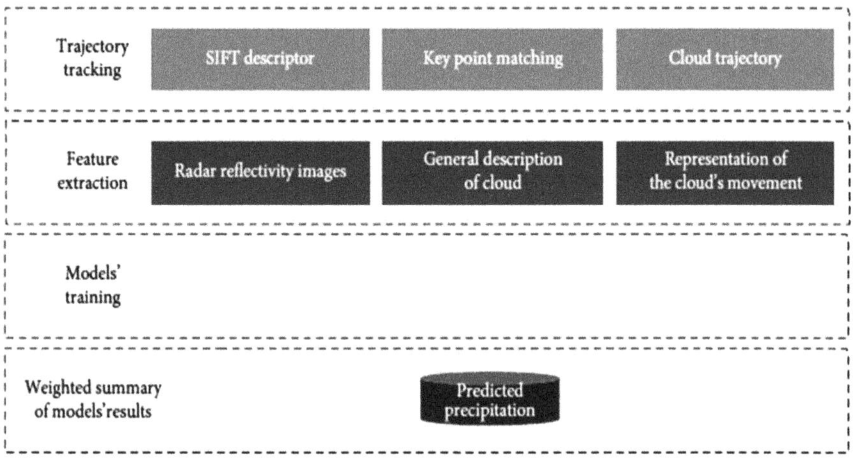

FIGURE 18.2 Precipitation nowcasting mechanism.

We propose an enhanced radar reflectivity procedure for not paired or accumulated elevation data using machine learning methods. The basic idea of our mechanism is to use features extracted from historical radar map series and equipment search patterns to predict hourly precipitation at a site 1–2 h into the future. Features include three types of radar reflectivity, and some non-imagery features include preferred cloud descriptions, cloud track maps, and ground track station heights. We had two goals. The original purpose was to predict the climatic splendor. We will show the final results using a selection tree regression for the month of August of a given year, and display the results related to the prevalence of humidity or heat. Another goal is to predict temperature using various procedures, including random forest regression, linear regression, and K nearest neighbor regression. The production cost should be a scalar cost based on some additional factors such as inhabitants' thickness and air quality [13].

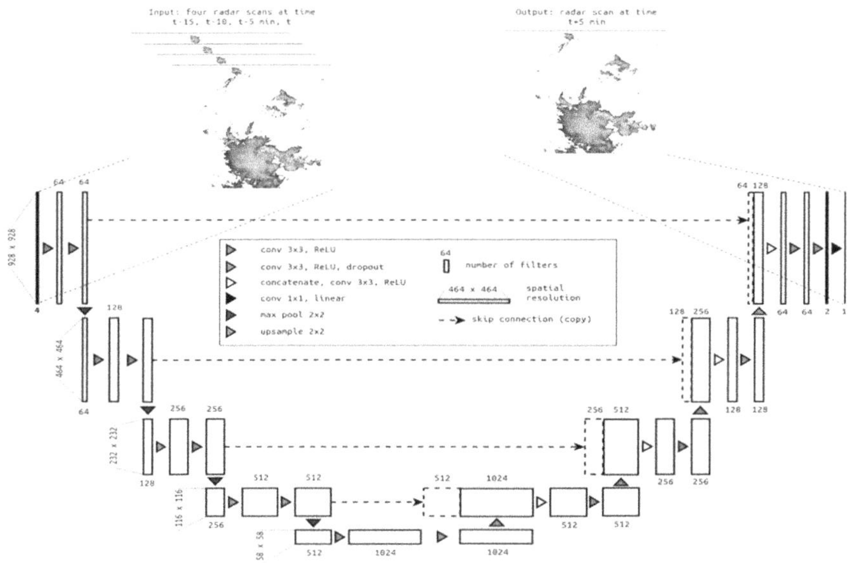

FIGURE 18.3 Proposed novel radar reflectivity mechanism.

18.4 RESULTS AND DISCUSSION

To execute the proposed work, first, we need to open the Pycharm tool related to python programming, and if we debug the GUI Script, we must detect a GUI pop-up. We can see the below-given screen. After, if we click on forecast, then it should estimate the weather relying on the data we offer. Behind selecting the forecast, we must be capable to visit the forecast, and the chart will be rendered.

The above figure gives us a comparison of the different parameters that are required in forecasting like day, month, year, maximum temperature, mean temperature, mean humidity, mean dew, pressure, heat, and wet.

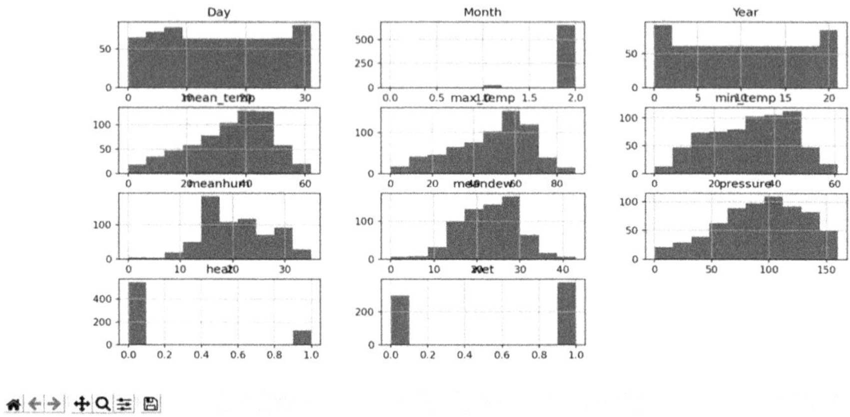

FIGURE 18.4 Forecasting on precipitation.

The graphs given below are Auto-Forecast Graphs. Here it gives us how mean temperature, mean humidity, maximum temperature, and mean dew vary day to day or in months.

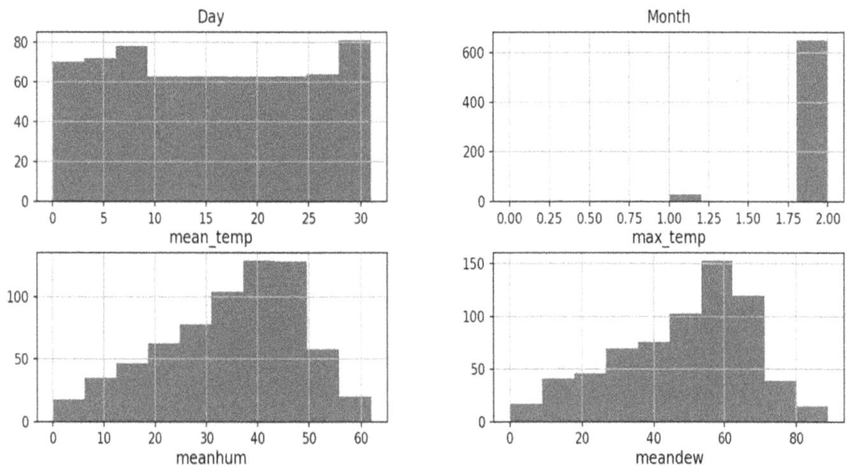

FIGURE 18.5 Auto-forecasting of precipitation.

The below graph gives us the comparison of Pressure versus Mean Temperature which gives us an idea of those parameters present how pressure impacts the mean temperature and the mean temperature is high at high pressures.

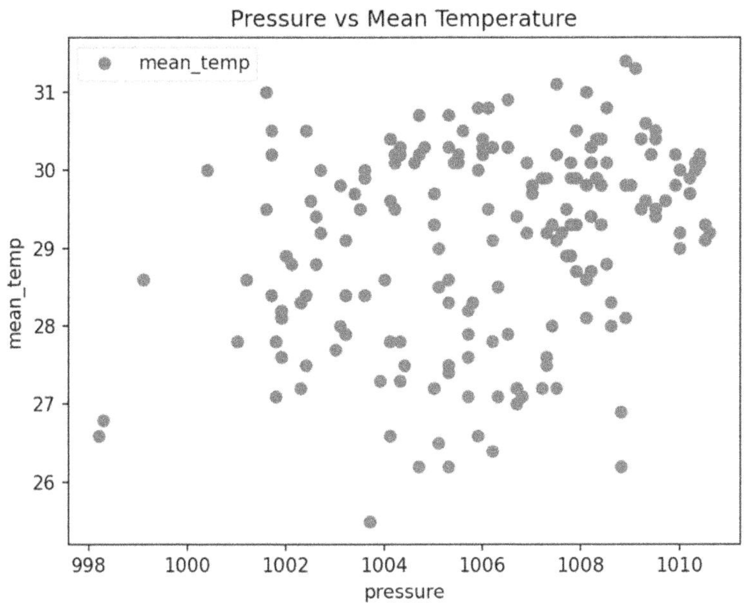

FIGURE 18.6 Pressure vs mean temperature.

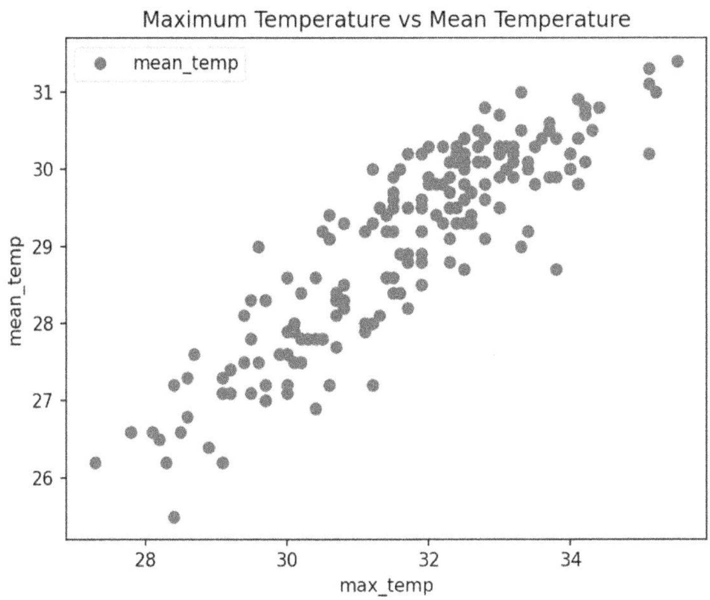

FIGURE 18.7 Maximum temperature vs mean temperature (MT).

The below graph gives us the comparison of Maximum Temperature versus Mean Temperature which gives us an idea of those parameters present and helps us in understanding the variation of temperature conditions. Mean Temperature and Maximum Temperature are directly proportional as we can see mean temperature increases as maximum temperature increases.

The below graph gives us a Regression plot of Actual Mean Temperature versus Predicted Mean Temperature Values which gives us an idea of those parameters present and the graph is even, that is, more accurate as the Actual Mean Temperature is similar to the predicted Mean Temperature.

The below graph gives us an idea of Actual Precipitation versus Prediction Precipitation that is more accurate as Actual Precipitation is similar to Predicted Precipitation. This graph predicts the precipitation and intense of rainfall which gives us a more comparative analysis that defines accuracy.

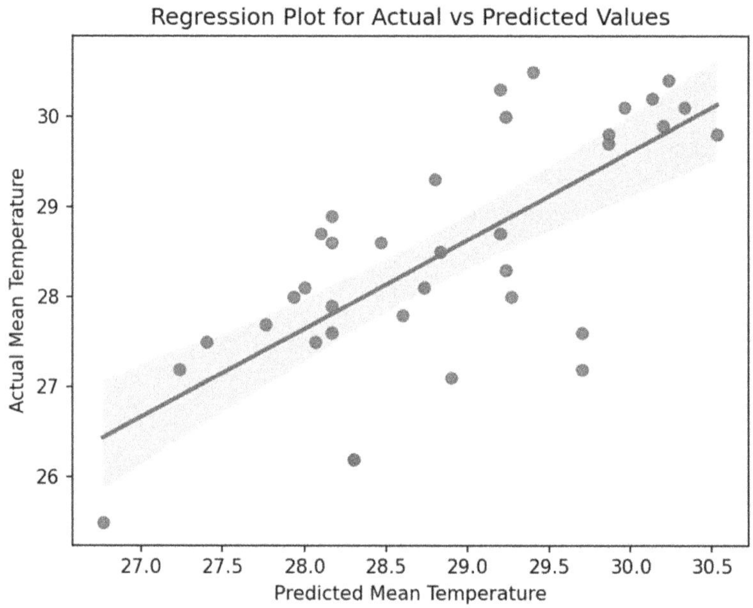

FIGURE 18.8 Actual mean temp vs predicted mean temp.

18.5 CONCLUSION

In this chapter, we have examined and implemented several pre-processing steps and translated their impact on the overall performance of our classifiers.

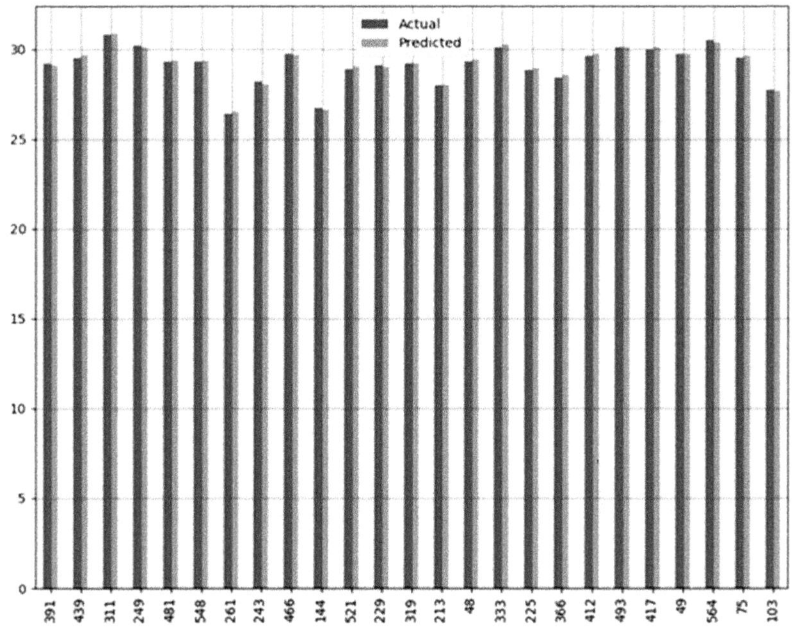

FIGURE 18.9 Actual vs prediction.

We also compared all classifier input information items and found how input statistics might affect model predictions. A common thought of our mechanism is to predict hourly rainfall for 1–2 h in the future at places utilizing features extracted from a series of recorded radar maps and the system gaining knowledge about patterns. Features included three types of radar reflective pixels and some non-pictured features along with general cloud description, cloud path diagram, and ground tracking station height. The representatives were CNN, NN, and GBDT. To acquire an average forecast from mode sequences, we allocate weights and summarize the effects of the three modes. Finally, we consider the whole bottom line as the last expected rainfall. The empirical outcomes demonstrate that our mechanism outperformed conventional lithography techniques. The accuracy of our mechanism should be limited to keeping the pattern type small in our statistical set. We thought that additional examples with exceptional hands-on scenes might want to donate to implementing our mechanism, which relies primarily on machine learning. In addition, features extracted from radar maps are closely related to fateful precipitation. The CNN performance worked particularly compared to the other two modes.

KEYWORDS

- **convolutional neural community**
- **nowcasting**
- **numerical weather prediction**
- **radar imagery**
- **RainNet**

REFERENCES

1. Kelleher, J. D., Mac Namee, B., & D'Arcy, A. (2015). *Fundamentals of machine learning for predictive data analytics: Algorithms worked examples, and case studies*. The MIT Press.
2. Witten, I. H., Frank, E., Hall, M. A., & Pal, C. J. (2016). *Data mining fourth edition: Practical machine learning tools and techniques*. San Francisco: Morgan Kaufmann Publishers Inc.
3. Stuart, R., & Norvig, P. (2016). *Artificial intelligence: A modern approach (Global edition)*. Harlow: Pearson.
4. Witten, I. H., & Frank, E. (2002). Data mining: Practical machine learning tools and techniques with Java implementations. *Acm Sigmod Record, 31*(1), 76–77.
5. Mitchell, T. M. (1997). *Machine learning*. The McGrawHill Companies, Inc.
6. Panhalkar, A., & Doye, D. (2017). An outlook in some aspects of hybrid decision tree classification approach: A survey. In *Proceedings of the International Conference on Data Engineering and Communication Technology: ICDECT 2016* (Vol. 2, pp. 85–95). Singapore: Springer.
7. Andersson, T., & Gustafsson, N. (1995). Coast of departure and coast of arrival: Two important concepts for the formation and structure of convective snowbands over seas and lakes. *Oceanographic Literature Review, 5*(42), 335–336.
8. Bocchieri, J. R. (1980). The objective use of upper air soundings to specify precipitation type. *Monthly Weather Review, 108*(5), 596–603.
9. Kishor Kumar Reddy, C., Anisha, P. R., Shastry, R., & Ramana Murthy, B. V. (2020). Comparative study on internet of things: Enablers and constraints. In *Data Engineering and Communication Technology: Proceedings of 3rd ICDECT-2K19* (pp. 677–684). Singapore: Springer.
10. Reddy, C. K. K., Anisha, P. R., & Prasad, L. N. (2016). A novel approach for detecting the bone cancer and its stage based on mean intensity and tumor size. *Recent Researches in Applied Computer Science, 20*(1), 162–171.
11. Huang, C., Newman, A. J., Clark, M. P., Wood, A. W., & Zheng, X. (2017). Evaluation of snow data assimilation using the ensemble Kalman filter for seasonal streamflow

prediction in the western United States. *Hydrology and Earth System Sciences*, *21*(1), 635–650.

12. Lafaysse, M., Cluzet, B., Dumont, M., Lejeune, Y., Vionnet, V., & Morin, S. (2017). A multiphysical ensemble system of numerical snow modelling. *The Cryosphere*, *11*(3), 1173–1198.
13. Cooper, S. J., Wood, N. B., & L'Ecuyer, T. S. (2017). A variational technique to estimate snowfall rate from coincident radar, snowflake, and fall-speed observations. *Atmospheric Measurement Techniques*, *10*(7), 2557–2571.

CHAPTER 19

SMART TRAFFIC MANAGEMENT AND EMERGENCY VEHICLE DETECTION SYSTEM USING IOT

T. TIRUPAL, B. UDAY KIRAN REDDY, K. SAI TEJA, UDAY KIRAN DHANE, and M. SIVA PRASAD

Department of ECE, G. Pullaiah College of Engineering and Technology, Kurnool, India

ABSTRACT

This chapter introduces the smart traffic management system which will help make roads less congested in big cities and give emergency vehicles a clear way. To prevent accidents and promote a smooth flow of traffic, the road dividers separate the road into two directions. A divider helps in preventing vehicles from colliding with each other. Due to traffic congestion, it will be difficult to clear the path and save time in cities. Using RFID it will provide a clear path for the ambulance or other emergency vehicle. Through the cloud, the traffic density at the dividers can be observed. The divider will be equipped with a buzzer and LCD display to indicate its movement automatically. The density of the vehicles can be counted using IR sensors. If a vehicle is found next to the divider, the ultrasonic sensor will detect it and emit a buzzing sound.

19.1 INTRODUCTION

Because of the increase in automobiles, countries all over the world are experiencing traffic problems [1, 2]. The current road infrastructure has

Cyber-Physical Systems: Applications, Challenges, and Research Directions. Rajanikanth Aluvalu, Samarjeet Borah, & Lalit Garg (Eds.)

DOI: 10.1201/9781779643421-19

not been updated to accommodate the growing number of cars, resulting in limited capacity to adapt to changes without encountering obstacles. Static roadway dividers with a predetermined number of lanes on each side of the street raise concerns. This necessitates making greater use of already-existing resources, such as the variety of available pathways. To solve this problem, we build a mobile roadway divider [3, 4] that moves in response to the flow of traffic. The microcontroller is capable of controlling the sensors to collect data from them and transmit it to the web via a Wi-Fi module. The Internet of Things gathers real-time data on vehicle traffic to analyze current traffic patterns and conditions. Infrared sensors will connect the Internet of Things to various traffic objects, including roadways and dividers [5].

19.2 PROPOSED METHOD

Additionally, the growing number of vehicles in cities leads to clogged roads. Every employee in the IT sector is working from morning until evening during that period. To reduce this traffic [6, 7], these methods are put into practice. There is a need to get to offices on time so that everyone will arrive at the same time. The proposed system is made of an Arduino Mega microcontroller, which manages the entire process, as shown in Figure 19.1. The transceiver, also referred to as an IR sensor [8], transmits light to count the density. The term “transducer” is also used to describe an ultrasonic sensor [9]. Before moving the partition, sound waves check to see if there are any vehicles nearby; if so, a buzzer sound may be produced. Emergency vehicles, such as ambulances, can be identified using RFID [10] modules, alerting onlookers to let them pass so that the ambulance can be reached. It can be updated in the cloud, which allows for remote monitoring.

19.2.1 OPERATION OF PROPOSED METHOD

People who are stuck in traffic feel stressed about reaching their respective destinations, such as offices and schools, on time, and also about various types of pollution like noise pollution, air pollution, and land pollution. Figure 19.2 demonstrates that traffic significantly affects students’ academic performance, primarily due to the stress they experience during their commute to college. By implementing smart traffic management and emergency vehicle detection, traffic can be controlled automatically by adjusting the dividers [11].

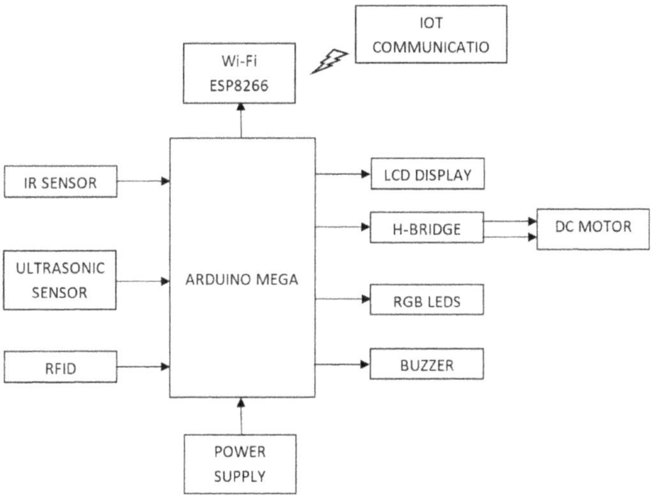

FIGURE 19.1 The proposed method is represented by a block diagram.

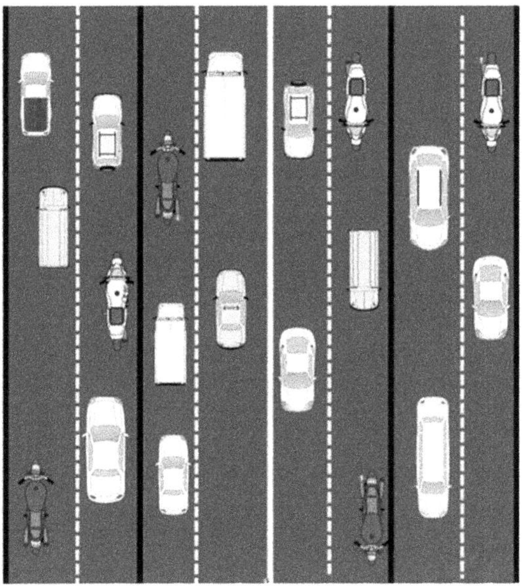

FIGURE 19.2 Moving the divider based on vehicle count.

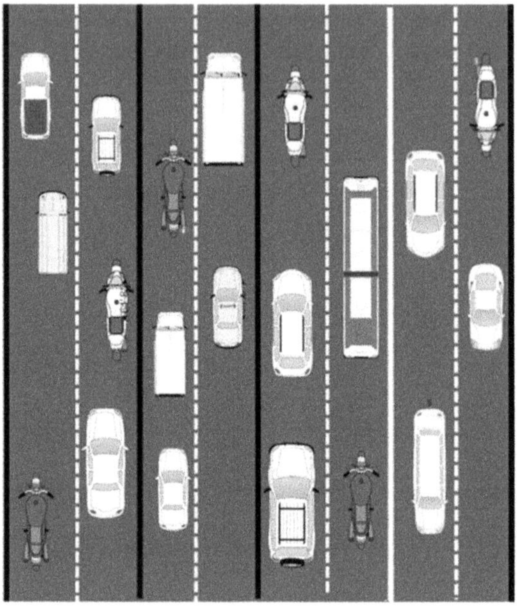

FIGURE 19.3 Divider's movement toward the lane with fewer vehicles.

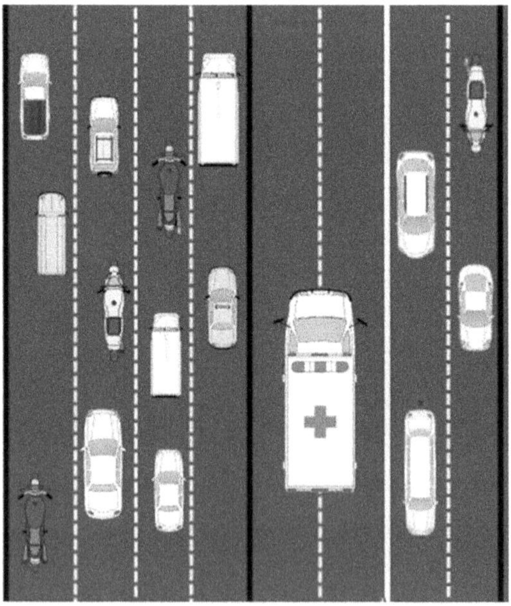

FIGURE 19.4 Controlling of traffic signals and ambulance detection.

By using these methods, dividers [12, 13] are moving from one side to another side based on the count of vehicles. When one side of the road has less number of vehicles and another road has a higher number of vehicles, a divider [14, 15] will automatically move based on the count of the vehicle, as shown in Figure 19.3.

When emergency vehicles [16, 17], such as ambulances, are stuck in traffic, lives are in danger for patients who are in the ambulance [19] and to save their lives, everyone should give way to the ambulance but there is no way in the road because of the fixed number of lanes. This chapter represents the smart movable road divider that moves automatically by controlling the traffic as shown in Figure 19.4.

19.2.2 ARDUINO MEGA 2560

Arduino is an open-source platform that encompasses both hardware and software components. The device is equipped with a 16 MHz artistic resonator, 6 primary data sources, a USB connection, a force jack, an ICSP header, and a reset button. The device features a total of 14 advanced input/output pins. It encompasses all the necessary components for microcontroller support. Figure 19.5 depicts an Arduino Mega 2560 microcontroller.

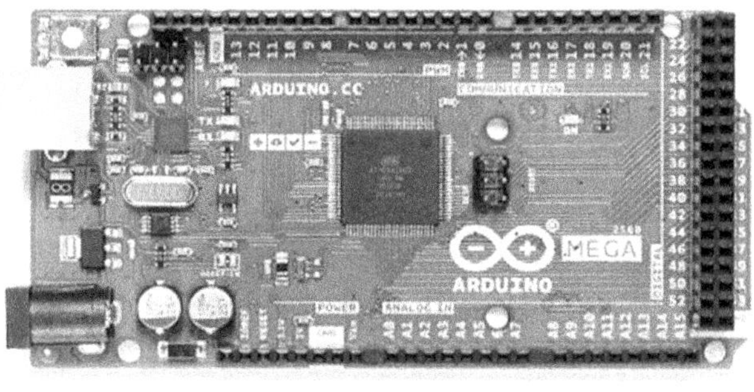

FIGURE 19.5 Arduino Mega 2560.

19.2.3 IR SENSOR

The IR sensor is an electronic device that emits light in order to detect objects within its surroundings. The emitter and receiver are the two components of an infrared sensor. As shown in Figure 19.6, an IR LED serves as the emitter and an IR photodiode serves as the receiver. Both the heat and motion of an object can be measured by an IR sensor. The operating voltage is 5 V DC. I/O pins are 3.3–5 V, and the measurement range is up to 20 cm.

FIGURE 19.6 IR sensor.

19.2.4 ULTRASONIC SENSOR

An ultrasonic sensor utilizes ultrasonic sound waves to measure the distance of a target object and converts the reflected sound into an electrical signal. Figure 19.7 illustrates a supply voltage of 5 V DC and a supply current of 15 mA. The distance is measured between 2 and 400 cm, with 0.3 cm accuracy.

FIGURE 19.7 Ultrasonic sensor.

19.2.5 ESP 0.1 WIFI MODULE

The ESP8266 ESP-01 is a module that enables serial to Wi-Fi communication. It includes an ARM microprocessor and has two GPIOs available for connection. The Serial WIFI Wireless Transceiver Module is a self-contained SOC that facilitates Wi-Fi network connectivity for any microcontroller. The device has 1 MB of Flash Memory, IEEE 802.11 b/g/n Wi-Fi, and operates on 3.3 V power. Figure 19.8 illustrates a 32-bit RISC Tensilica Xtensa LX Processor operating at a frequency of 80 MHz.

FIGURE 19.8 ESP 0.1 Wi-Fi module.

19.2.6 L293D BRIDGE

The L293D Bridge is an electronic circuit. The L293D Bridge allows for the application of voltage across a load in both directions. The same IC can be utilized to drive a DC motor shown in Figure 19.9. The maximum peak current of the motor is 1.2 A. The maximum continuous current of the motor is 600 mA. The motor voltage Vcc (Vs) is between 4.5 V and 3.6 V.

FIGURE 19.9 L293D bridge.

19.2.7 RFID

RFID is an integrated MF RC522 13.56 MHz contactless communication card chip. The RFID tags are small chips that contain a unique ID and can be read using an RFID reader. These tags can be embedded in products and documents to automate the tracking process shown in Figure 19.10.

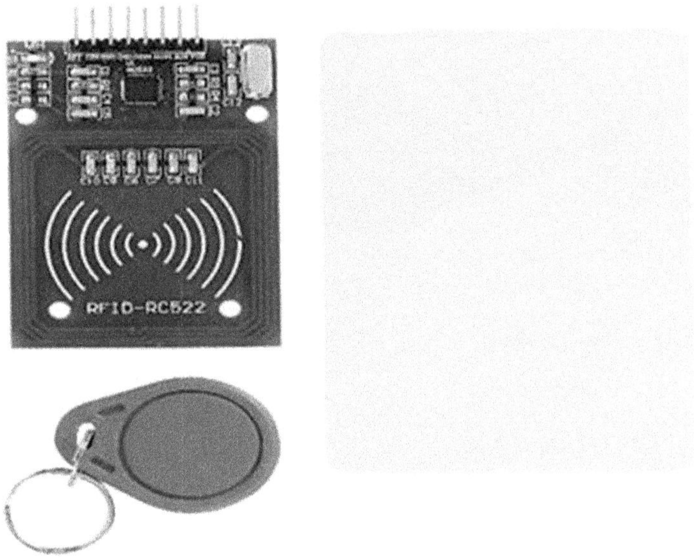

FIGURE 19.10 RFID.

19.2.8 DC MOTOR

A DC motor converts electrical power from direct current into mechanical power. This DC motor voltage rating is 12 V DC Motor with 200 RPM and can be used in all-terrain robots and a wide range of robotic applications. Figure 19.11 illustrates that for better coupling shaft has a hole.

19.2.9 LIQUID CRYSTAL DISPLAY

LCD is an acronym for liquid crystal display. The LCD is a type of flat panel display technology that utilizes liquid crystals. Figure 19.12

illustrates that LCD has 2 rows and each row produces 16 characters. The display light puts less strain on the eyes. The LCD has operating voltage rangeing from 4.7 V to 5.3 V.

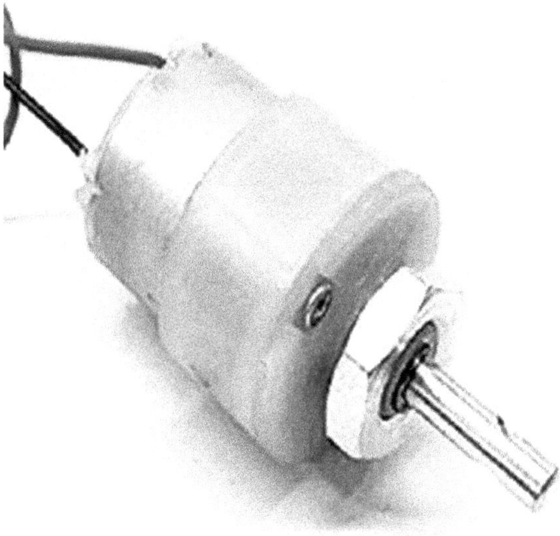

FIGURE 19.11 DC motor.

FIGURE 19.12 Liquid crystal display.

19.2.10 BUZZER

A Buzzer is a signaling device that emits audible sounds which are mechanical, piezoelectric, or electromechanical. Figure 19.13 shows that the Buzzer is a simple effective component to add sound feature to our project. The frequency range spans 3300 Hz. The operating voltage spans from 4 V to 9 V DC.

FIGURE 19.13 Buzzer.

19.3 RESULTS AND DISCUSSION

After integrating all modules, the resulting structure aims to minimize traffic congestion and facilitate the movement of road vehicles. We have designed and developed a demonstration and working of our project. But as this is only a prototype, there is space for improvement when it is implemented as a product type as shown in Figure 19.14.

19.3.1 COMPARISON OF THE EXISTING MODEL AND PROPOSED MODEL

19.3.1.1 Vehicles Counting Using IR Sensor

In the annals of past history of vehicle system and navigation, serial monitor was the only medium of display of the vehicle count (density of road). After several researches, we have come up with a unique solution where the vehicle count will be constantly monitored and updated, and additionally, in this project, the problem of serial monitor was eradicated and replaced with mobile phones. Now, the vehicle count can be constantly viewed on mobile phones.

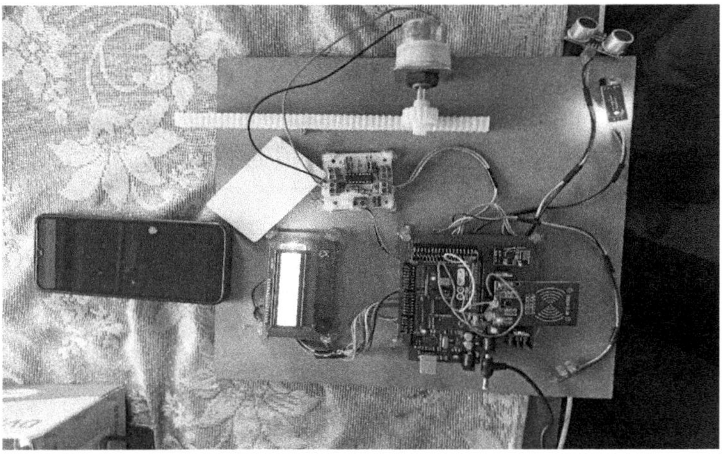

FIGURE 19.14 Working of the proposed system.

19.3.1.2 IDENTIFYING OBSTACLES USING ULTRASONIC SENSOR

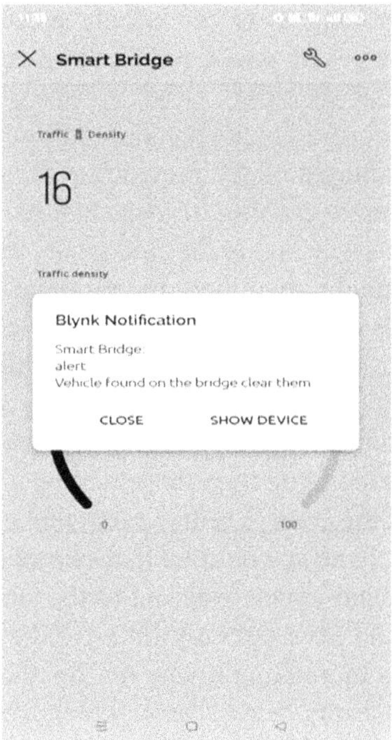

FIGURE 19.15 Vehicle is detected at the divider using an ultrasonic sensor.

In earlier times, when this sort of project was implemented, there were no proper precautions and due to that it led to many accidents. The main cause was that when the bridge was divided into two, there were no proper alert systems and that led to accidents in so many cases.

But now in our project, our system will notify the control system to clear the vehicles on a bridge before the bridge is getting divided. And, moreover, in our project, a buzzer is also included which gives a high-frequency beep sound that will alert the drivers of vehicles on the bridge to make their further move. This way our project is very effective in terms of handling things and minimizing the danger cases as shown in Figure 19.15.

19.3.1.3 AMBULANCE OR EMERGENCY VEHICLE DETECTION USING RFID

In earlier days, if there was any ambulance present in the traffic, there was no alert system and if there are any emergency vehicles in the traffic, nobody would even get to know about that vehicle and apparently it would lead to miscellaneous dangers to the people present in the emergency vehicles or ambulances. This was a major disadvantage in the previous system.

Our project consists of a unique solution that will be a remedy for all the mentioned disadvantages of the previous system. The solution which is presented in this chapter consists of ideas that will apparently control a traffic signal and make a way out for the emergency vehicles or ambulances.

There might be a doubt about how the problem of noticing the emergency vehicles ambulances is done? Our chapter consists of the idea of where RFID tags will be given to emergency vehicles and ambulances. RFID tags help in noticing them and accordingly vehicles in the traffic will make a way for the emergency vehicles and ambulances and even if the vehicles are not responding then there will be a message delivered to the control center and the control center personnel ensure that emergency vehicles are available. Traffic would be maneuvered and accordingly, the bridge will be moved and pave a way out to the emergency vehicles and ambulances. This would save not only time for the people in emergency vehicles but also lives by making a way out for the ambulances in risky traffic. Our project gives the best solution which saves time for people in emergency vehicles.

19.3.2 APPLICATIONS

19.3.2.1 USED FOR SMART TRAFFIC MANAGEMENT SYSTEM

Implementing measures to mitigate traffic congestion and minimize the occurrence of road accidents. Facilitating expeditious authorization for emergency vehicles. Enabling the implementation of measures to enhance the safety and efficiency of commuting, resulting in reduced travel durations. Implementing strategies to mitigate traffic congestion and optimize energy consumption at intersections.

19.3.2.2 USED FOR RESCUING THE PATIENTS

Hospitals will establish communication with ambulances using ultrasonic sensors in order to assess their availability for attending to the victim. In addition, the ambulance's path would involve the utilization of GSM communication to control the traffic light signals. This will optimize the response time of the ambulance to arrive at the medical facility.

19.3.2.3 CAN BE USED FOR FIRE FIGHTING VEHICLES

The smart traffic control solution is implemented by utilizing visual sensors, QR code detection technology, and ESP modules. Real-time modifications: The traffic signal system undergoes real-time adjustments, prioritizing emergency vehicles and facilitating their swift and secure passage through intersections.

19.4 CONCLUSION

The proposed model aims to alleviate traffic congestion at junctions and facilitate the passage of emergency vehicles, including ambulances and VIP vehicles. It will aid in making smart road arrangements and the traffic density is updated in the cloud via the blynk application, and the moment of the divider is indicated via buzzer sound and LCD display. These projects adhere to traffic regulations. Also, as a future enhancement of the project, it is possible to employ speech alarms to indicate the various controlling devices' operational conditions. The introduction of time-controlled gadgets for usage in commercial settings is a further planned development.

KEYWORDS

- **buzzer**
- **IR sensor**
- **RFID**
- **smart traffic**
- **ultrasonic sensor**

REFERENCES

1. Keerthan, A. J., Likhith, H. D., Chetan, H. S., Kiran, K. P., & Sumaiya, M. N. (2020). Automatic movable smart road dividers-IOT based solution to traffic congestion problems. *International Journal of Innovative Science and Research Technology, 5*, 2456–2165.
2. Naveen, N., & Sowmya, C. N. (2019). IoT deployed automatic movable smart road divider to avoid traffic problems. *International Journal of Computer Science Trends and Technology (IJCST), 7*, 23–27.
3. Sri, B. D., Nirosha, K., & Gouse, S. (2017). Design and implementation of smart movable road divider using IOT. In *2017 International Conference on Intelligent Sustainable Systems (ICISS)* (pp. 1145–1148). IEEE. doi: 10.1109/ISS1.2017.8389364.
4. Sowjanya, K. N., Nikhil J. H. M., Unnimaya, K., & Bhavana, G. (2021). Automatic movable smart road divider using IOT. *International Advanced Research Journal in Science, Engineering and Technology, 8*, 324–329.
5. Ravish, R., Gupta, V. R., Nagesh, K. J., Karnam, A., & Rangaswamy, S. (2019). Software implementation of an automatic movable road barrier. In *2019 International Carnahan Conference on Security Technology (ICCST)* (pp. 1–6). IEEE. doi: 10.1109/CCST.2019.8888417.
6. Neha, A., Deshmukh, K., Khan, S., Katre, A., Gajbhiye, S., Bansod, A., Meshram, R., & Sewatkar, R. (2021). Controlling of traffic using movable road dividers. *International Journal of Innovative Science and Research Technology, 6*, 340–342.
7. Rashmi, C., Roopa, T. N., Samrudh, R., & Sandhya, M. (2020). Movable Road Dividers. *International Research Journal of Engineering and Technology, 7*, 4621–4623.

8. Basavaraju, A., Doddigarla, S., Naidu, N., & Malgatti, S. (2014). Vehicle density sensor system to manage traffic. *IJRET: International Journal of Research in Engineering and Technology eISSN, 3*, 2319–1163.
9. Jo, Y., Choi, J., & Jung, I. (2014). Traffic information acquisition system with ultrasonic sensors in wireless sensor networks. *International Journal of Distributed Sensor Networks, 10*(5), 961073.
10. Arvind, D. S., Poornima, L., Sai Pavan Kiran, T., Sai Soundarya, K., Hema Sunder, K., *et al.* (2022). Adjustable road divider using IOT. *Journal of Science, Computing and Engineering Research, 3*, 222–225.
11. Rao, V. S., Rao, M. S., & Rani, K. P. (2017). Shrewd street dividers driven by IOT technology. *International Journal of Civil Engineering and Technology (IJCIET), 8*(7), 385–389.
12. Agrawal, S., & Maheshwari, P. (2021). Controlling of smart movable road divider and clearance ambulance path using IOT cloud. In *2021 International Conference on Computer Communication and Informatics (ICCCI)* (pp. 1–4). IEEE. doi: 10.1109/ICCCI50826.2021.9402497.
13. Vastava, S. S. S., Vandana, B., Bhavana, M., & Gongati, R. (2023). Automatic movable road divider using Arduino UNO with node micro controller unit (MCU). *Materials Today: Proceedings, 80*, 1842–1845.
14. Dinesh, M. (2021). Movable road dividers for road traffic control with automated light solutions in embedded. *Turkish Journal of Computer and Mathematics Education (TURCOMAT), 12*(12), 2187–2191.
15. Immadi, G., Puthiyaveettil, A. A., Ramakanth, A., Mannuru, S., & Surya Kumari, T. (2020). Traffic density management using movable divider and RFID. *International Research Journal of Engineering and Technology, 5*, 6130–6134.
16. Lai, Y. L., Chou, Y. H., & Chang, L. C. (2018). An intelligent IoT emergency vehicle warning system using RFID and Wi-Fi technologies for emergency medical services. *Technology and Health Care, 26*(1), 43–55.
17. Amir, S., Kamal, M. S., Khan, S. S., & Salam, K. A. (2017). PLC based traffic control system with emergency vehicle detection and management. In *2017 International Conference on Intelligent Computing, Instrumentation and Control Technologies (ICICICT)* (pp. 1467–1472). IEEE.
18. Naik, T., Roopalakshmi, R., Ravi, N. D., Jain, P., & Sowmya, B. H. (2018). RFID-based smart traffic control framework for emergency vehicles. In *2018 Second International Conference on Inventive Communication and Computational Technologies (ICICCT)* (pp. 398–401). IEEE.
19. Saradha, B. J., Vijayshri, G., & Subha, T. (2017). Intelligent traffic signal control system for ambulance using RFID and cloud. In *2017 2nd International Conference on Computing and Communications Technologies (ICCCT)* (pp. 90–96). IEEE.

CHAPTER 20

DIGITAL CONTROLLED NOTICE BOARD USING IOT

N. RAMAMURTHY, KURUBA ANKITHA, NANDIVARGAM SWAPNA, and GADIGE SHIRISHA

Department of Electronics and Communication Engineering, G. Pullaiah College of Engineering and Technology, Kurnool, India

ABSTRACT

Digital notice boards have grown in significance in institutions, businesses, and numerous public locations such as bus stops, train stations, and hospitals. These electronic message boards cut down on paper, labor, and wastage of time. The main goal of this project is to create a wireless, digitally controlled notice board using Internet of Things (IoT) in conjunction with NodeMCU and ESP8266 WiFi module technologies. The ESP8266 WiFi module technology-based communication offers long-range, dependable, and secure communication. There are numerous wireless communication methods, such as ZigBee and others. This project uses an LED display to show data that we send from a mobile device, and we may remotely modify the notification using the web.

20.1 INTRODUCTION

We place a lot of emphasis on information accessibility in today's age of intense connectivity. We love to keep ourselves informed and up-to-date, thanks to the Internet and news publications. The main feature of any

Cyber-Physical Systems: Applications, Challenges, and Research Directions. Rajanikanth Aluvalu, Samarjeet Borah, & Lalit Garg (Eds.)

DOI: 10.1201/9781779643421-20

institution or public space, including bus stops, train stations, colleges, shopping centers, and so on, is a notice board. Papers or cable notice board displays are now sufficient for the purpose. The responsibility for publishing the notification is given to a certain individual. This not only necessitates additional labor costs but also takes a lot of time. Faster message delivery is essential in the modern era, where accessibility has always occupied the top spot in the development hierarchy. This project makes it possible to communicate without even needing to know someone's contact information or having an account with the system, making it superior to all currently offered Internet services. Its greatest strength is the range it offers. It can serve a whole organization that is based in one location by using routers as repeaters. By having all the tabs for the courses in a single app, additional work may be done to simplify complexity and deliver separate messages to several classrooms. Therefore, the basis of this research is an intelligent, rather than exciting, method of communicating with peers or the public by using a wireless electronic board that is synced using an Android app.

20.2 LITERATURE SURVEY

A Bluetooth-enabled wireless e-notice board, as described by Ramya R, Priyanka M, and Bavithra N, is discussed in this essay. The electronic board is explained using Bluetooth technology [1]. This chapter describes a novel and incredibly intriguing way of using a Bluetooth-synchronized wireless electronic display board to interact with the audience. As a result, it is possible for us to deliver any message by SMS almost immediately and without waiting, which is an improvement over the antiquated traditional approach of posting the message on a notice board.

Dharmendra Kumar Sharma and Vineet Tiwari described a small and medium-sized wireless electronic notice board using Bluetooth and ZigBee technologies [25]. Data security and authentication become more crucial when using networks for communication [2]. This chapter introduces a low-cost, portable, wireless electronic notice board utilizing an Atmel ATmega32 microcontroller and various wireless technologies such as Bluetooth and ZigBee. Performance analysis is conducted based on variables including range, bit error rate (BER), received signal strength indicator (RSSI), signal attenuation, and power consumption. We have created a Bluetooth and ZigBee wireless standard communication receiver for notice boards. We used an I2 × 64 graphical LCD based on the KS0108 as a display component.

SubbuLakshmi et al. presented a work on GSM-based wireless notice board using Arduino [3]. This study uses GSM technology to construct a notice board. The GSM-based notice board is intended for use in schools and institutions to broadcast current events continuously or periodically throughout business hours. Due to the GSM base, it can display announcements or news via flash rather than through a programmed mechanism. It comprises an embedded microcontroller system to create a GSM-free notice board whose content may be modified via SMS. It is straightforward, user-friendly, simple to install, and capable of receiving and displaying notice in a highly exact manner. Thanks to GSM, SMS technology messages can be shown on a notice board using a user's mobile phone.

Since they are so portable and have such a wide variety of purposes, smartphones in daily life are getting more and more significant. As a result, an Android-powered smartphone can be used for this. On the user's smartphone, an Android app is used to install the broadcast. The receiver end uses an expensive microcontroller board (Arduino Uno), which may display and receive messages in any of the four mentioned communication types. The methods that have been used to accomplish two unique applications: wireless person contacting and putting messages on a remote digital notice board. To save time, money, and paper, the developed system will try to communicate information to its target audience without the use of wires. Similar tasks can be found in the literature [4–24].

20.3 EXISTING SYSTEM

Many financial institutions, public spaces, and educational institutions have access to this innovative messages board. A microcontroller and a Bluetooth connection are used to accomplish this. Users can access data on LCD by sending messages as the existing system uses a Bluetooth-based notice board.

20.4 PROPOSED SYSTEM

We have created an IoT-based smart notice board for this project. We use a Node MCU, an ESP8266, and a dot matrix LED control display module to display the data in this scenario. A cheap open-source platform is Node MCU. We may access the WiFi network using Node MCU. The user can enter the information they want to display on the notice board in the ThingSpeak cloud. The end user can obtain these data in dot matrix LED control display module to display after writing the data. These display units are more useful in places of learning, banks, and public transportation like trains and airplanes.

20.4.1 PERFORMANCE

Figure 20.1 describes the digitally controlled notice board.

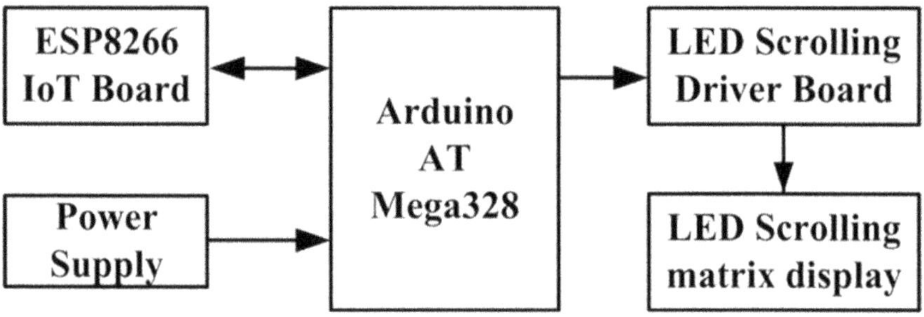

FIGURE 20.1 Venn diagram of digitally controlled notice board.

20.4.2 SOFTWARE DESCRIPTION

20.4.2.1 ARDUINO IDE

Arduino is a platform for both hardware and software prototyping, which is free and open source. By using an Arduino board to record inputs like light on a sensor and finger on a button, anything can be posted online, including a motor that can be started and an LED display that can be switched on. The Arduino Software (IDE), which is based on processing and wiring-based Arduino programming language can be used to accomplish this. Over the years, a variety of projects, from basic household items to complex scientific equipment, have employed Arduino as their brain.

FIGURE 20.2 Arduino IDE.

20.4.2.2 EMBEDDED C

Embedded system, commonly known as "Firmware," is a combination of hardware and software. A specific purpose computer system known as an embedded system is one that is entirely contained by the object it controls. The system is computer-driven. A specialized system that is a component of a bigger system or equipment is called an embedded system. Electronic devices that use microprocessors as part of their implementations are a part of a bigger system.

The major goals of microprocessors are to increase flexibility and streamline system design. The software for embedded devices is frequently kept in a read-only memory (RAM) chip. The monitoring of the analogue environment by receiving data from sensors and operating actuators is one of the key functions that embedded systems do.

20.4.3 HARDWARE DESCRIPTION

20.4.3.1 ARDUINO UNO

The AT mega328p is the brains of the Arduino Uno microcontroller board, with six analogues inputs, 15 digital input/output ports, an ICSP header,

USB connections, power jack, and reset button. Moreover, there is a 16MHZ ceramic resonator (CSTCE16MOV53-R0). The microcontroller comes with everything you need to operate it; all you need to power it with is a USB cable, an AC-to-DC converter, or a battery (Italian name "nattery"). The MOV53-oneme IDE 1.0 designates the release of the Arduino software. On the basis of the Uno board and Arduino software (IDE) version 1.0., the later versions of Arduino were created. The Uno board, the first of a line of USB Arduino boards, is the platform for default design. To view a complete list of all the current, previous, and outdated boards, go to the Arduino index of boards.

FIGURE 20.3 Arduino UNO.

20.4.3.2 ESP 8266 WIFI MODULE

A microcontroller known as the ESP8266 was created by Espressif Systems. The Esp8266 is a self-contained Wi-Fi networking system that can execute standalone programming and serves as a bridge between Wi-Fi and preexisting microcontrollers. The ESP8266 connects the microcontroller with Wi-Fi and creates a simple TCP/IP relationship, all while being incredibly simple to use. It is capable of offloading or enabling the entire Wi-Fi capability to several CPUs.

FIGURE 20.4 ESP 8266 WiFi module.

Both a built-in USB connection and wide range of pinouts are included with this module. Like how Arduino is easily flashed, you may connect the NodeMCU deviate to your laptop using a micro-USB connection. Furthermore, it is right away breadboard friendly.

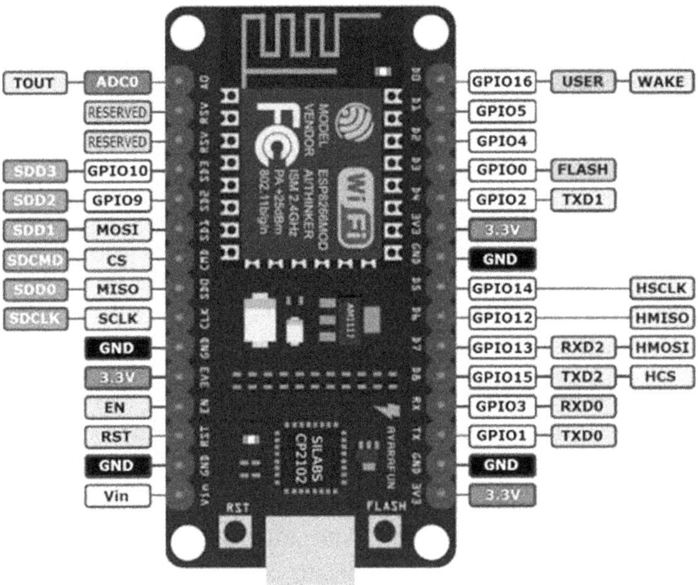

FIGURE 20.5 Pins of node MCU.

20.4.3.3 LIGHT-EMITTING DIODE

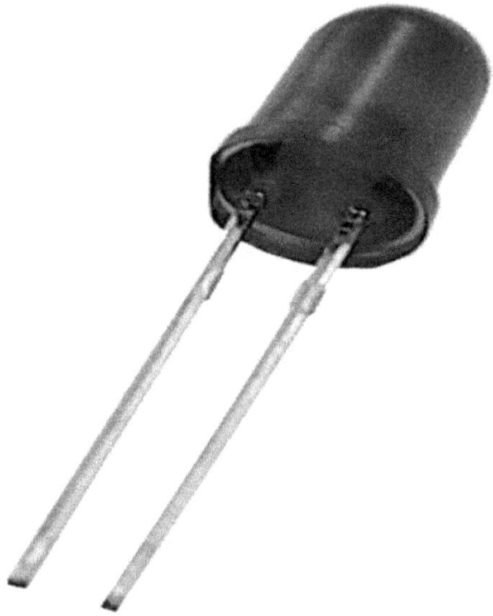

FIGURE 20.6 Light-emitting diode.

Special diodes known as LEDs produce light when linked to a circuit. In electronic appliances, they are widely employed as "pilot" lights to show whether the circuit is closed or not. The semiconductor chip in the center of an LED was enclosed in a clear (or frequently colored) epoxy shell. LEDs must be wired in the proper direction; the schematic may be marked with a "a" or "+" symbol for the cathode, and k or for the anode. There are two ways to identify an LED lead's negative side:

1. By the bulb
2. By using the cable that extends farthest from the LED

The cathode is the bigger electrode inside the LED. You can see it although this is not a recognized method of identification. The adverse lead should be positioned as described here and is identified using the bulb recognition method. Only when a significant voltage is applied to the semiconductor chip, an electron can pass across the junction and enter the *p* region, allowing current to flow. The junction functions as an electric potential barrier to the passage of electrons when significant voltage is applied to the semiconductor (voltage) across the LED leads.

FIGURE 20.7 LED view.

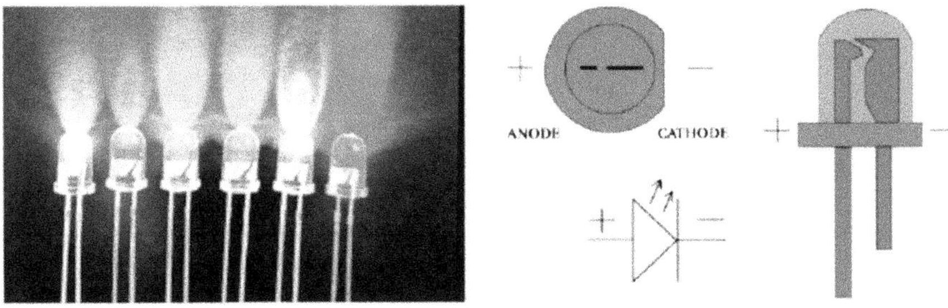

FIGURE 20.8 Light-emitting diode.

20.4.3.4 POWER SUPPLY

The Arduino Uno requires a power source to function and can be fueled in a variety of ways. As is customary, a USB cable can be used to connect the board directly to your computer. If you want to make your project portable, use a 9 V battery pack. A power supply is the component that provides electricity to the boards; it may be a battery, USB cable, AC adaptor, or another sort of controlled power source.

FIGURE 20.9 Power supply.

20.4.3.5 DOT MATRIX

FIGURE 20.10 Dot matrix.

Any messages entered by the user are shown on an LED dot matrix display. An 8 × 8 matrix and two blocks make up the LED dot matrix that is used in this project. The position of the least significant bit (LSB) and most significant bit (MSB) are illuminated. The second column comes after it, and the first column is turned off. Until the final column, the procedure will continue. Then, in response to a user request, the procedure is repeated to create a character or word. Because the LEDs turn off so quickly, they appear to be all lighting up at once. The latch was primarily responsible for organizing the data into columns and rows for the LED. Software created in assembly language for the microcontroller MC68HC11A1 regulates this procedure. For instance, the microcontroller must receive a sequence of hex number instructions for the rows and columns to display in the dot matrix. For rows, the integers are $B5, $CB, and for columns, $C3. The LED on the dot matrix will turn on, thanks to logic0.

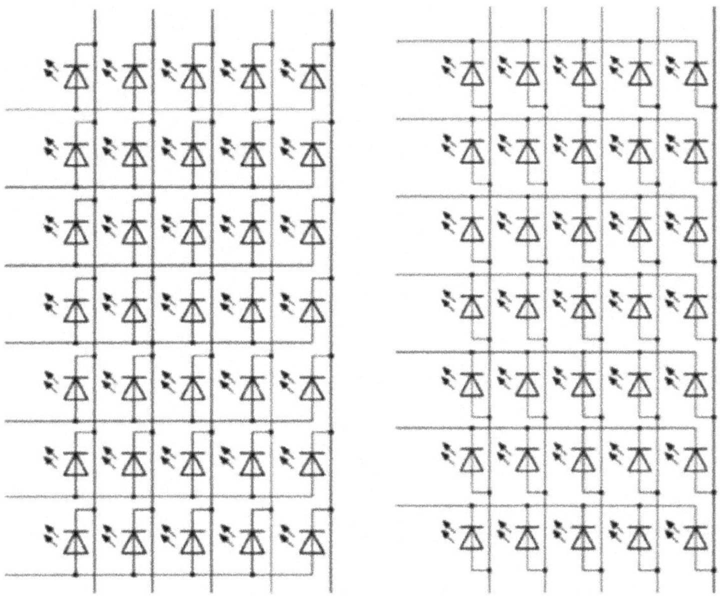

FIGURE 20.11 LED dot matrix configurations—CCAR and ACCR.

20.4.4 ANDROID APPLICATION

Google first made an open-source web program available which is called App Inventor for Android, and the Massachusetts Institute of Technology is still maintaining it today (MIT). It enables anyone, regardless of background, to produce software for the Android operating system (OS). Users can drag and drop visual components to develop an app that can run on Android handsets using a graphical user interface that is comparable to Star Logo TNG and scratch. Google drew on a lot of historical expertise in educational computers as well as internal work on online development environments for developing App Inventor.

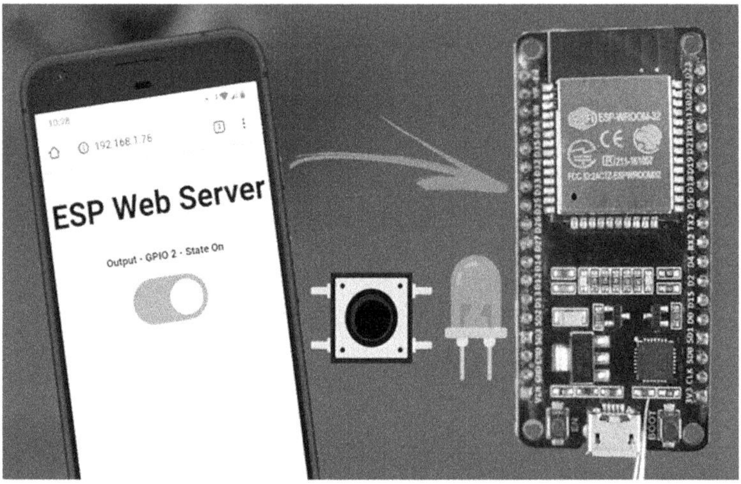

FIGURE 20.12 Android application.

Constructionist learning theories, which emphasize that programming may be a tool for promoting the active learning of significant concepts, informing App Investor and the projects on which it is built. The Firebase Database plugin now supports the MIT App Investor. People can now save data on Google Firebase thanks to this.

20.4.5 WORKING OF THE PROJECT

The ESPWI-FI.h and MD-MAX72xx.h files allow the NodeMCU to build a web server over a WiFi network and manage a LED matrix, respectively. Before continuing, take some time to familiarize yourself with this. After the code has been uploaded to the NodeMCU, the URL can be checked from the serial monitor and used to access the web page for managing the text that is displayed on the MAX LED display. On this website, we can enter a message, and it will be shown on the grid.

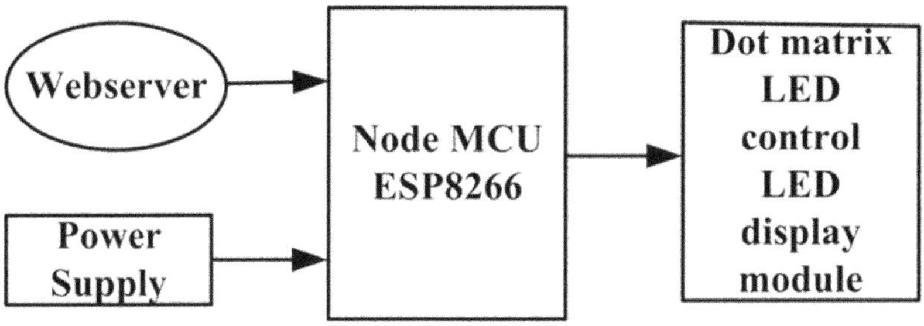

FIGURE 20.13 Diagram of the front end of the app.

In this block diagram, NodeMCU ESP8266 is used as a main controller. Web server cloud is used to write the data in the display. The dot matrix LED control display module is connected as the output device which is used to show the data to the end users.

20.4.6 *FUTURE SCOPE*

The implemented system can be used for dynamically updating notices and messaging at bus stops and retail centers. The suggested system can display scanned copies of notices as well. To this design, a speech capability can be added to make announcements or display advertisements in public areas.

20.5 CONCLUSION AND RESULTS

The front end of the Android application is depicted in Figure 20.14. There are two text boxes in it. One box is used to type the local host's IP address in order to connect, and the other is used to type text that the user wants to display on a matrix display. Figure 20.4 demonstrates how a straightforward word was presented in dot matrix, reading from left to right. Here, the user had input the message that would appear in the app's text box.

The figure shows a hot spot is shared by mobile devices and hardware equipment. Here, the SPI protocol was used for message transport. Figure 20.5010 demonstrates how a digit was presented in a dot matrix from left to right. Here, the user input message that would be displayed in the app's text field, as seen in the figure. Here, a hot spot is shared by mobile devices and hardware equipment. Here, the SPI protocol was used for message transport and shows how text was shown in a dot matrix, left to right. Here, the user inputs the message that would be displayed in the app's text field, as seen in the figure. Here, a hot spot is shared by mobile devices and hardware equipment. Here, the SPI protocol was used for message transport Wi-Fi.

FIGURE 20.14 Front end of the app.

FIGURE 20.15 Front end of the app.

ACKNOWLEDGMENT

We would like to take this time to thank everyone who helped make this capstone project a success. We deeply appreciate Dr. N. Ramamurthy, our project guide, for his unwavering support, knowledgeable direction, encouragement, and courteous cooperation throughout the microproject work. We owe a great deal for his assistance with the capstone project.

FIGURE 20.16 Experiment results.

KEYWORDS

- **ESP8266**
- **IoT**
- **LED display**
- **notice board**
- **Wi-Fi module**

REFERENCES

1. Kumar, P., Bhrdwaj, V., Pal, K., Rathor, N. S., & Mishra, A. (2012). GSM based e-notice board: wireless communication. *International Journal of Soft Computing and Engineering, 2*(3), 601–605.
2. Memon, A. R., Chowdhry, B. S., Shah, S. M. S., Memon, T. R., & Shah, S. M. A. (2015). An electronic information desk system for information dissemination in

educational institutions. In *2015 2nd International Conference on Computing for Sustainable Global Development (INDIACom)* (pp. 1275–1280). IEEE.
3. http://www.pcmag.com/encyclopedia/term/54 272/webapplication
4. Power, M. (2011). Mobile web apps: a briefing paper. *Centre for Educational Technology & Interoperability Standards, March. Mark Power@ CETIS.*
5. Persada, S. F., Razif, M., Lin, S. C., & Nadlifatin, R. (2014). Toward paperless public announcement on environmental impact assessment (EIA) through SMS gateway in Indonesia. *Procedia Environmental Sciences, 20*, 271–279.
6. Sharma, D. K., Tiwari, V., Kumar, K., Botre, B. A., & Akbar, S. A. (2015). Small and medium range wireless electronic notice board using Bluetooth and ZigBee. In *2015 Annual IEEE India Conference (INDICON)* (pp. 1–5). IEEE.
7. Khera, N., Shukla, D., & Awasthi, S. (2016). Development of simple and low cost Android based wireless notice board. In *2016 5th International Conference on Reliability, Infocom Technologies and Optimization (Trends and Future Directions) (ICRITO)* (pp. 630–633). IEEE.
8. Simha, K., Kumar, C., Parinitha, C., & Tantry, S. (2016). Electronic notice board with multiple output display. In *2016 International Conference on Signal Processing, Communication, Power and Embedded System (SCOPES)* (pp. 1558–1561). IEEE.
9. Arun, M., Monika, P., Lavanya, G., Arun, M., Monika, P., & Lavanya, G. (2017). Raspberry Pi controlled smart e-notice board using arduino. *IJCAT*.
10. AbduAllah, Z. M., Mahmood, O. T., & AL-Naib, A. M. I. (2014). Photovoltaic battery charging system based on PIC16F877A microcontroller. *International Journal of Engineering and Advanced Technology (IJEAT), 3*(4), 27–31.
11. Sunehra, D., & Rajasri, S. (2017). Automatic street light control system using wireless sensor networks. In *2017 IEEE International Conference on Power, Control, Signals and Instrumentation Engineering (ICPCSI)* (pp. 2915–2919). IEEE.
12. Kamdar, F., Malhotra, A., & Mahadik, P. (2013). Display message on notice board using GSM. *Advance in Electronic and Electric Engineering, 3*(7), 827–832.
13. Kamdar, F., Malhotra, A., & Mahadik, P. (2013). Display message on notice board using GSM. *Advance in Electronic and Electric Engineering, 3*(7), 827–832.
14. Gaikwad, A., Kapadia, T., Lakhani, M., & Karia, D. (2013). Wireless electronic notice board. *International Journal on Advanced Computer Theory and Engineering (IJACTE), 2*.
15. RaspberryPi. https://www.raspberrypi.org/documentation/hardware/raspberrypi/power/README.md (accessed November 02, 2017).
16. Raspberry Pi. https://www.raspberrypi.org/documentation/hardware/raspberrypi/bcm2835/README.md (accessed November 03).
17. Swapna, M., Viswanadhula, U. M., Aluvalu, R., Vardharajan, V., & Kotecha, K. (2022). Bio-signals in medical applications and challenges using artificial intelligence. *Journal of Sensor and Actuator Networks, 11*(1), 17.
18. Ramana, K., Aluvalu, R., Gunjan, V. K., Singh, N., & Prasadhu, M. N. (2022). Multipath transmission control protocol for live virtual machine migration in the cloud environment. *Wireless Communications and Mobile Computing, 2022*, 2060875.
19. Dugyala, R., Reddy, N. H., Maheswari, V. U., Mohammad, G. B., Alenezi, F., & Polat, K. (2022). Analysis of malware detection and signature generation using a novel hybrid approach. *Mathematical Problems in Engineering, 2022*, 1–13.

20. Polkowski, Z., Mishra, S. K., Mishra, B. K., Borah, S., & Mohanty, A. (2023). Impact of internet of things in social and agricultural domains in rural sector: a case study, *International Journal of Cloud Computing, 12*, 90–105, https://doi.org/10.1504/IJCC.2023.129776.
21. Nath, M. P., Priyadarshini, S. B. B., Mishra, D., Borah, S. (2020). A comprehensive study of contemporary IoT technologies and varied machine learning (ML) schemes, In *Soft computing techniques and applications* (pp. 623–634). Springer Nature. https://doi.org/10.1007/978-981-15-7394-1_1.
22. Mishra, J. P., Borah, S., & Mishra, S. K. (2021). Performance of scalable data linked to internet of things: a case study. In S. Borah, R. Pradhan, N. Dey, P. Gupta (Eds.), *Soft computing techniques and applications. Advances in intelligent systems and computing* (Vol. 1248). Springer. https://doi.org/10.1007/978-981-15-7394-1_54.
23. Sourav Prasad, S., Mishra, J. P., Mishra, S. K., & Borah, S. (2020). *Challenges toward blockchain and renewable energy linked to IoT: a survey, advances in energy technology* (pp. 49–55), Springer. https://doi.org/10.1007/978-981-15-8700-9.
24. Verma, U., Sohani, M., Borah, S., Nagwanshi, K. K., & Pathak, S. (2022), Role of fog-assisted internet of things-enabled system for managing the impact of COVID-19. In S. Borah, S. K. Mishra, V. E. Balas, & Z. Polkowski (Eds.), *Advances in data science and management. Lecture notes on data engineering and communications technologies* (Vol. 86). Springer. https://doi.org/10.1007/978-981-16-5685-9_39.
25. Sharma, D. K., Tiwari, V., Kumar, K., Botre, B. A. and Akbar, S. A. Small and medium range wireless electronic notice board using Bluetooth and ZigBee, 2015 Annual IEEE India Conference (INDICON), New Delhi, India, 2015, pp. 1–5, doi: 10.1109/INDICON.2015.7443394.

CHAPTER 21

IOT COMPONENTS FOR FUTURE SMART CITIES WITH 6G COMMUNICATIONS

SWATI SUCHARITA ROY, BHARAT JYOTI RANJAN SAHU, and SHATARUPA DASH

Department of Computer Science & Engineering, ITER, Siksha 'O' Anusandhan University, Bhubaneswar, India

ABSTRACT

Rapid urbanization urges further advancement of information and communication technology (ICT), a digital and sustainable environment, and smart governance to improve resident's quality of life (QoL). The Internet of Things (IoT) is a framework of interconnected digital objects that can configure themselves to carry out assigned tasks autonomously. The IoT has been a crucial driving force in establishing the smart city since the last decade. Understanding the city dweller's behavior is one of the important parts of smart city management. Thanks to the integration of machine learning (ML) for the analysis and prediction of IoT operations which further accelerates the advancement in smart city development. This chapter discusses various aspects of smart cities, such as transportation, energy, agriculture, industry, and healthcare management, with a systematic review that can provide a new direction and helps to improve the services provided by smart cities. The primary focus of this work is to explore the impact of IoT on smart city development with the integration of ML, metaheuristic algorithms, and 6G communication technology. Furthermore, we have applied a metaheuristic algorithm for smart parking space management.

Cyber-Physical Systems: Applications, Challenges, and Research Directions. Rajanikanth Aluvalu, Samarjeet Borah, & Lalit Garg (Eds.)

DOI: 10.1201/9781779643421-21

21.1 INTRODUCTION

The rapid advancement of technology accelerates improvement in urban area size and better lifestyle. Recent studies show that urbanization is the key to economic advancement. It is essential to understand the behavior of city dwellers in managing urban spaces. With the Internet of Things (IoT), tools like information and communication technology (ICT), Big data, and machine learning (ML) are of utmost importance to managing extensive physical infrastructure. IoT gathers and accumulates essential data autonomously, analyzes those data for future decision-making, and improves the system's overall performance. The existing traditional system has several flaws which need to be handled. ML and other advanced computational techniques have demonstrated their ability to deal with big data in the field of smart cities. Various consequences are incorporated, as well as conflicts caused by the adoption of IoT and ML in the smart city system. IoT in smart cities operates in a variety of industries autonomously, that is, without human intervention, including agriculture, government, retail, manufacturing, transportation, and healthcare. Remote monitoring, predictive maintenance, facility management, manufacturing efficiency, connected products, and a variety of other tasks are all performed by IoT.

The main focus of the smart city is that the city can be faster, friendlier, energy efficient (i.e., green), safe, and services can be efficient, flexible, and sustainable. Various smart city components are smart buildings, smart transportation, smart public spaces, green energy, e-healthcare, e-education, and smart public infrastructure [1].

In this chapter, first, we discuss various aspects of IoT and its different application scenarios. Then, this work focuses on the effect of IoT, ML, metaheuristic algorithms, and 6G in smart cities lastly talks about different components in smart cities and the essential features of smart cities like energy consumption, end-to-end delay, reliability, scalability, and latency. Finally, we have demonstrated the use of a metaheuristic algorithm in the smart parking use case.

21.1.1 COMMUNICATION ARCHITECTURE IN IOT

An IoT-based system consists of various components that provide services such as sensing, monitoring, communication, identification, and management. Smart objects in IoT perform various functions such as sensing, monitoring, and controlling. Through various communication technologies, devices in an IoT system can share data with other applications and devices connected to the system. Figure 21.2 represents the overall communication in IoT. IoT devices can exchange collected data and process it locally or send it to a server or cloud for processing. IoT devices use various interfaces to communicate with other devices, for example, storage interfaces, I/O interfaces, and audio/video communication interfaces. Device modeling, monitoring, controlling, and communication are all standard IoT services. The IoTs system includes several modules for analyzing data and making decisions. The most common technologies in IoT are its hardware platforms, which include processors, clock speeds, bus width, and various communication technologies for use in various applications [2].

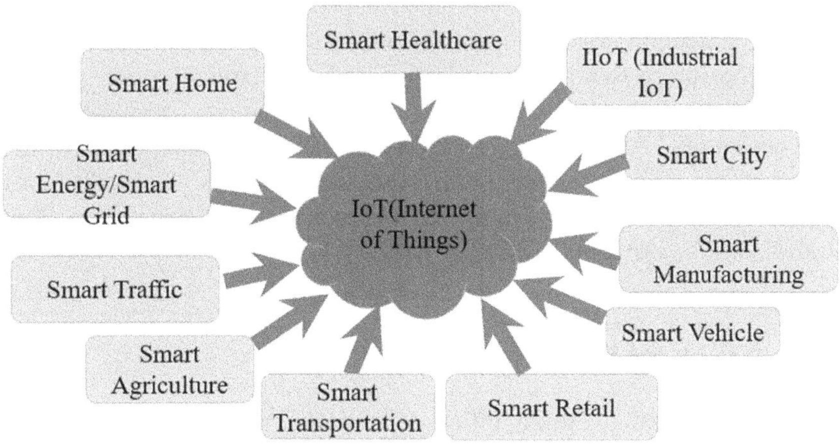

FIGURE 21.1 Applications of IoT.

The IoT umbrella has heterogeneous devices communicating with each other with different service requirements. Smarter IoT devices, objects, or things can communicate with one another. IoT processing is also smarter because the system performs autonomously based on data from the environment and stores the data for analysis and prediction, allowing it to make smart decisions that increase efficiency, productivity, and performance and reduce overall cost. The IoTs have a wide range of implications for various applications, industries, and organizations.

Ploennigs et al. [3] propose that integrating blockchain and AI can improve prediction and decision-making and that incorporating some features of cloud computing, edge computing, and fog computing can improve IoT efficiency. Perera et al. [4] discuss the effect of IoT in our day-to-day life and sense to find out the challenges in IoT. Mohanty et al. [1] focus on IoT in the smart city. The authors have given a broad overview of various application areas in IoT, focusing on enhancements in each component of smart cities. Additionally, managing IoT infrastructure should be emphasized to provide more convenience to society. Sodhro et al. [5] focus on quality of experience (QoE) optimization and propose a QJEEO algorithm for QoE based on energy and entropy optimization. It connects QoS and QoE. Examine the relationship between test scores (i.e., survey data) and performance metrics in terms of mobility on the speed of portable devices.

Improved cloud computing and big data management can increase the impact of IoT in smart cities. Kirimtat et al. [8] discuss smart city IoT infrastructure in general. IoT, Big Data, and cloud computing are mentioned to discuss the future trend of smart cities. IoT infrastructure can be more efficient through the utilization of cloud computing, fog computing, and edge computing. Ahmed et al. [9] focus on the various necessities of IoT environments like communication type, technologies, network types, objectives of the infrastructure, and characteristics. Furthermore, they discuss various application areas of IoT.

21.2 IOT IN SMART CITIES

The establishment of smart cities depends on their required features, components, and requirements. Smart city infrastructure includes various components such as healthcare, education, smart vehicle, smart transportation, security and surveillance, smart home, smart building, smart industry, smart business, smart energy/smart grid, smart bins, and so forth. IoT influences each corner of the cities, improves operation and services, and affects the daily life of citizens, giving convenience to society. Smart cities include smart infrastructure, smart buildings, smart homes, smart traffic, transportation, energy, healthcare, and smart industry.

Advancement of technology encourages merging the cities. However, it requires physical infrastructure, ICT, Big data, and IoT. IoT improves the effectiveness of cities. In recent decades, the world population has been significantly increasing, and by the end of 2050, 70% of the population will reside in urban areas, which will consume more energy and generate greenhouse gases. Thus, it adversely affects the environment and necessitates the smart city concept [6]. Adaptation of advancement in communication and interconnection can improve efficiency in the smart city concept and impact millions of lives. Cano et al. [7] focus on the advancement of the IoT, specifically in industry, and discuss various communication advancement for interconnection and communication. Broad knowledge about IoT architecture, including communication and methodologies used in various applications, encourages the advancement of communication to work in the IoT environment establishing smart cities effectively depending on their required features, components, and requirements.

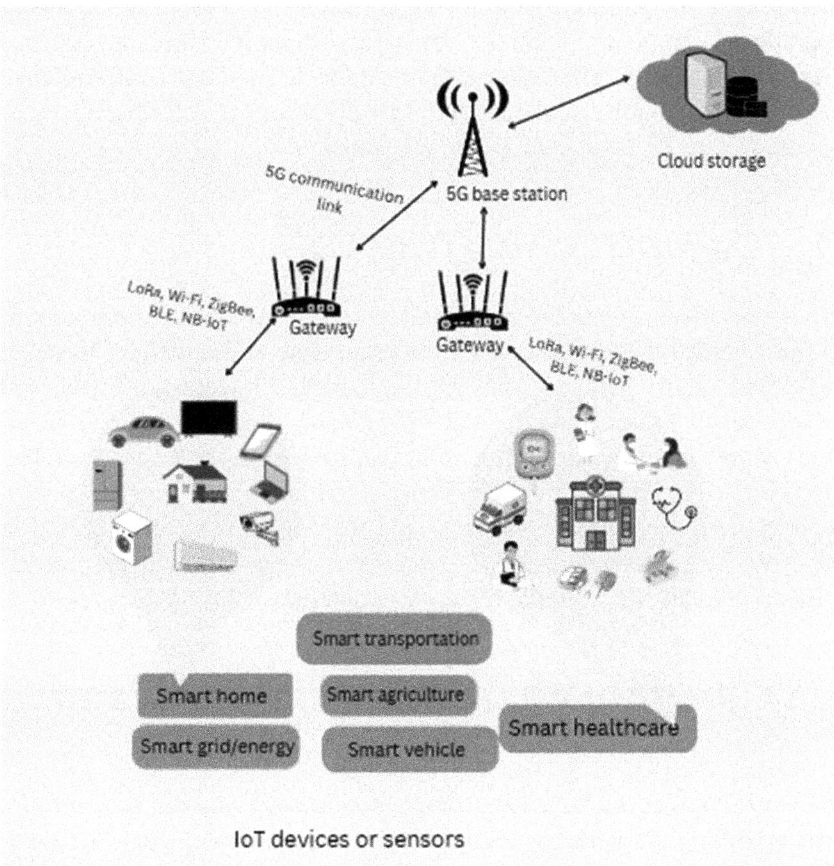

FIGURE 21.2 The communication architecture of IoT.

21.2.1 BENEFITS OF IOT IN SMART CITIES

IoT in smart cities has many benefits, such as improving the QoL, increasing the efficiency of urban operations and services, influencing competitiveness, reducing energy consumption, depreciating water consumption, decreasing carbon emission, regulating transportation requirements, managing traffic, and managing city waste.

21.2.2 CHALLENGES OF SMART CITIES

With several benefits, IoT also has some challenges for the advancement in smart cities, such as configuration cost, infrastructure establishment cost, operational efficiency, reduction of carbon emission, disaster management, data security, efficient communication technology (full coverage, low latency, reliable, scalable), public safety, city waste management, system failure, and water supply management.

21.2.3 SECURITY IN SMART CITIES

Privacy and security are the main challenges in an IoT-based environment, as massive devices are connected for processing and communication. In IoT-based smart cities, security can be managed by standardization and various technologies. Violette et al. [10] discuss standardization improvement, work on leak path finding, and focus on security. Kranz et al. [11] discuss the standardization problems regarding data transfer and its security, IoT's effects on business, and the financial benefits of IoT in business. Mosenia and Jha's [12] security is the primary concern in IoT infrastructure as it is a global network of heterogeneous devices.

21.2.4 IOT COMMUNICATION TECHNOLOGIES FOR SMART CITIES

IoT devices collect and exchange data autonomously and can be controlled remotely through various communication technologies such as Wi-Fi (Wireless Fidelity), Zigbee, Z-wave, NB-IoT (Narrowband IoT), Sigfox, Lora (Long range radio), 3G, 4G, 5G, and 6G. Devices in IoT infrastructure are smarter, the processing is more intelligent, and communication is more informative.

IoT, the internet of a significant massive number of devices (or objects or things), renders countless services. However, the interconnection of billions of smart devices and exchanging confidential data over the internet increases security challenges [12]. IoT influences a better lifestyle and new industrial opportunities; however, challenges come with opportunities. Existing technologies for connectivity, like Wi-Fi, suffer from collision and congestion; 5G wireless networks can handle these issues by adopting new methodologies [18]. Ray [2] broadly discusses the IoT architecture, focuses on various communication technologies, and compares existing methodologies. Arzo et al. [13] intend to discuss various communication methods, Network Function Virtualization (NFV), and Software Defined Networks (SDN) and discuss various standardization methods. 6G, ML, blockchain, and edge computing handle communication among massive IoT devices and across the global IoT network [6]. 6G, cellular IoT, integrates ECC (Energy, Computation and Communication) and beam forming to reduce energy consumption [14]. It is necessary to work on QoE to improve IoT infrastructure and explore energy harvesting in industrial applications over edge intelligent 6G networks. Agiwal et al. [15] discuss the 5G communication technology for IoT devices and its future requirements. Zanella et al. [16] intend to focus on transparent and seamless digital services. This chapter's main focus is on urban IoT, which supports advancement in communication technologies. It includes constrained application protocol (CoAP), 6lowPAN.

21.2.5 NEED OF 6G

Nowadays a massive number of devices are connected, and many unmanageable applications come into the picture, such as augmented reality/virtual reality, online games, smart vehicles, and autonomous driving, making everything smarter and focusing on computation-intensive, data-intensive, and delay sensitivity. To manage the requirement, specific applications and today's huge demand encourage the advancement in communication technology. 5G is insufficient to support today's colossal demand; it encourages the use of 6G, which supports a high data rate, widespread coverage, very low latency, accurate localization, and maintains privacy and security against unauthorized users [17]. To overcome the challenges or limitations in 5G, 6G communication technology can be adopted. Saxena et al. [18] broadly discuss communication in IoT infrastructure. As day by day, a huge number of devices are connected in IoT infrastructure, enhancement of communication in the IoT environment is a challenging task in 5G.

21.3 COMPONENTS OF SMART CITIES

There are various components included in smart cities. IoT has become popular in making cities smarter using various technologies in almost every field; it affects society deeply, provides convenience to the citizens, and makes devices smarter compared to the traditional system. IoT can work for large industries and can perform household work smartly. In IoT, the "smart" word is usually used as it makes the system and individuals operate autonomously, that is, without giving external instruction all the time. Figure 21.3 represents the overall configuration of smart cities, which includes their components, features, technologies, infrastructure, and challenges. Choudhary et al. [19] focus on the utilization of IoT in smart city urbanization and discuss various components of smart cities. Ianuale et al. [20] focus on advancement in smart cities and discuss big data in smart cities system.

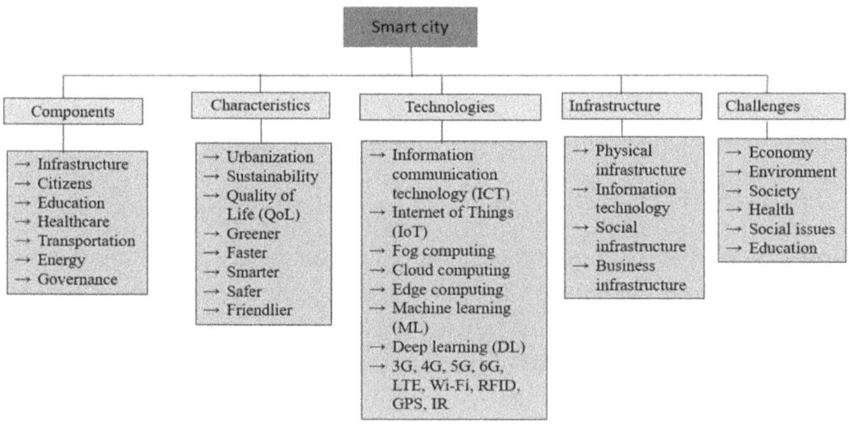

FIGURE 21.3 Overall configuration of smart cities.

The development of the smart city focuses on traffic congestion, power consumption, public safety, smart city service optimization, real-time operations, waste management, transport management, optimum utilization of resources, and effective interaction among citizens and inhabitants. Hurrison et al. [21] focus on making a city smarter through the advancement of various technologies. Various components of smart cities are smart homes, buildings, healthcare, industry, agriculture, manufacturing, grid, and transportation. Figure 21.4 shows various aspects of a smart city and its various components.

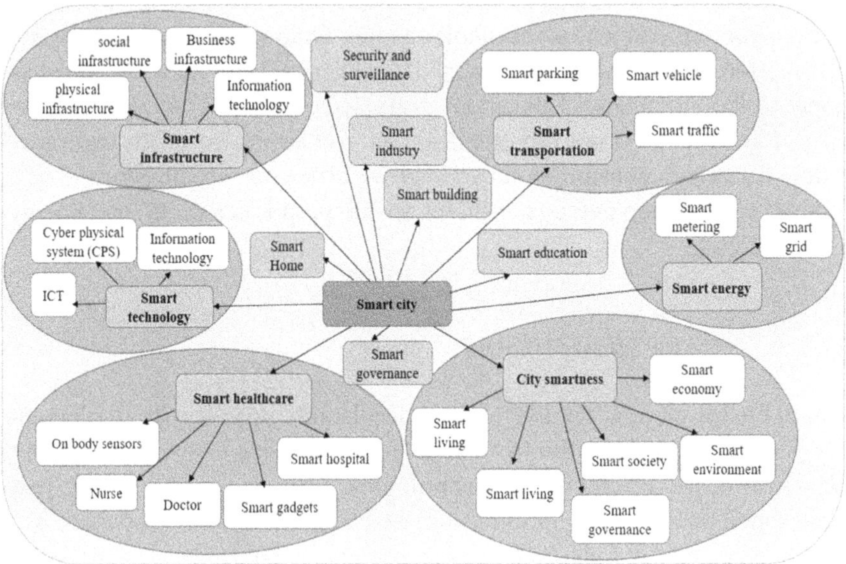

FIGURE 21.4 Elements of smart city components.

21.3.1 SMART TRANSPORTATION

Earlier, traditional modes of transportation such as road, rail, water, and airline transport were available. Smart transportation uses ICT and real-time data processing to make global airway hubs, intelligent road transport, protected cycle routes, secure pedestrian paths, and an intercity railway network a safer, faster, more reliable, and cost-effective transportation system. It increases the efficiency of vehicle utilization; citizens can use or select different modes of transportation based on their needs at a low cost and with minimal effort. More advanced parking in a smart city can improve efficiency and make a smart city more advanced. Tandon and

Gupta [22] include fog computing in smart parking systems to improve the efficiency of managing parking, reduce congestion, and avoid the negative impacts of unmanaged parking.

21.3.2 SMART GRID/ENERGY

A smart grid has various components like a smart meter, operational control mechanism, load balancing, and fault-tolerant mechanism, which are responsible for efficient and reliable manner from power delivery. Energy distribution, optimum consumption of energy, and efficient management improve the efficiency of the smart grid. Heck et al. [23] focus on energy or grid systems in smart city applications. Advancement in the multilevel model for energy consumption can improve efficiency. Rostirolla et al. [24] focused on reducing energy consumption and proposed a multilevel energy-saving model.

21.3.3 SMART HEALTHCARE

Smart health care is an IoT system for health services. It implements various technologies to improve healthcare services in society. It adopts new healthcare services and technologies, provides better health care, makes healthcare more mobile, and ensures quick and easy access to health services for kids and old-age society. Aldahiri et al. [28] intend to overview ML technologies for analyzing healthcare datasets for disease prediction and decision-making by including the benefits of IoT for generating medical data. People in remote areas face challenges with healthcare issues; they need the facilities for diagnosis and timely treatment. Hameed et al. [29] introduce a modern low-cost healthcare system to provide healthcare-related facilities based on IoT and ML, which sense the patient's health-related data and process it for prediction and decision-making about serious health issues and treatment. Sharma et al. [30] intend to predict the patient's body condition and disease by analyzing the data generated from the IoT-based healthcare environment. The system collects the patient's body temperature, heart rate, and oxygen label through sensors and sends it to the Raspberry Pi and then exchanges it to the cloud server using a ML technique to predict the patient's diseases and health conditions. Here data is generated from different sources and healthcare systems. Store the generated healthcare data in the cloud server. Through mobile applications, users can get patient information easily, and also the system maintains privacy by using cloud virtual machines. Jeong et al. [31] intend to adopt IoT for smart healthcare systems to gather data through various sensing technologies and then communicate the collected data to the server for analysis. Mohamed et al. [32] intend to highlight the benefits of IoT in monitoring the behavior of patients with critical conditions like neurological illness, mental disorder, and mental trauma, and then use deep belief neural network to analyze data collected from the IoT-based infrastructure.

21.3.4 SMART HOME

Various everyday used devices like AC, coffee machines, microwaves, water heaters, water sprinklers, and security alarms can be controlled through our mobile phones remotely. The IoT creates an environment where everyday objects, such as air conditioning, refrigerators, microwaves, cars, coffee machines, water heaters, and security alarms, are connected to the internet, making users' lives more productive and convenient [4].

21.3.5 SMART AGRICULTURE

IoT supports smart agriculture for the advancement of smart agriculture by using various technologies for crop monitoring, livestock monitoring, irrigation management, smart pest control (i.e., detection of pesticide presences), fertilizer management, weather forecasting, and harvesting automation. Its primary objective is to enhance the quality and productivity of the crop with less human effort.

21.3.6 SMART MANUFACTURING

Smart manufacturing is a system consisting of various methods to upgrade manufacturing, such that productivity can be maximized and also cost, and effort can be reduced.

21.3.7 SMART VEHICLE

The smart car allocates the shortest and less crowded path or the most suitable path for our journey, which can manage our time and safety.

IoT provides tremendous advantages in various application areas like smart cities, that is, providing convenience to the citizens in the respective city, healthcare, and industry by improving production, manufacturing, collaboration with other companies, and in many more essential analyses, prediction, and decision-making things, transportation sectors, agriculture by improving traditional farming techniques, seed selection, farming methods, and various necessary steps for improvement of production, smart grid by utilizing energy supply smartly. IoT acts as a key player in the modernization and advancement of society and also reduces human effort with improved and quality work within stipulated time with more efficiency.

The connected devices in the IoT system gather and exchange data through sensors. These IoT devices are connected through gateways. Wired or 3G, 4G, and 5G wireless technologies are also used to pass the gathered information by connecting the devices. Wi-Fi, Bluetooth, ZigBee, and Z-wave provide connectivity and maintain communication protocols. IoT helps to perform autonomously instead of manually, increasing efficiency, reducing cost, and reducing user efforts. Figure 21.5 shows the communication architecture of the IoT infrastructure by considering the smart home.

All appliances can be connected to the internet to perform automatically and can be controlled remotely. In 2018, there were 23 billion devices connected; now, more devices are connected and exchange information. IoT also connects people to communicate with each other for effective communication. Let us consider a smart home where the alarm goes off in the morning, the IoT opens the window blind, switches on the coffee machine, and switches on the water heater for the bath.

21.4 USE OF MACHINE LEARNING

Deep learning requires more computational power and takes more time for model training. However, more efficient hardware and training algorithms improve the efficiency of analysis and prediction. Like ML, deep learning has two main categories of learning methods: unsupervised and supervised learning. Restricted Boltzmann machine (RBM) and Autoencoder are some unsupervised learning, and Convolutional Neural Networks (CNN) and Recurrent Neural Networks (RNN) are some supervised learning algorithms. In the healthcare system, deep learning is essential to disease prediction, early diagnosis, treatment, and medication. Deep learning has a significant effect on analysis and prediction. However, deep learning has many challenges and complexities and is unsuitable for all scenarios. In smart cities, ML, deep learning, IoT, big data, 5G, and 6G are used as per their requirements. For example, metaheuristic algorithms can be used in parking management for the nearest parking slot allotment. In agriculture, ML can help in future prediction to do farming smartly.

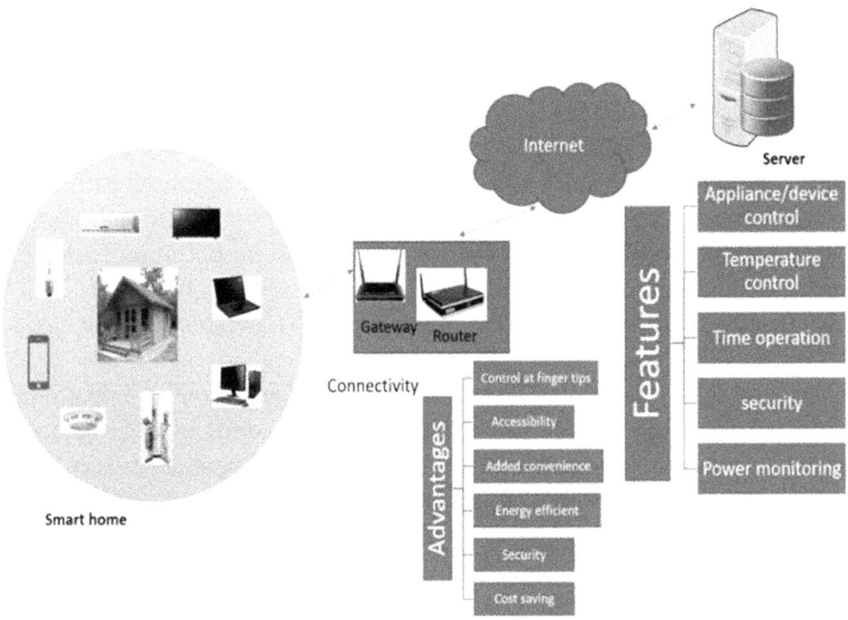

FIGURE 21.5 The communication architecture of IoT in smart home.

21.5 ROLE OF METAHEURISTIC ALGORITHM IN IOT

There are various metaheuristic algorithms available that can be used for various purposes, such as finding the shortest path, reducing energy consumption, lowering latency, and optimizing operations. Metaheuristic algorithms are categorized into different types, such as swarm-based algorithms, nature-inspired algorithms, biogeographic stimulated algorithms, evolutionary algorithms, and physics-based algorithms. Ant Colony Optimization (ACO), Artificial Bee Colony (ABC), Fish Swarm Algorithm (FSA), and Particle Swarm Optimization (PSO) are some examples of swarm-based algorithms. Bat Algorithms (BA), Cuckoo Search Algorithms (CSA), and Flower Pollination Algorithms (FPA) are some natures inspired algorithms. Spotted hyena optimizer (SHO), Gray wolf optimizer (GWO) are some biogeographic stimulated algorithms. Genetic Algorithm (GA) and Differential Evolution (DE) are some evolutionary algorithms. Black hole algorithm (BHA) and Harmony Search (HS) are some physics-based algorithms. Among all these metaheuristic algorithms we can adopt any one or can use hybrid algorithms for various purposes on IoT.

ACO is one of the nature-inspired metaheuristic algorithms. ACO is motivated by ant colonies to search for food, that is, the foraging activity of ants; they use a pheromone to find the shortest path. It is a population-based nature-inspired method for the prediction of the optimum path. ACO is utilized to detect estimated solutions to complex optimization issues. In ACO, artificial ants, that is, software agents, explore optimum solutions, such as searching for the best path based on the user's requirements. ACO, for example, can be used to find the best path to a viable parking slot in smart city infrastructure.

Maeda et al. [30] focus on technological advancement to make cities smarter, like big data processing, coordination among heterogeneous devices, and cutting-edge ICT for system control. Advancements in load-balancing technologies may reduce energy consumption and system failure in the smart city. Khan et al. [31] focus on energy management in smart city infrastructure using fog computing and a metaheuristic algorithm. Butt et al. [32] focus on energy consumption in IoT-based smart city infrastructure with cloud and fog computing using metaheuristic algorithms. A hybrid metaheuristic algorithm can be used to reduce energy consumption, enhance green communication, and enhance network life span. It is required to enhance the network life span, focus on green communication, and reduce energy consumption for the IoT environment. Verma et al. [17] proposed hybrid whale spotted hyena optimization by integrating the whale optimizer algorithm and SHO to enhance green communication in 6G.

21.6 SMART PARKING SYSTEM USE CASE

In smart cities, parking vehicle during pick hours in an overcrowded place is a very challenging task; it wastes gasoline, wastes the valuable time of the citizens, and some people suffer due to the unavailability of parking slots and busy schedules or for emergency work and some people park their vehicles here and there which affects the traffic and transportation management and makes the place overcrowded and unmanageable which may cause an unhealthy situation or may cause accidents. To avoid these undesirable situations parking system should be managed smartly. IoT plays a vital role in managing parking systems in smart cities. Roy et al. [33] focus on energy efficiency in the IoT environment and discuss routing techniques to lower the latency, reduce energy consumption, and reduce end-to-delay. IoT builds an infrastructure to collect information on parking lots in the respective city and the people's requests for parking vehicles in the cloud. The parking system is observed or monitored to maintain the operation and improve efficiency. Tang et al. [35] proposed a smart parking system based on fog computing, and the main focus is on real-time processing.

21.6.1 ISSUES IN TRADITIONAL PARKING SYSTEM

The rapid growth of vehicles for transportation is challenging to handle the parking system properly and the best utilization of the limited parking space. Thus, it is required to handle the issues or challenges in traditional parking systems through the advancement of technologies. There are various issues in the existing parking system such as inefficient use of parking space, lack of security to the vehicles, inefficient management of the operation of the parking system, high operating cost, manual payment system, low speed, manual processing, and less operational efficiency [37–44].

21.6.2 MOTIVATION OF PARKING SYSTEM

Due to the day-by-day faster population growth, the requirement for vehicles for comfortable transportation also increases rapidly. Thus, it is essential to manage the parking system to maintain proper transportation and reduce traffic congestion. Efficient parking system management can improve society's lifestyle and reduce unnecessary trouble for citizens in Tandon and Gupta [34]. Overall energy consumption compared to throughput should be reduced to improve efficiency, which can be handled by implementing IoT in the parking system. Various factors motivate the establishment of smarter parking systems, such as faster growth of the number of vehicles, limited parking space, parking and traffic jam, energy consumption, and environmental pollution.

21.6.3 PROPOSED FEATURES OF SMART PARKING SYSTEM

There are various strategies to make the city's parking system smarter. Various measures are considered for the smart parking system. Here various proposed features are included to improve the efficiency of the traditional parking system, such as the availability of slots in the nearest parking lot, parking space utilization, environment-friendly, no manual intervention, energy-efficient communication, low communication overhead, and local management, cloud data offload.

A city has many parking lots, each with a different capacity for parking purposes, that is, it has many parking slots with various capacities. In this proposal, each parking lot in the respective city is connected with different fogs, that is, fog node (Fogi) allocated to the parking lot (Parkloti) and all the fogs nodes (Fog1, Fog2, Fog3, ..., Fogn) of respective parking lots (Parklot1, Parklot2, Parklot3, ..., Parklotn) are connected with the cloud. Fog locally handles the operations in each parking lot and sends the essential data to the cloud. The use of fog can reduce the load, increase the speed of operation, reduce congestion, and also lower the security and privacy risk as it operates locally. Implementation of 5G communication technology improves communication efficiency compared to other existing communication technologies. Soon, 6G will come into the picture with a high data rate and low end-to-end delay, which will significantly impact building the smart city concept. 6G will support technological advancement and a massive no. of devices. For lightweight devices, there is a requirement to advance methodologies that can reduce overall energy consumption. Various metaheuristic algorithms can be used to reduce energy consumption and load balancing. Figure 21.6 represents the overview of the smart parking system; it shows the cloud and fog implementation in the parking system and the use of 5G communication technology in the parking system for efficient and faster communication. In [36], focus on allocating energy intelligently to the plug-in hybrid electric vehicle using PSO. Figure 21.7 shows the flow chart for finding the nearest parking lot for the requesting vehicle using PSO. Smart parking system reduces manual intervention, autonomously performs operations, provides services, and lowers communication overhead.

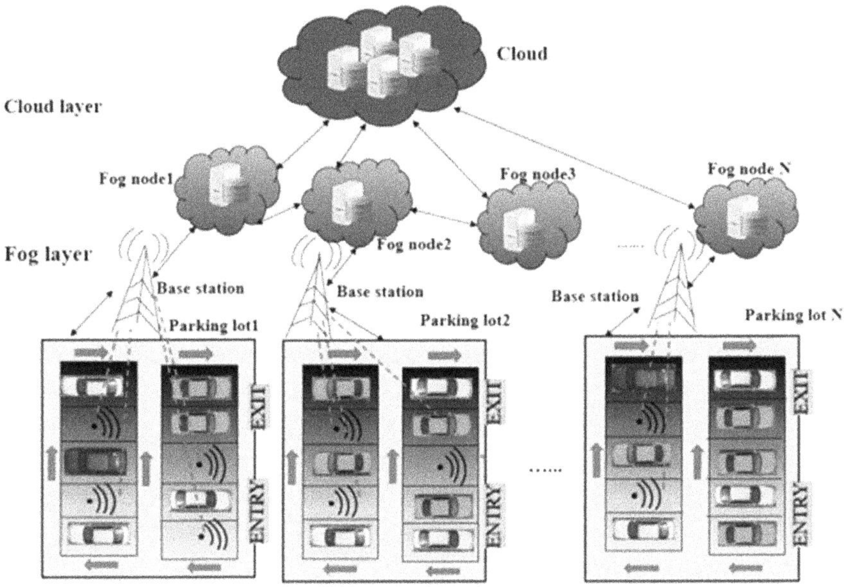

FIGURE 21.6 Configuration of smart parking system with fog and cloud.

In the proposed smart parking management system, when fog node Fogi receives a parking request from vehicle Vj, Fogi sends an acknowledgment message to the vehicle, and searches for an unoccupied parking slot in the respective parking lot (Parkloti). Here some cases may arise, such as,

Case 1: If a parking slot is available in the respective lot, it will immediately send a slot availability response message to the requesting vehicle and, after receiving the confirmation update the parking slot status.

Case 2: If a parking slot is not available, then respective fog node Fogi checks the availability of the slot in the near future (be_unoccuped_slot[j]) with time stamp Tk (i.e., maximum time vehicle should wait, which can vary based on the vehicle waiting_status). Furthermore, if a slot can be available within Tk, in that case, it will send a response message for waiting with time duration Tk and update the waiting status after receiving the confirmation from vehicle Vi for waiting.

Case 3: If the previous two cases are not satisfied, the fog forward the request to the cloud to find the nearest parking lot with vacant parking slot information. Here cloud uses PSO algorithm to find the nearest parking lot with the specified requirements of requesting vehicle Vi (type of vehicle, vehicle destination, parking duration, and reservation or on-spot parking info) and traffic conditions.

21.6.4 EXPERIMENTAL RESULT

The experimental result consists of various parameters such as energy efficiency or overall energy consumption (energy consumption due to transfer of more control packets, transfer of data to more distance, network traffic congestion, data loss, and retransmission), throughput, latency (low latency is required of efficient communication). When all the parking requests are directly communicated to the cloud for process, it will increase latency or response time.

In the proposed model, one fog is allocated for each parking lot. When a vehicle requests a vacant slot, the respective fog processes the request and provides the vacant slot information. However, when all slots are occupied and there is no vacant slot in the requested parking lot, the respective fog node forwards the request to the cloud to find the nearest parking lot. Here data or request packet and control packet from fog to cloud and response packet and control packet from cloud to fog communication will occur. It consumes more energy and increases the control packet and packet dropping rate.

In the proposed model, packet overhead and request to cloud occurred in exceptional cases when the requested parking lot does not have a vacant slot; otherwise, it handles locally. Furthermore, the cloud finds the nearest parking lot using PSO to update the status. But the existing model always sends the request to the cloud for parking slot allocation. In this article, the simulated result compares the proposed and existing models to determine the energy consumption and latency difference. Here, we have taken 50 vehicles initially, then increased it by 50 more each time, and compared the response time. Figure 21.8 shows the comparison of the response time of the existing and proposed model.

Algorithm 1: Parking slot allocation

1. Procedure:
2. Fog_i receive a parking request (park_Req)
3. Check unoccupied parking space
4. if $fog\_Available_i > 0$
   a. Allocate parking slot Park[i] & Update Park_ST
5. else if $fog\_Available_i == 0$
   a. Send a Waiting response message & Update Wait_ST
6. else
   a. if Wait_ST> T2(i.e. maximum no. of vehicle can wait)
      i. Send Parking requests to the cloud
      ii. Find the nearest parking (Fog_j) lot for the vehicle using PSO
      iii. if reservation needed
         1. Generate reservation_duration with location
      iv. else
      Allocate parking space with the location
      endif
   b. endif
7. endif

21.7 CONCLUSION

The smart city infrastructure is very complex based on responsibility, timeliness, and dedication, that is, proper planning, timely decisions, proper management of citizens' needs, and availability of required

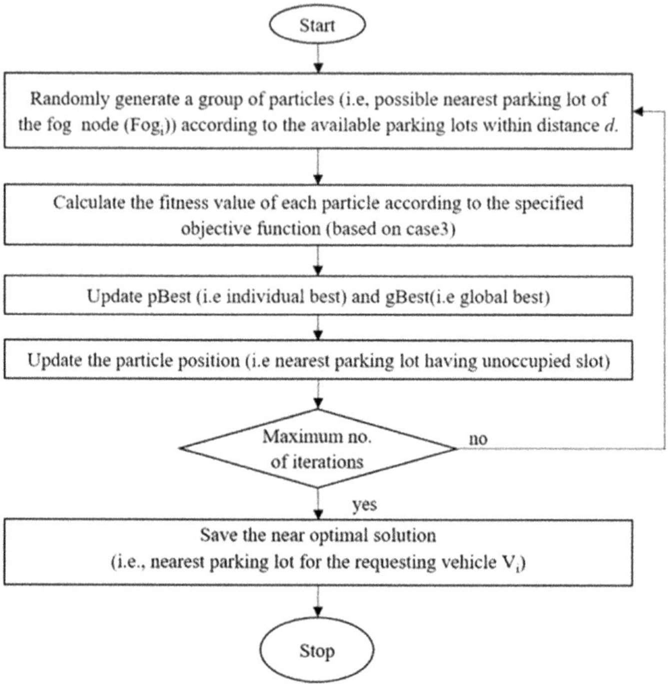

FIGURE 21.7 Flow chart for finding the nearest parking lot using PSO.

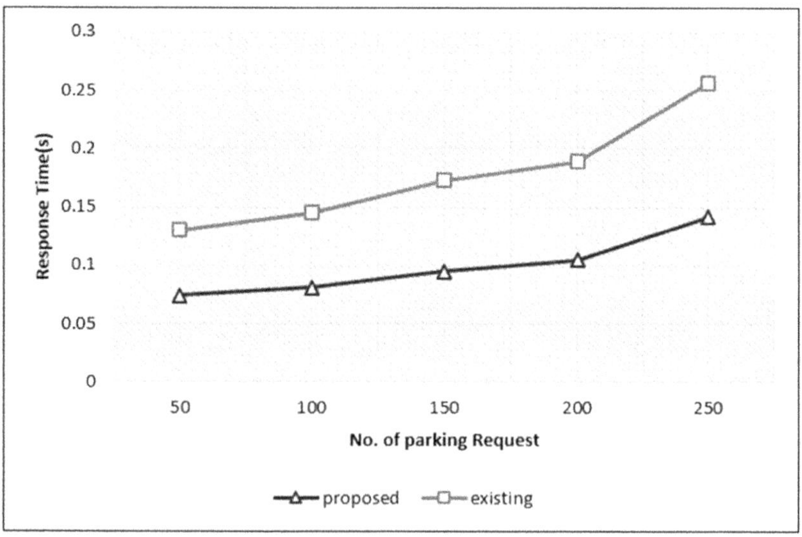

FIGURE 21.8 Response time comparison among existing and proposed model.

facilities with the convenience of the citizens. There are various categories of citizens staying in cities, and they have different levels of requirements in various areas such as healthcare needs, education needs, transport, business, and job finding. Thus, the advancement of technologies and efficient management of IoT-based smart city infrastructure, as also prediction and decision-making through various improved methods, can handle these critical situations. Here, the main focus is on using IoT technologies for autonomous configuration, and management of various components of smart city infrastructure for traffic management, transport management, advancement of the education system, treatment, diagnosis, and patient monitoring, in healthcare, surveillance, system management, and database management again applies ML methods to do analysis and prediction based on the collected data for the IoT-based smart city environment. IoT automates the smart city environment and provides convenience to society and again for prediction, integration of ML, and also metaheuristic algorithms to improve the efficiency of the smart cities. IoT and ML together give better solutions to the difficulties of the traditional system. In the future, advancement in IoT technologies and ML methodologies shall be a key driver for the smart city system, which could generate more revenue, sustainability, and convenience for city dwellers.

KEYWORDS

- **5G**
- **6G**
- **IoT**
- **machine learning**
- **metaheuristic algorithm**

REFERENCES

1. Mohanty, S. P., Choppali, U., & Kougianos, E. (2016). Everything you wanted to know about smart cities: The internet of things is the backbone. *IEEE Consumer Electronics Magazine, 5*(3).

2. Ray, P. P. (2018). A survey on internet of things architectures. *Journal of King Saud University-Computer and Information Sciences, 30*(3), 291–319.
3. Ploennigs, J., Cohn, J., & Stanford-Clark, A. (2018).The future of IoT. *IEEE Internet of Things Magazine*, *1*(1), 28–33.
4. Perera, C., Barhamgi, M., De, S., Baarslag, T., Vecchio, M., & Choo, K-K. R. (2018). Designing the sensing as a service ecosystem for the internet of things. *IEEE Internet of Things Magazine, 1*(2), 18–23.
5. Sodhro, A. H., Pirbhulal, S., Luo, Z., Muhammad, K., & Zahid, N. Z. (2020). Toward 6G architecture for energy-efficient communication in IoT-enabled smart automation systems. *IEEE Internet of Things Journal, PP*(99), 1–1.
6. Guo, F., Yu, F. R., Zhang, H., Li, X., Ji, H., & Leung, V. C. M. (2021). Enabling massive IoT toward 6G: A comprehensive survey. *IEEE Internet of Things Journal, 8*(15), 11891–11915.
7. Cano, J. C., Berrios, V., Garcia, B., & Toh, C. K. (2018). Evolution of IoT: An industry perspective. *IEEE Internet of Things Magazine, 1*(2), 12–17.
8. Kirimtat, A., Krejcar, O., Kertesz, A., &Tasgetiren, M. F. (2020). Future trends and current state of smart city concepts: A survey. *IEEE Access, PP*(99), 1–1.
9. Ahmed, E., Yaqoob, I., Gani, A., Imran, M., & Guizani, M. (2016). Internet- of-things-based smart environments: State of the art, taxonomy, and open research challenges. *IEEE Wireless Communications, 23*(5).
10. Violette, M. (2018). IoT standards. *IEEE Internet of Things Magazine, 1*(1), 6–7.
11. Kranz, M. (2018). Why industry needs to accelerate IoT standards. *IEEE Internet of Things Magazine, 1*(1), 14–18.
12. Mosenia, A., &Jha, N. K. (2016). A comprehensive study of security of internet-of-things. *IEEE Transactions on Emerging Topics in Computing, 5*(4), 586–602.
13. Arzo, S. T., Naiga, C., Granelli, F., Bassoli, R., Devetsikiotis M., & Fitzek, F. H. P. (2021). A theoretical discussion and survey of network automation for IoT: Challenges and opportunity. *IEEE Internet of Things Journal, 8*(15), 12021–12045.
14. Qi, Q., Chen, X., Zhong, C., & Zhang, Z. (2020). Integration of energy, computation and communication in 6G cellular internet of things. *IEEE Communications Letters, 24*(6), 1333–1337.
15. Agiwal, M., Roy, A., & Saxena, N. (2016). Next generation 5G wireless networks: A comprehensive survey. *IEEE Communications Surveys & Tutorials, 18*(3), 1617–1655.
16. Zanella, A., Bui, N., Castellani, A., Vangelista, L., & Zorzi, M. (2014). Internet of things for smart cities. *IEEE Internet of Things Journal, 1*(1), 22–32.
17. Verma, S., Kaur, S., Khan, M. A., & Sehdev, P. S. (2021). Toward green communication in 6G-enabled massive internet of things. *IEEE Internet of Things Journal, 8*(7), 5408–5415.
18. Saxena, N., Roy, A., Sahu, B. J. R., & Kim, H. (2017). Efficient IoT gateway over 5G wireless: A new design with prototype and implementation results. *IEEE Communications Magazine, 55*(2), 97–105.
19. Choudhary, P., Bhargava, L., Suhag, A. K., Choudhary, M., & Singh, S. (2021). An era of internet of things leads to smart cities initiatives towards urbanization. In *Digital Cities Roadmap: IoT-Based Architecture and Sustainable Buildings.*
20. Ianuale, N., Schiavon, D., & Capobianco, E. (2015). Smart cities, big data, and communities: Reasoning from the viewpoint of attractors. *IEEE Access, 4,* 41–47.

21. Harrison, C., Eckman, B., Hamilton, R., Hartswick, P., Kalagnanam, J., Paraszczak, J., & Williams, P. (2010). Foundations for smarter cities. *IBM Journal of Research and Development, 54*(4), 1–16.
22. Tandon, R., & Gupta, P. K. (2019). Optimizing smart parking system by using fog computing. In *International Conference on Advances in Computing and Data Sciences* (pp. 724–737). Springer, Singapore.
23. Heck, G. C., Hexsel, R., Gomes, V. B., Iantorno, L., Junior, L. L., & Santana, T. (2021). GRID-CITY: A framework to share smart grids communication with Smart City Applications. In *2021 IEEE International Smart Cities Conference (ISC2)* (pp. 1–4). IEEE.
24. Rostirolla, G., da Rosa Righi, R., Barbosa, J. L. V., & da Costa, C. A. (2017). ElCity: An elastic multilevel energy saving model for smart cities. *IEEE Transactions on Sustainable Computing, 3*(1), 30–43.
25. Aldahiri, A., Alrashed, B., & Hussain, W. (2021). Trends in using IoT with machine learning in health prediction system. *Forecasting, 3*(1), 181–206.
26. Hameed, K., Bajwa, I. S., Ramzan, S., Anwar, W., Khan, A. (2020). An intelligent IoT based healthcare system using fuzzy neural networks. *Scientific Programming.*
27. Sharma, Y. K., & Sudam, K. S. (2019). Health care patient monitoring using IoT and machine learning. *IOSR Journal of Engineering (IOSR JEN)-,* ISSN (e).
28. Jeong, J.-S., Han, O., & You, Y. Y. (2016). A design characteristics of smart healthcare system as the IoT application. *Indian Journal of Science and Technology, 9*(37).
29. Mohamed, R. M., Shahin, O. R., Hamed, N. O., Zahran, H. Y., & Abdellattif, M. H. (2022). Analyzing the patient behavior for improving the medical treatment using smart healthcare and IoT-based deep belief network. *Journal of Healthcare Engineering, 2022*, 6389069.
30. Maeda, A. (2012). Technology innovations for smart cities. In *2012 Symposium on VLSI Circuits (VLSIC)* (pp. 6–9). IEEE.
31. Khan, Z. A., Butt, A. A., Alghamdi, T. A., Fatima, A., Akbar, M., Ramzan, M., & Javaid, N. (2019). Energy management in smart sectors using fog based environment and meta-heuristic algorithms. *IEEE Access, 7*, 157254–157267.
32. Butt, A. A., Khan, S., Ashfaq, T., Javaid, S., Sattar, N. A., & Javaid, N. (2019). A cloud and fog based architecture for energy management of smart city by using meta-heuristic techniques. In *2019 15th International Wireless Communications & Mobile Computing Conference (IWCMC)* (pp. 1588–1593). IEEE.
33. Roy, S. S., Puthal, D., Sharma, S., Mohanty, S. P., & Zomaya, A. Y. (2018). Building a sustainable internet of things: Energy-efficient routing using low-power sensors will meet the need. *IEEE Consumer Electronics Magazine, 7*(2), 42–49.
34. Tandon, R., & Gupta, P. K. (2019). Optimizing smart parking system by using fog computing. In *International Conference on Advances in Computing and Data Sciences*. Springer, Singapore.
35. Tang, C., Wei, X., Zhu, C., Chen, W., & Rodrigues, J. J. P. C. (2018). Towards smart parking based on fog computing. *IEEE Access, 6*, 70172–70185.
36. Su, W., & Chow, M. -Y. (2011). Performance evaluation of a PHEV parking station using Particle Swarm Optimization. In *2011 IEEE Power and Energy Society General Meeting*, (pp. 1–6).

37. Mallick, P. K., Mishra, D., Patnaik, S., & Shaw, K. (2016). A semi-supervised rough set and random forest approach for pattern classification of gene expression data. *International Journal of Reasoning-based Intelligent Systems, 8*(3-4), 155–167.
38. Mallick, P. K., Satapathy, B. S., Mohanty, M. N., & Kumar, S. S. (2015, February). Intelligent technique for CT brain image segmentation. In *2015 2nd International Conference on Electronics and Communication Systems (ICECS)* (pp. 1269–1277). IEEE.
39. Rautaray, B., & Swain, D. K. (2021). Intellectual structure and publication pattern in journal of money laundering control from 2014 to 2020: A bibliometric study.
40. Prasad, S. S., Mishra, J. P., Mishra, S. K., & Borah, S. (2020). Challenges toward blockchain and renewable energy linked to IoT: A survey. In *Advances in Energy Technology* (pp. 49–55). Springer Nature. DOI 10.1007/978-981-15-8700-9.
41. Nath, M. P., Priyadarshini, S. B. B., Mishra, D., & Borah, S. (2020). A comprehensive study of contemporary IoT technologies and varied machine learning (ML) schemes, soft computing techniques and applications. *Advances in Intelligent Systems and Computing Soft Computing Techniques and Applications*, 623–634. DOI 10.1007/978-981-15-7394-1_1.
42. Mishra, J. P., Borah, S., & Mishra, S. K. (2021). Performance of scalable data linked to internet of things: A case study. In S. Borah, R. Pradhan, N. Dey & P. Gupta (Eds.), *Soft Computing Techniques and Applications. Advances in Intelligent Systems and Computing* (Vol. 1248, pp. 601–610). Springer, Singapore. https://doi.org/10.1007/978-981-15-7394-1_54
43. Polkowski, Z., Mishra, S. K., Mishra, B. K., Borah, S., & Mohanty, A. (2023). Impact of internet of things in social and agricultural domains in rural sector: A case study. *International Journal of Cloud Computing, 12*(1), 90–105. ISSN: 2043-9997. DOI: https://doi.org/10.1504/IJCC.2023.129776
44. Verma, U., Sohani, M., Borah, S., Nagwanshi, K. K., & Pathak, S. (2022). Role of fog-assisted internet of things-enabled system for managing the impact of COVID-19. In S. Borah, S. K. Mishra, B. K., Mishra,V. E. Balas, & Z. Polkowski (Eds.), *Advances in Data Science and Management. Lecture Notes on Data Engineering and Communications Technologies* (Vol. 86). Springer, Singapore. https://doi.org/10.1007/978-981-16-5685-9_39

CHAPTER 22

IOT-BASED BREATH ANALYSIS FOR THE PREDICTION OF RHEUMATISM USING BI-LSTM

NILAKSHI MARUTI MULE[1] and DIPTI DURGESH PATIL[2]

[1]*Department of Computer Engineering, Smt. Kashibai Navale College of Engineering, Savitribai Phule Pune University, Pune, Maharashtra, India*

[2]*Department of Information Technology, MKSSS's Cummins College of Engineering for Women, Karve Nagar, Savitribai Phule Pune University, Pune, Maharashtra, India*

ABSTRACT

This chapter presents the design and development of an IoT-based breath analyzer to detect and monitor respiratory, malignant, neurodegenerative, and metabolic diseases using exhaled breath. The primary goal of this research is to find out if exhaled breath analysis can distinguish between individuals with inflammatory joint disorders and healthy controls. This system has assessed a group of rheumatoid arthritis (RA) patients and healthy individuals. It is designed for both wired and wireless modules. Its primary function is to obtain the exhaled breath data coupled with a sensory system composed of a gas sensor. Here, the breath sample is conditioned for the initial stage data collected through a wired module using CoolTerm software. After data collection, a deep learning algorithm is employed to carry out the characteristic analysis of the disease. That results in a clear

Cyber-Physical Systems: Applications, Challenges, and Research Directions. Rajanikanth Aluvalu, Samarjeet Borah, & Lalit Garg (Eds.)

DOI: 10.1201/9781779643421-22

distinction between the samples of rheumatism and healthy samples by obtaining 84% accuracy. With the aid of this system, a nontransmissible or transmissible disease could be identified early, preventing infection in other people.

22.1 INTRODUCTION

The human breath samples contain thousands of volatile organic compounds (VOCs). Many compounds are endogenous, and others are exogenous [1, 2]. The data collected through the breath analyzer contains various sources of variance. These sources include the interest level and the noise level. Thus, the data generated by the breath is large and complex. The research aims to extract useful information from the data by analyzing exhaled breath [3].

Various analytical methods are used to perform this kind of analysis. Mass spectrometry and gas chromatography (GC-MS) are commonly used to analyze breath air. This sensitive and reliable technique can be used in offline breath analysis. It necessitates capillary separation and sample preconcentration, which takes time [4]. Online real-time breath analysis is based on mass spectrometry (MS) such as SESI-MS, SIFT-MS, and PTR-MS. SIFT-MS is an analytical technique for quantifying real-time trace gases in air or breath samples. This method can detect disease states early and quickly [5, 27].

PTR-MS is an ionization molecule used in online analysis. It is equipped with a quadrupole mass analyzer with high sensitivity [6]. SESI-MS is an ambient ionization method that is sensitive, robust, and reliable. However, calibration is needed for this method, and absolute quantification is impossible [7]. E-nose is used for a short analysis time. The mixtures do not have to be separated, and the portable version is available. It has high sensitivity and reproducibility, but it is possible that poisoning and moisture-sensitive sensors suffer from ageing [8, 26].

The primary goal of this research is to find out if exhaled breath analysis can distinguish between individuals with inflammatory joint disorders and healthy controls. Rheumatoid arthritis (RA) is the most prevalent chronic systemic inflammatory disease and chronic inflammatory arthritis [9]. Rheumatoid is an autoimmune and inflammatory disease that attacks healthy cells, causing severe inflammation (swelling) in the affected body areas [10]. Approximately 1% of the global population is affected by

RA, with women two to three times more likely than males [11, 12]. This system has assessed a group of RA patients and healthy individuals.

In 1987, the American Rheumatism Association proposed diagnostic standards for RA. These requirements include morning stiffness lasting at least an hour, simultaneous RA in three or more joint locations, rheumatoid hand disease, symmetric joint involvement, subcutaneous rheumatoid nodules, presence of rheumatoid factor, and radiographic erosion or bone decalcification of the affected joints [13].

Many patients are diagnosed without fulfilling these requirements. The evolution of risk factors and molecular markers is anticipated to affect the diagnostic standards for RA [14]. In addition to RA's effects on the joints, lung involvement frequently develops and contributes significantly to the disease's morbidity and death. Lung illness directly related to RA (RA-LD) is the most common consequence, even though pulmonary infection, drug toxicity, or both are frequently encountered [15]. Interstitial lung disease (ILD), airways disease, Rheumatoid nodules, and pleural fluid are the most common types of lung disease [16].

RA-LD includes ILDs like typical interstitial pneumonia, nonspecific interstitial pneumonia, organizing pneumonia, widespread alveolar damage, pleuritis, pulmonary vascular disease, and rheumatoid nodules [17].

Tsuchiya et al. [18] evaluated RA patients with RA-LD, noting changes in clinical characteristics among different disease types, assessing prognostic implications, and examining reasons for mortality.

The prognostic significance of the interstitial pneumonia pattern observed on high-resolution computed tomography (HRCT) in RA-associated ILD is examined by Kim et al. [19]. HRCT is more accessible, reliable in the hands of skilled radiologists, less expensive, and less risky than surgical lung biopsy. ILD, according to Cavagna et al.'s [20] review, is a significant extra-articular manifestation of RA that can develop either early on or as a side effect of a chronic condition. The prognosis of these patients' quoad vitam drastically changes due to RA-ILD [20]. Robles-Perez et al. [21] look at the prevalence of asymptomatic lung abnormalities in early RA patients and any possible links between lung involvement and positive RA blood-reactive biomolecules.

A cohort of patients with early RA who had no respiratory symptoms and had undergone complete pulmonary function tests and a baseline chest radiograph (CR) as part of a lung disease screening programme were the subjects of a prospective observational study. HRCT was done on patients with lung abnormalities on the CR or PFTs. Forty patients were a part of them. Eighteen of these patients had altered PFTs found. In these circumstances, the forced vital capacity did not significantly decrease, and the diffusion lung transfer capacity of carbon monoxide (DLCO) was about 80% of what was predicted [21]. To study the association between systemic disease activity measures, medications for RA, and smoking with lung abnormalities, Robles-Perez et al. [21] explain the characteristics of lung involvement within the first 5 years following RA diagnosis. 32 of the 40 enrolled patients completed the 5-year follow-up. Three of the 13 individuals with lung involvement in the first 5 years following an RA diagnosis had ILD.

To evaluate the correlations between clinical features, lung involvement, and frequency of pulmonary involvement in RA [23], they prospectively assessed 52 RA patients (8 men, 44 women, mean age 53.6 years) using HRCT and PFT. The HRCT was abnormal in 35 individuals (67.3%), with reticulonodular patterns (found in 22 patients; 62.9%), ground-glass attenuation (20%), and bronchiectasis (found in 17%) being the most common abnormalities. Six patients, two of whom had ILD, observed that their diffusion lung transfer capacity of carbon monoxide significantly declined with time [22].

Thirteen (17%) of this group's patients had normal PFT findings. Results indicated that clinical symptoms and PFT have poor diagnostic specificity for detecting pulmonary involvement in RA at an early stage. They discovered correlations between RA severity and lung involvement, a significant component of RA's systemic disease. They also imply that the severity of RA is an essential predictor of HRCT abnormalities indicative of lung involvement [23]. Marigliano et al. [24] aim to describe how lung injury affects CTDs. They primarily concentrate on RA, which, in contrast to other rheumatic diseases, shares several characteristics with idiopathic pulmonary fibrosis (IPF) [24].

The TGS-822, TGS-2600, TGS-2602, MQ-7, and MQ-9 gas sensors are used in this research work. These sensors are compact, have a comprehensive dynamic range, are inexpensive, simple to implement automatic measurement, and are used online and in continuous detection [25].

22.1.1 RESEARCH OBJECTIVES

1. This study aims to identify the gas that RA patients exhale. The goal is to create a technique that can efficiently gather breath samples.

2. The study attempts to address the issue of how to store the gathered breath samples. The goal is to suggest an IoT-based breath analyzer that can first collect breath when the breath buffer is full before passing that breath to the sensor module.
3. The planned research will examine the variations in exhaled breath gases between people with rheumatism and healthy individuals.
4. One of the research's goals is to assess the breath analyzer device's performance using an IoT-based long-short memory algorithm.

22.1.2 MAJOR CONTRIBUTIONS

1. To create and develop an IoT-based breath analyzer.
2. To create a real-time data set or to gather breath samples from people with rheumatism and healthy people to detect differences in exhaled gases.
3. Using deep learning techniques makes it possible to differentiate between exhaled gases.

22.1.3 THE STRUCTURAL FLOW OF THE CHAPTER

The chapter is organized as follows. Section 22.1 provides introduction to the research area, related work, and major contributions of this chapter. Section 22.2 gives a description of the data acquisition process and methodology. Section 22.3 describes detailed results and analysis. Section 22.4 includes discussion on results. Section 22.5 concludes the study.

The following section explains the proposed methods.

22.2 MATERIALS AND METHODS

22.2.1 GAS SENSORS

Using sensor technologies to diagnose and detect diseases through the breath involves studying and analyzing VOCs. The presence of VOCs in the breath indicates the existence of diseases. The sensors in the device are sensitive to those mixtures of volatile compounds. Five commercial gas sensors are used to create portable measurement equipment. Table 22.1 summarizes each sensor's properties taken from its datasheet.

TABLE 22.1 Specification of the Gas Sensor.

Sensors	Application	Gas Identity	Material
TGS-822	Breath alcohol detector Solvent detector, dry cleaner, and semiconductor Gas leakage detector	Organic solvent Volatile vapors Carbon monoxide	Tin dioxide semiconductor (SnO2)
TGS-2600	Air cleaner Ventilation control Air quality monitor	Air contaminants (Hydrogen, Carbon Monoxide)	Metal oxide semiconductor
TGS-2602	Air cleaner Ventilation control Air quality monitor VoCs and odor monitor	Air contaminants (VoCs, ammonia, hydrogen sulfide)	Metal oxide semiconductor
MQ-7	Gas-detecting equipment for carbon monoxide (CO.)	Carbon monoxide	SnO2
MQ-9	Gas-detecting equipment for carbon monoxide, methane, LPG	Carbon monoxide, methane	SnO2

22.2.2 SYSTEM STRUCTURE

The design of the sampling device of the breath analyzer system consists of a blower, a hand compressor to push air from the sampling chamber, an analyzer, and an IoT interface. The outline of the sampling device is shown in Figure 22.1.

The blower is initially utilized to collect gas inlets from the patient. The gas then travels through the ensuing chambers after being blown. To prevent air reverse, the non-return valve, by default, closes. The gas sample is held in the air sac in the sampling chamber until it passes through the gas analyzer. The air valve is closed until gas from the sampling chamber does not travel through the gas analyzer. A containt analyzer is a type of gas analyzer that combines several gas sensors into one chamber or has various chambers with various gas sensors inside of them. It depends on the sensors' ability to perform. The gas analyzer has numerous gas sensors to examine the contents of the sample taken from the patient.

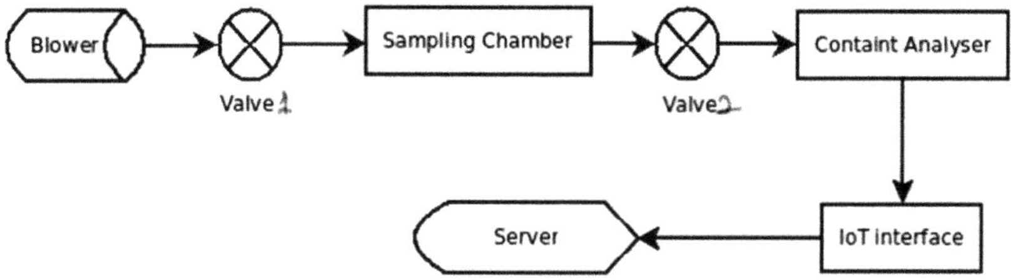

FIGURE 22.1 Outline of IoT-based breath analyzer prototype.

22.2.3 *WIRED DATA COLLECTION*

IoT-based breath analyzer collects data in two ways: one is wired using a USB cable, and another uses a wireless module. The data was primarily obtained by a wired method in the present research. Wired or wireless modules carry out the data acquisition. An application called CoolTerm for the serial port is used to collect the data. It exchanges text and other data between connected serial ports. It helps to record received data in text file format, and displays received data in textual or hexadecimal format. Figure 22.2 shows the samples of collected data.

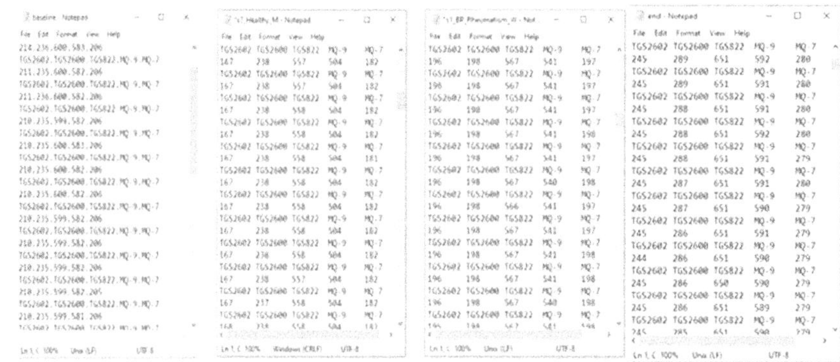

FIGURE 22.2 Collected breath samples.

22.3 DATA ANALYSIS

The tests were performed on individuals diagnosed with rheumatism and healthy persons with prior permissions. Table 22.2 shows the details of the number of samples.

TABLE 22.2 Number of Samples.

Class	Number	Age	Male	Female
Rheumatism	13	25–65	02	11
Healthy	8	26–40	03	05

Every device component is removable rather than having a sensor module; the sampling phase is completed concurrently with device assembly. After assembling all parts of the device, the reading of baseline air samples of the sensors is collected to identify the changes in breath concentration before the collection of breath samples, during and after collecting the samples.

After ensuring the device's proper functioning, the patient starts the exhalation by inserting the mouthpiece into the mouth to begin the process. Once this is completed, the sac is filled with the patient's breathing air. Then, the collected air is passed into the sensor module until the breath buffer is empty. Then, the device starts acquiring the data. After completing data collection using a hand compressor, the breathing air from the sensor module is removed using an exhaust valve. An indicator indicates whether the air is removed from the sensor module so that we can repeat the process. A graphical representation in Figure 22.3 shows the response of the gas sensors to a sample of exhaled breath of a patient with rheumatism, and Figure 22.4 shows a sample of a healthy person.

Evidence of the difference between the breath samples is found in the graph analysis. Primarily, the response of the sensors MQ-3, MQ-7, TGS-822, TGS-2600, and TGS-2605 shows that the sensors' reply to a rheumatism sample has a higher concentration than that of a healthy person. This suggests a chemical imbalance in the body when a condition like rheumatism exists.

22.4 RESULT AND DISCUSSION

High computation has been used in healthcare to improve patient outcomes and cut costs in two ways: artificial intelligence and machine learning. These technologies have been used to develop prediction models, automate repetitive tasks, and help decision-making [28]. When processing sequential data, time series analysis, weather forecasting, image partitioning, and text generation tasks, long-short-term memory (LSTM) network variations

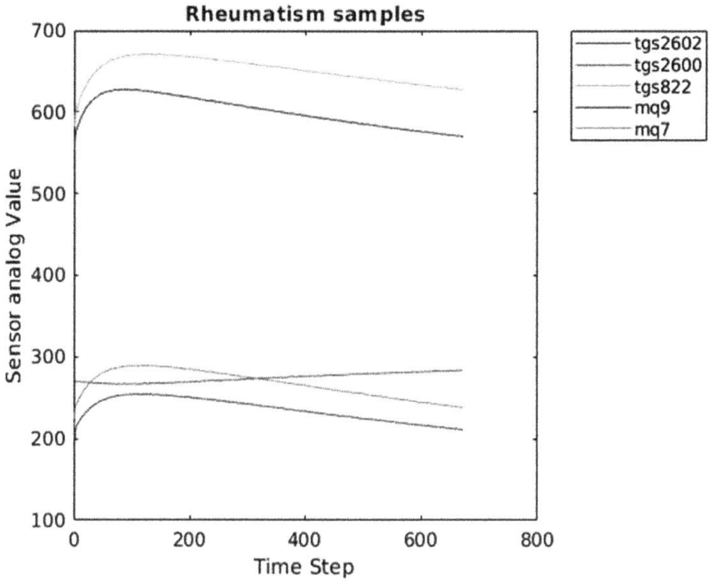

FIGURE 22.3 Samples of rheumatism patient.

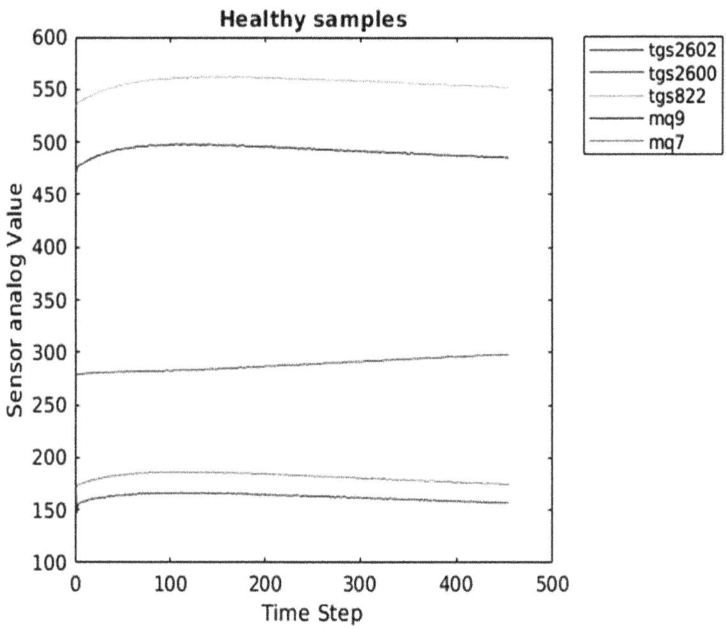

FIGURE 22.4 Samples of healthy persons.

of RNN deliver outstanding results. The lengthier sequences network recursively processes the LSTM sequences using a gating mechanism, which also helps it remember previous data. An effective model for managing textual information has been demonstrated by the bidirectional-long-short-term memory (Bi-LSTM) network, which has expanded variants of the LSTM network [29]. Bi-LSTM is used in artificial intelligence and deep learning for processing the collected data. Bi-LSTMs are predominately used to learn processes and classify sequential data. These networks identify long-term dependencies between time steps of data. The data collected is categorized into baseline (reading before collecting breath sample) and end (after collecting breath sample) of a healthy and rheumatism-diseased person. Then, the data is separated for each sensor. The columns are the responses of the sensor's analogue values for each breath sample captured, and the rows are the time steps shown in Figures 22.3 and 22.4.

The data set is divided into two parts for training and testing purposes. These two parts are the training part of the first sequence X (Xtrain) and the training part of the second sequence Y (Ytrain). 75% is given to the training part, and 25% to the testing part in the Bi-LSTM model. First, the sequence lengths for each observation are declared. The number of observations is taken in the Xtrain variable. The sequence is extracted, and a sequence length is declared according to the number of observations. The sequence length changes the size on every loop iteration.

The Bi-LSTM network architecture is explained here. First, set the input size to 5 and output the final element of the sequence while specifying the Bi-LSTM layer to contain 60 hidden units. Include a fully connected layer with a size that depends on the type of disease, followed by a softmax layer and a classification layer to identify classes. List the available training options. Set "adam" for the solver and "Gradient Threshold" at 1 and set the maximum epochs to 500 and the mini-batch size to 25. The CPU is better suited for training due to the mini-batches and short sequences. Turn the "Execution Environment" to the "CPU" set. Use the designated training choices to train the Bi-LSTM network.

$$\text{Net} = \text{train Network (Xtrain, Ytrain, layers, options)} \quad (22.1)$$

Load the information of the collected breath sample to test data. Load the test set and classify the sequences into Xtest and Ytest. The Xtest system is a cell array that stores sequences of dimension 25 of various lengths. Ytest is a category vector containing labels related to the five sensors. These labels are TGS-822, TGS-2600, TGS-2602, MQ-7, and MQ-9. The Bi-LSTM network is educated with the use of mini-batches of sequences that are of the same length. Classify the test data and specify the same mini-batch size used for training.

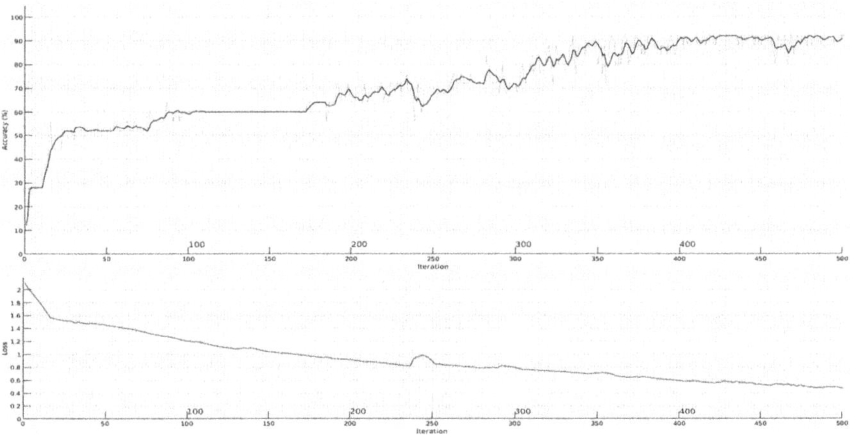

FIGURE 22.5 Training and testing of the samples.

```
layers =

  5x1 Layer array with layers:

     1   ''   Sequence Input            Sequence input with 5 dimensions
     2   ''   BiLSTM                    BiLSTM with 60 hidden units
     3   ''   Fully Connected           8 fully connected layer
     4   ''   Softmax                   softmax
     5   ''   Classification Output     crossentropyex
```

Epoch	Iteration	Time Elapsed (hh:mm:ss)	Mini-batch Accuracy	Mini-batch Loss	Base Learning Rate
1	1	00:00:02	4.00%	2.3177	0.0010
50	50	00:00:27	56.00%	1.3909	0.0010
100	100	00:00:50	60.00%	1.3146	0.0010
150	150	00:01:15	60.00%	1.2352	0.0010
200	200	00:01:39	60.00%	1.0910	0.0010
250	250	00:02:03	64.00%	0.9389	0.0010
300	300	00:02:28	68.00%	0.9063	0.0010
350	350	00:02:52	76.00%	0.8480	0.0010
400	400	00:03:18	80.00%	0.7560	0.0010
450	450	00:03:43	84.00%	0.6911	0.0010
500	500	00:04:07	84.00%	0.6512	0.0010

FIGURE 22.6 5 × 1 layers array.

The breath samples from patients with rheumatism (previously confirmed their diagnosis through clinical history) are obtained to compare with healthy people to measure how well the system works. The procedure described above is followed to acquire these breath samples. Figures 22.3 and 22.4 show the discrimination resulting from rheumatism and healthy samples. The figure shows that the exhaled gas from rheumatism patients has Ammonia, Ethanol, Acetone, Carbon Monoxide, Methane, Hydrogen, and Isobutene.

TGS-822, TGS-2600, TGS2602, MQ-7, and MQ-9 sensors generate a high degree of relevant information to rheumatism exhaled gas. As seen in both Figures 22.7 and 22.8, the system obtained a clear discrimination of the baseline and end reading of rheumatism patients and healthy people. For baseline results, all sensors responded due to the absence of the gas in the initial reading; it shows a straight line in the graph.

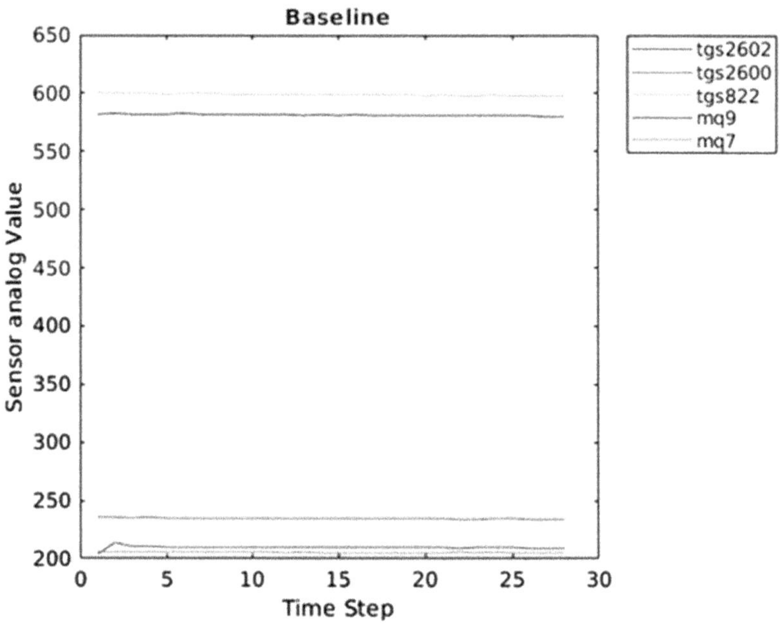

FIGURE 22.7 Before collecting breath sample (baseline) result.

22.4.1 *LIMITATION*

1. The small sample size can make it harder to generalize the results and conduct further research on larger datasets. The currently used

datasets are not very large. It would be helpful to evaluate the effectiveness of the suggested strategy using more diverse datasets.

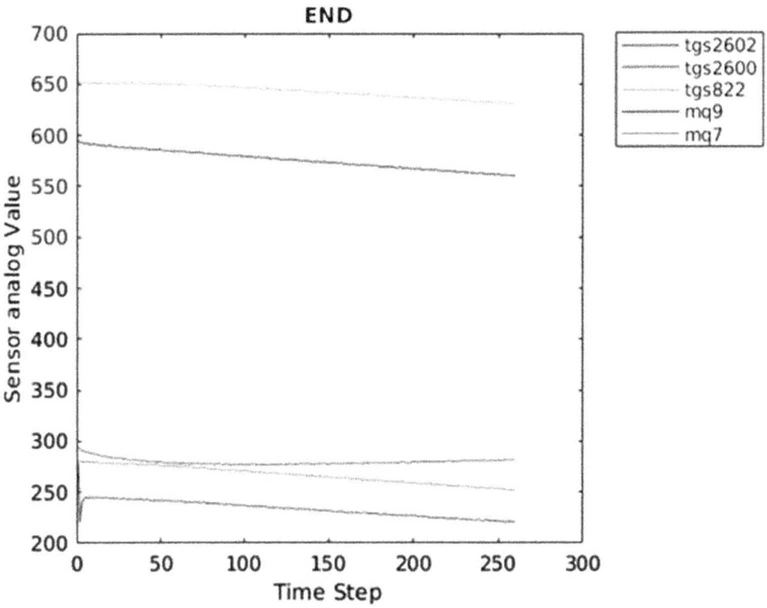

FIGURE 22.8 After collecting breath sample (end) result.

2. More training data are needed for the Bi-LSTM to learn appropriately. In future studies, to overcome the drawbacks of Bi-LSTM, more diverse approaches can be tested for exhaled breath prediction.
3. It is essential to give careful thought to and further research the diagnostic accuracy.
4. Evaluating the method's generalizability and effectiveness in more actual clinical contexts is crucial.

22.4.2 *PRACTICAL IMPLICATIONS*

22.4.2.1 EARLY DIAGNOSIS OF RHEUMATOID ARTHRITIS

The suggested approach, "IoT-based breath analysis for prediction of rheumatism using Bi-LSTM," may help with the early identification and diagnosis of rheumatism diseases. Healthcare personnel can start prompt interventions and treatment plans, improving patient outcomes and possibly saving lives by recognizing problems early.

22.4.2.2 *ENHANCED DIAGNOSTIC PRECISION*

The study results imply that the noninvasive strategy can produce higher classification accuracy levels.

22.4.2.3 *USEFUL APPLICATIONS FOR REHABILITATION*

The suggested method might make the breath collection process more straightforward, and the breath buffer element of the sample collection bag is reusable and detachable.

22.4.2.4 *POSSIBILITY OF SOFT COMPUTING TECHNIQUES*

The study emphasizes the usage of Bi-LSTM, an artificial neural network utilized in deep learning and artificial intelligence. It indicates the more significant potential of these clever approaches in rheumatology.

22.5 CONCLUSION

The IoT-based breath analyzer system has recorded the most typical VOCs and detected the distinctive pattern of each group of exhaled breath samples observed in the rheumatism patient testing and control subjects. The results prove the discriminative capacities of the patient. The IoT-based breath analyzer makes diagnosing diseases in the early stage possible. It supports the purpose of this research, which claims that diseases alter the body's chemistry. It can be observed by comparing the sensors' reactions to a sample of a rheumatoid person's breath to a sample of a healthy person's breath. In the future, the wireless system can be improved further, and more data can be collected to identify the most representative biomarkers of transmittable or nontransmittable diseases. It will enable the proposed system to be trained more effectively, leading to a more precise diagnosis, which will be helpful as a support tool in medical diagnosis at healthcare centers.

KEYWORDS

- **Bi-LSTM**
- **breath analysis**
- **healthcare**
- **internet of things**
- **rheumatism**

REFERENCES

1. David, Z., Dongmin, G., & Ke, Y. (2017). *Breath analysis for medical applications*. Springer Nature Singapore Pte Ltd.
2. Todor, A. (2011). Human exhaled breath analysis: CME review. *Annals of Allergy, Asthma & Immunology, 106*, 451–456.
3. Smolinska, A., Hauschild, A. C., Dallinga, J. W. (2014). Current breathomics – A review on data preprocessing techniques and machine learning in metabolomics breath analysis. *Journal of Breath Research, 8*(2), 027105.
4. Kim, K.-H., Ara, J. S., Ehsanul, K. (2012). A review of breath analysis for diagnosis of human health. *Trends Analysis Chemistry, 33*, 1–8.
5. Moorhead, K. T., & et al. (2008). Classifying algorithms for SIFT-MS technology and medical diagnosis. *Computer Methods Programs in Biomedicine, 89*, 226–238.
6. Sun, Y., Chen, Y., Sun, C., Liu, H., Wanga, Y., & Jiang, X. (2019). Analysis of volatile organic compounds from patients and cell lines for the validation of lung cancer biomarkers by proton-transfer-reaction mass spectrometry, *Analytical Methods, 11*, 3188–3197.
7. Ho, C. S., Lam, C. W. K., Chan, M. H. M., Cheung, R. C. K., Law, L. K., Lit, L. C. W., Ng, K. F., Suen, M. W. M., & Tai, H. L. (2003). Electrospray ionisation mass spectrometry: Principles and clinical applications. *Clinival Biochemist Review, 24*(1), 3–12.
8. Wardencki, W., Chmiel, T., & Dymerski, T. (2013). Gas chromatography-olfactometry (GC-O), electronic noses (e-noses) and electronic tongues (e-tongues) for in vivo food flavour measurement. In *Instrumental assessment of food sensory quality. A pract guide woodhead publish series in food science, technology and nutrition* (pp. 195–229).
9. Groner, L. K., Green, D. B., Weisman, S. V., Legasto, A. C., Toy, D., Gruden, J. F., & Escalon, J. G. (2021). Thoracic manifestations of rheumatoid arthritis. *RSNA Journal, 41*(9).

10. Bogale, Z., Feleke, Y. (2022). Prevalence, clinical manifestations, and treatment patterns of patients with rheumatoid arthritis attending the rheumatology clinic at Tikur Anbessa Specialized Hospital, Ethiopia: A cross-sectional study, *Open Access Rheumatology, 14*, 221–229.
11. Brekelmans, M. P., Fens, N., Brinkman, P., Bos, L. D., Sterk, P. J., Tak, P. P., Gerlag, D. M. (2016). Smelling the diagnosis: The electronic nose as diagnostic tool in inflammatory arthritis. A case-reference study. *PLoS One, 11*(3), e0151715.| DOI:10.1371/journal.pone.0151715
12. Shaw, M., Collins, B. F., Ho, L. A., & Raghu, G. (2015). Rheumatoid arthritis-associated lung disease. *European Respiratory Review, 24*, 1–16. | DOI: 10.1183/09059180.00008014.
13. Arnett, F. C., Edworthy, S. M., Bloch, D. A., Mcshane, D. J., Fries, J. F., Cooper, N. S., Healey, L. A., Kaplan, S. R., Liang, M. H., Luthra, H. S., Medsger, T. A., Mitchell, D. J., Neustadt, D. H., Pinals, R. S., Schaller, J. G., Sharp, J. T., Wilder, R. L., & Hunder, G. (1988). The American Rheumatism Association 1987 Revised criteria for the classification of rheumatoid arthritis. *Arthritis and Rheumatism*, *31*(3), 315–324.
14. Ozerkis, D. A., Evans, J., Rubinowitz, A., Homer, R. J., Matthay, R. A. (2010). Pulmonary manifestations of rheumatoid arthritis. *Clinics in Chest Medicine, 31*, 451–478. DOI: 10.1016/J.Ccm.2010.04.003.
15. Brown, K. K. (2007). Rheumatoid lung disease. *Proceedings of the American Thoracic Society*, *4*(5), 443–448.
16. Gabriel, S. E., & Michaud, K. (2009). Review: Epidemiological studies in incidence, prevalence, mortality, and comorbidity of the rheumatic diseases. *Arthritis Research & Therapy*, *11*(3), 229.
17. Yousem, S. A., Colby, T. V., & Carrington, C. B. (1985). Lung biopsy in rheumatoid arthritis. *American Review of Respiratory Disease, 131*(5), 770–777.
18. Tsuchiya, Y., Takayanagi, N., Sugiura, H., Miyahara, Y., Tokunaga, D., Kawabata, Y., & Sugita, Y. (2011). Lung diseases directly associated with rheumatoid arthritis and their relationship to outcome. *European Respiratory Journal*, *37*(6), 1411–1417.
19. Kim, E. J., Elicker, B. M., Maldonado, F., Webb, W. R., Ryu, J. H., Van Uden, J. H., Lee, J. S., King Jr., T. E., & Collard, H.R. (2010). Usual interstitial pneumonia in rheumatoid arthritis-associated interstitial lung disease. *European Respiratory Journal, 35*(6), 1322–1328.
20. Cavagna, L., Monti, S., Grosso, V., Boffini, N., Scorletti, E., Crepaldi, G., & Caporali, R. (2013). The multifaceted aspects of interstitial lung disease in rheumatoid arthritis. *Biomed Research International, 2013*, 759760.
21. Robles-Perez, A., Luburich, P., Sanchon, B. R., Dorca, J., Nolla, J., Molina, M. M., & Garcia, J. N. (2016). Preclinical lung disease in early rheumatoid arthritis. *Chronic Respiratory Disease, 13*(1), 75–81.
22. Pérez, A. R., Luburich, P., Bolivar, S., Dorca, J., Nolla, J. M., Molina-Molina, M., & Narváez, J. (2020). A prospective study of lung disease in a cohort of early rheumatoid arthritis patients. *Scientific Reports, 10*, 15640.
23. Bilgici, A., Ulusoy, H., Kuru, O., Celenk, C., Unsal, M., & Danacı, M. (2005). Pulmonary involvement in rheumatoid arthritis. *Rheumatology International, 25*, 429–435.

24. Marigliano, B., Soriano, A., Margiotta, D., Vadacca, M., & Afeltra, A. Review lung involvement in connective tissue diseases: A comprehensive review and a focus on rheumatoid arthritis. *Autoimmunity Reviews, 12*, 1076–1084.
25. Rydosz, A. (2018). Review sensors for enhanced detection of acetone as a potential tool for non-invasive diabetes monitoring. *Sensors*, *18*(7), 2298.
26. Meshram, D. A., & Patil, D. D. (2021). Digital representation techniques for olfactory features. *Information Technology*, *9*(1).
27. Mule, N. M., Patil, D. D., & Kaur, M. (2021). A comprehensive survey on investigation techniques of exhaled breath (EB) for diagnosis of diseases in human body. *Informatics in Medicine Unlocked*, *26*, 100715.
28. Laxmikant, K., Bhuvaneswari, R., & Natarajan, B. (2023). An efficient approach to detect diabetes using XGBoost classifier. In *Winter Summit on Smart Computing and Networks (WiSSCoN), Chennai, India* (pp. 1–8). DOI: 10.1109/WiSSCoN56857.2023.10133854.
29. Natarajan, B., Elakkiya, R., Bhuvaneswari, R., Saleem, K., Chaudhary, D., & Samsudeen, S. H. (2023). Creathing alert messages based on wild animal activity using hybrid deep neural network. *IEEE Access, 11*, 67308–67321.

CHAPTER 23

APPLICATION DOMAINS AND ENABLING TECHNOLOGIES OF INTERNET OF THINGS: A SURVEY

SIKHINAM NAGAMANI and N. RENUGADEVI

Department of Computer Science Engineering, Indian Institute of Information Technology, Tiruchirappalli, Tamil Nadu, India

ABSTRACT

The Internet of Things (IoT) is a revolutionary paradigm that connects things and devices to the Internet, allowing for seamless data exchange and communication that boosts automation and efficiency across a variety of industries. The transformational promise of IoT rests in its capacity to collect, process, and act on massive volumes of real-time data from connected devices, enabling organizations and people to make wise decisions, streamline procedures, and enhance the quality of life in general. The widespread adoption of IoT has revolutionized industries and daily life, empowering businesses to streamline operations, optimize resource utilization, and enhance productivity through real-time data insights and intelligent automation. Businesses may optimize resource allocation, minimize downtime, and make choices based on real-time data by utilizing IoT's capabilities, while people can benefit from increased convenience, increased safety, and a more connected lifestyle. With the proliferation of IoT-enabled devices and applications, the seamless integration of the physical and digital realms has unlocked opportunities for innovative solutions, creating a smarter, more connected world that fosters efficiency,

Cyber-Physical Systems: Applications, Challenges, and Research Directions. Rajanikanth Aluvalu, Samarjeet Borah, & Lalit Garg (Eds.)

DOI: 10.1201/9781779643421-23

convenience, and personalized experiences for users. This chapter provides an overview of the IoT, the enabling technologies, and its applications. This chapter also reviews some of the existing research papers related to the application domain of IoT.

23.1 INTRODUCTION

The Internet of Things (IoT) involves different activities, such as streamlining data collection to reduce asset costs, assuring quick processing of data quantities, establishing secure data transfer, and improving data storage capacities [1]. IoT connects everything to the Internet and it involves incredible resources. IoT requires a special focus on the improvement of transmission protocols to lower energy consumption [2]. IoT refers to all the devices and entities that are connected and communicate with each other via the Internet without any human interaction. It is an expanded service provided by the Internet [3]. The main objective of the IoT paradigm is the integration of various technologies and standards. The objects associated with digital functionalities can communicate via the Internet. IoT is the best solution for many problems related to the world of communication and smart objects [4, 5].

The world has witnessed four industrial revolutions, that is, Industry 1.0, Industry 2.0, Industry 3.0, and Industry 4.0. Industry 1.0 was introduced in the late 18th century and was specific to mechanical assistance, basically dealing with advancements related to water and steam. The next phase of Industry 2.0 was introduced in the early 20th century that created mass production facilities. Industry 3.0 was introduced in the early 1970s for the automation of products. Later, it was revised by electronics and information technology. Finally, Industry 4.0 was introduced in recent years wherein sensors are integrated with different technologies in an interconnected way [6]. Emerging IoT provides great opportunities to create, enhance, and develop huge services for the flexibility of users. Over time, IoT has considerably expanded and evolved in a diversity of applications [7]. IoT offers a wide range of platforms and various technologies that are connected by various devices and automated by using sensors [8]. The future world will be all about connecting various interconnected things, which sense the data, actuation real-time monitoring, and controlling everything by automation through intelligent systems [9]. These entire essential stages of the evolution of IoT are shown in Figure 23.1.

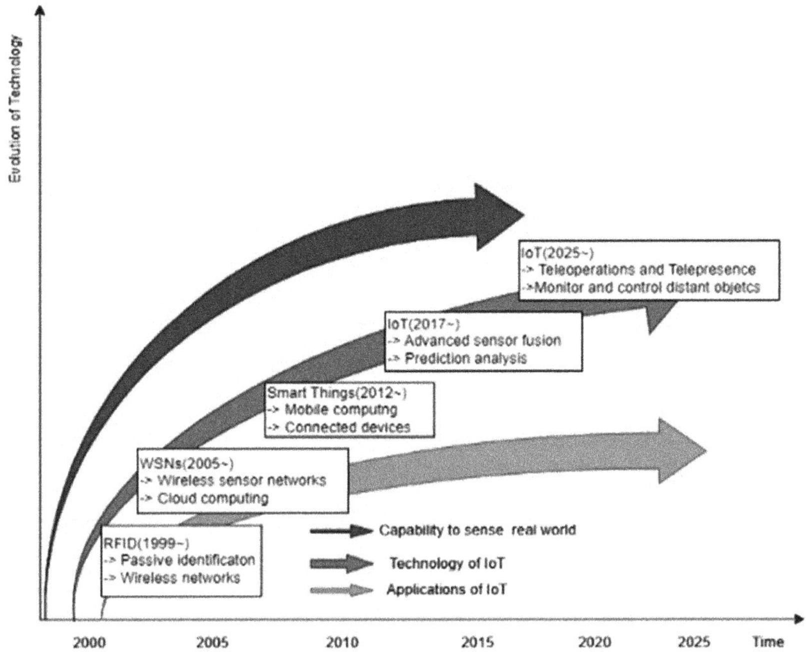

FIGURE 23.1 Evolution of IoT.

Kevin Ashton, the father of IoT, proposed the term "IoT" in 1999 when he was a member of the Radio Frequency Identification Development (RFID) community [10]. Recently, RFID become more popular due to the tremendous usage of mobile phones, ubiquitous computing, embedded devices, and data analytics. In 2005, wireless sensor networks were introduced for processing of cloud computing applications and Web 2.0. It supports low energy consumption of devices. In 2012, smart things were proposed such as mobile computing, where cooperation among objects is required. In 2017, IoT came into the picture for faster connectivity, faster data transmissions, and predictive analysis [11].

Every day, technology is changing dynamically. There are lots of enhancements, extensions, and revisions to the existing technology, which gives motivation to researchers, eminent academicians, and industrialists [12]. During the past 10 years, there have been a lot of advancements in the field of IoT, that is, integration of Artificial Intelligence and Machine Learning (ML) with IoT, integration of Big data and Cloud Computing, Edge and Fog Computing, and so forth [13]. IoT has numerous applications in our everyday life. IoT offers various commercial openings in this digitized world [14]. This chapter introduces the enabling technologies of IoT. In addition to that, some IoT applications are reviewed namely smart cities, smart farming, smart healthcare, smart weather monitoring, and smart crime analysis. Further, this chapter examines their objectives, implemented methodologies, metrics considered, challenges, advantages, and so forth. This study gives an overview of current research requirements and their revisions in the field of IoT.

This chapter is organized as follows. Section 23.11 provides an overview of IoT, while Section 23.2 describes the enabling technologies of IoT. Section 23.3 explores a literature review of various application areas of IoT, Section 23.4 summarizes the research work done related to some smart application domains of IoT, Section 23.5 gives challenges and future directions, and Section 6 concludes this chapter.

23.2 ENABLING TECHNOLOGIES

In recent years, IoT has grown exponentially with the advancements in microelectronics as well as the Internet. However, the evolution of IoT has happened over a long period with the very first contributions taking place in the early 19th century, and the IoT revolution has gained a lot of momentum in the last decade. In the real world, the implementation of the IoT concept is feasible over the combination of various enabling tools. IoT allows smart intelligent decision-making systems [15]. The role of enabling technologies in IoT is discussed in this section.

23.2.1 CLOUD COMPUTING

To establish a strong connection between objects and people, a communication system is required. The backbone of the communication system is IoT which merges various things via the Internet. Cloud is one of the crucial components of IoT that provides specific IoT services all over the world. IoT cloud platform solves several issues related to application development and management of systems, heterogeneity and data management, storage of data, visualization, deployment, analyzing tools, monitoring, and research [16]. Cloud computing is a leading business model and has unlimited storage and processing capabilities in the evolution of information technology. We are highly dependent upon cloud services in our day-to-day life, for example, storing and processing of data, document writing, banking services, E-commerce services, e-mail, data backup, infrastructure services, social media, and so forth. Cloud has four key components, namely infrastructure, data center (network), platform, and application; its services are categorized into three, namely, software as a service (SaaS), Infrastructure as a Service (IaaS), and Platform as a Service (PaaS). The applications that run on cloud environments can be delivered by SaaS through the Internet. PaaS is used to develop and deploy software frameworks and applications, manage operating system support, and so forth. IaaS is responsible for storing and processing data, instantaneous computing infrastructure, network-related resources, and so forth [17]. Virtual private cloud is to access the features of the public cloud via private cloud (own). By using virtual private network technology, it addresses the problems associated with public and private clouds such as security and topology [18].

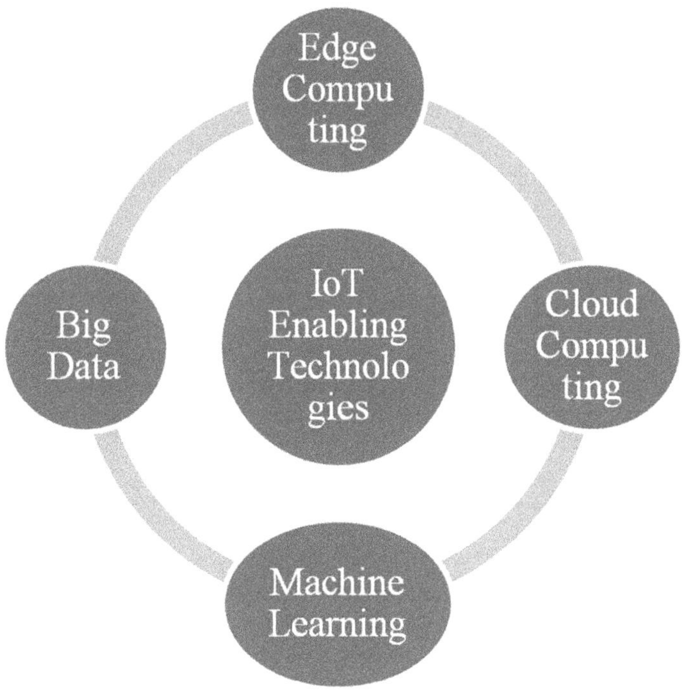

FIGURE 23.2 IoT-enabling technologies.

23.2.2 FOG COMPUTING

It deals with the conversion of simple sensor recordings into actionable industrial insights by using ML. It helps to manage high volumes of data and automation opportunities for predictive analysis. This kind of architecture is implemented in areas like e-health and military applications. However, Fog computing provides limited communication, storage, and network services. The necessity of IoT computing framework is to process data. For data processing, the frameworks like fog and cloud computing are required. Instant access to data and network architecture supports faster and better computing [15].

23.2.3 BIG DATA

The term Big Data is universally employed for large sets of data. Big data contains a huge volume of heterogeneous data that can be structured, unstructured, or semistructured. Mainly, IoT systems consist of Internet-equipped devices containing sensors and communication components to store data, forward, and perform on the extracted data. Any IoT device that shares the data with the cloud to analyze it uses smart devices as major components of IoT. These devices generate huge amounts of data, and then explore this data which is investigated in real time. This is the key point where Big Data Analytics comes into play [13].

23.2.4 MACHINE LEARNING

IoT and ML are buzzwords in the digital world. ML takes huge amounts of data for processing which is critical to minimize costs and produce a better experience for the customer. The goal of ML is to find minor changes in real-time applications by using different technologies within a fraction of a second. With traditional data analysis, we can generate general reports or models of what previously existed and what is happening today. With data analytics, data can be tracked, enable smart decisions, and measure success over time. ML is used in applications of IoT which helps save cost in industrial applications because predictive analysis is a more useful component in industry. A large number of sensors are interconnected for communication over the Internet. By taking data from multiple sensors on machines, the sensed data is transmitted to servers, after which stored data in the servers are compared to the existing data for analysis [19]. The sensor data is uploaded to the cloud for analysis. ML splits the information into data which is used for training and verification. ML model looks at hundreds and thousands of data records for anomalies, correlation, and projections to come up with a hypothesis. Once the hypothesis has been created, it is tested and validated; then the validated result is published as an executable endpoint [20].

TABLE 23.1 Industrial IoT-Enabled Technologies.

Technology	Objective	Result	Applications	Remarks
Cloud computing [49]	To integrate VR simulator into AR in neurosurgery in virtual teaching techniques	AR simulator gives better surgical skills than VR simulator	Neurosurgery simulation	Pulse and blood pressure are not yet given accurate simulation results
Edge computing [50]	To provide offload service without any delay	To resolve scheduling problems in edge computing	Computation offloading	Due to regional discrepancies in network traffic, the algorithm has complications
Bigdata [51]	To concentrate resource optimization and service guarantee in IIoT	Lower bandwidth usage	Service assurance and optimizing routes	In the future, it could be extended by considering scalability and efficiency on the control plane
Machine learning [52]	To replace manual procedures in critical activities using Machine Learning	Explores manufacturing and service industry intelligence	Supply chain and agile solutions	Machine Learning may remove hidden IoT data patterns with advanced algorithms

Table 23.1 describes various technologies, their objectives, results, and remarks on several applications. Integration of all these technologies gives better results in industrial applications of IoT.

23.3 LITERATURE REVIEW

In this section, several research works related to IoT are identified and discussed. Many researchers have contributed insights into the development of IoT as an emerging engineering technology. After examining the various existing research contributions, the findings are encapsulated in separate sections for better and clearer understanding. Such application-specific information will be useful to identify the relative developments in each area of application and gain significant perspective for future needs.

23.3.1 SMART CITIES

The main goal of smart cities is to resolve civic problems effectively by using modern technologies and infrastructure. The objectives of smart cities are to offer quality services to citizens, to improve the usage of public resources, and so forth. To accomplish these objectives, the smart city needs to set up an infrastructure that offers a simple and flexible environment.

As shown in Figure 23.3, a typical application of a smart city offers a well-connected network of technologies, facilities/amenities, and people. Usually, the amenities include smart parking, road traffic, public transport, street lighting, smart cities in hyper-connected cities for the ease of lining-walkable streets, parks, waterfronts, smart governance-command, and control centers, mobile applications, connected communities-smart schools, heritage, arts and culture, urban resilience-solar and wind energy, water and wastewater, and waste to energy. Largely, a smart city application covers smart roads, traffic management, and smart parking as road maintenance and parking management are major challenges of smart cities.

In 2012, Hammi et al. [11] proposed a framework for a smart city by taking into consideration the practical needs of government professionals. The smart city initiative framework is aimed at building the IT infrastructure for maintaining data security and protecting the privacy of citizens at a lower operational cost. This initiative promised to increase sustainability and ensure better management of natural resources so that the IT infrastructure does not cause any destruction as it is done when physical infrastructure is used in cities [11]. In 2015, Harmon et al. [20] projected the implementation of an IoT system in a smart city context, which deals with the foundational development of smart cities. An IoT architecture was proposed to ensure integrity, security policies, and data storage. This architecture is useful in transportation, retail, and energy [20].

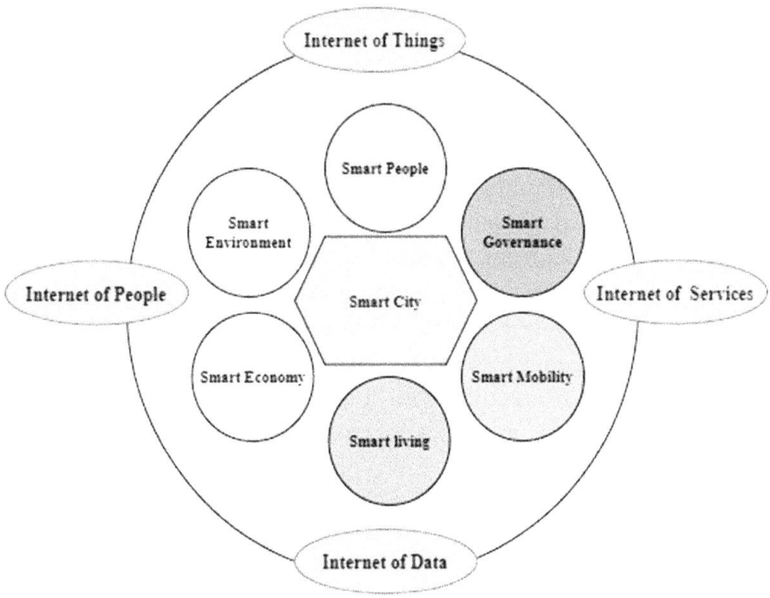

FIGURE 23.3 Smart city applications.

In 2016, Pribadi et al. [21] described an open-source intelligent IoT platform based on intelligent unmanned aerial vehicles (UAV). In this smart city platform, the authors have considered the elements of data, connectivity, devices, and services for creating a smart platform. It addressed issues such as scalability, integration, and interoperability. This platform is best suitable for crime analysis, agriculture, and military applications [21]. In 2020, Kirimtat et al. [22] described the challenges faced by cities by using information and communication technologies (ICT) in collaborative recommenders. For measuring smart city complexities, they have considered parameters like quality ranking, precision and recall, and rating of prediction error using mean absolute error (MAE) and root mean squared error (RMSE). The biggest advantage of this method is found to be the promised flexibility and comfort to the citizens [22].

23.3.2 SMART FARMING

Smart farming is an evolving concept for managing farms and seeks to optimize human effort and enhance the quantity as well as the quality of products. To achieve these goals, it addresses issues such as high population, climate conditions, labor, planting, and watering of crops. To produce good crops, the proper utilization of the soil is a key component that IoT can address by providing ample information about crops and the condition of the soil. Prathibha et al. [23] proposed IoT-based smart agriculture based on temperature and humidity levels through a CC3200 chip. Formerly, a methodology named Climate Smart Agriculture (CSA) was introduced after analyzing the current agricultural trends, human food wastage, and changes in climatic conditions. A comprehensive layout of CSA is presented to cater to both its aspects, that is, climate change mitigation and socioeconomic policies. Lipper et al. [24] proposed a taxonomy that establishes a five-layer architecture with a data reception layer, middleware layer, optimization layer, analytics layer, and application layer.

As observed from Figure 23.4, a typical application of smart farming offers a well-connected network of technologies, control tools, and human effort. The soil condition and climatic properties are examined by using sensors. The collected data is used for smart analysis and planning at a remote location using a cloud-based data management system. Then, the analyzed data is distilled to generate actionable inputs for the smart control devices.

In 2016, Zamora-Izquierdo et al. [25] proposed a remote monitoring system. It collects real-time data on agriculture for easy access to agriculture facilities through a short message service. The metrics used for monitoring fields are temperature, humidity, light, and intensity. By receiving prior alert messages, the farmers can protect the crops in a better manner [25]. In 2018, an IoT-based polyhouse was proposed by Kirimtat et al. [26]. Polyhouses use various sensors such as temperature, humidity, soil, and water. The proposed method is useful for creating better crop management, better resource management, reducing cost, and improving the quality of the field [26].

In 2018, an IoT-based smart farming was proposed by Deepa et al. [27]. In smart farming temperature and humidity sensors are used to capture images of the field, and it sends multimedia messages to farmers so that crops can be protected from animals, birds, and so forth [27]. In 2018, Bauer and Aschen Bruck [28] proposed an IoT-based agriculture monitoring system that was implemented by using wireless sensor networks. In this, the Leaf area index (the ratio of the leaf area to a given unit of land area) was continuously assessed by sensor networks for continuous monitoring of crop growth [28]. In 2019, Bothe et al. [29] proposed an RFID system-based authentication system and it is used for farming applications. In Farm Management Information Systems (FMISs), an interconnection of agriculture machines was used to continuously authenticate the user for higher security.

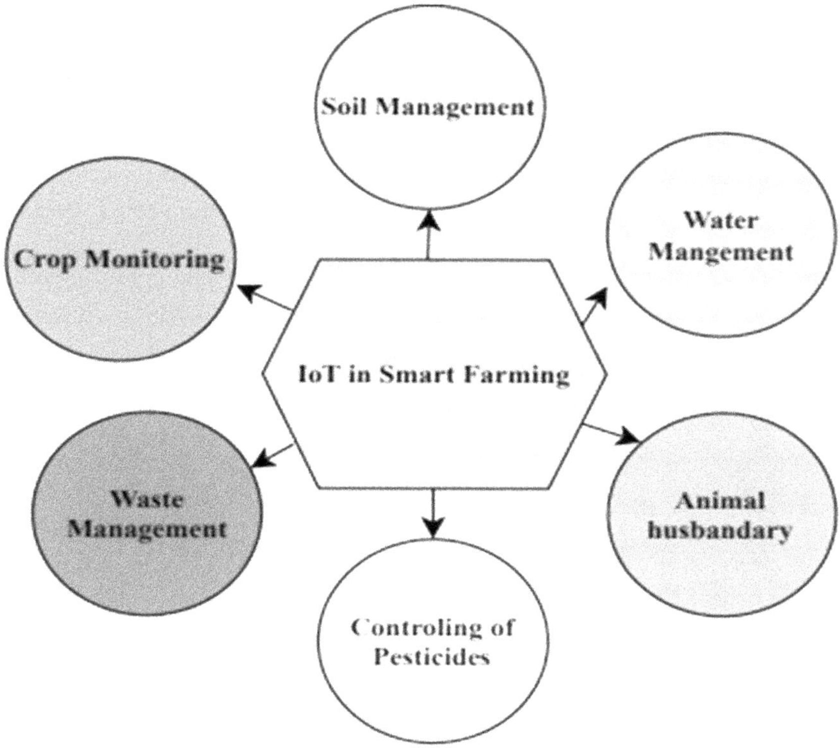

FIGURE 23.4 Smart farming.

23.3.3 SMART HEALTHCARE

IoT acts as a structure chain system that is mainly responsible for collecting and transmitting information and transmitted to interconnected smart devices. These interconnected devices communicate with each other via the Internet. In recent times, the healthcare industry has been using digital health monitoring systems for tracking patient's fitness using smart wearable devices. IoT in the health care system involves various technologies, that is, Cloud Computing, Wi-Fi, Bluetooth, RFID, wireless medical sensors, and so forth [30]. By combining IoT with healthcare systems, the eminence of human life can be enhanced. It controls the real-time tracking of patients' health and improves the interaction between patients and healthcare workers or caregivers. The evolution of the IoT healthcare sector has given a boost to the traditional healthcare system [31].

As shown in Figure 23.5, a smart healthcare system combines various smart devices in an interconnected network. A smart device uses sensors attached to a smartphone or a wearable gadget and monitors the vital health parameters of a patient. Then the data is sent via cloud to other workstations for analysis by the healthcare system. Internet Healthcare Network (IoTHNet) as a key feature is used for transmitting and collecting information [32].

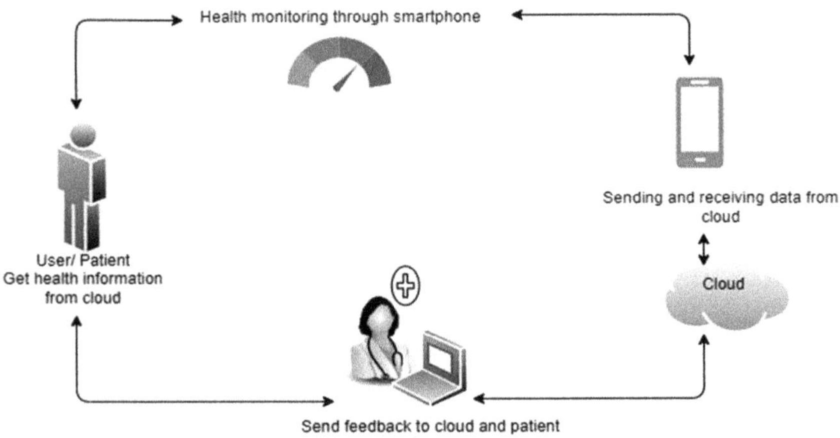

FIGURE 23.5 Smart healthcare system.

In 2016, Islam et al. [33] proposed wearable IoT devices based on wireless sensor networks. By using wearable devices patients can be monitored remotely in an effective manner. In 2017, Vippalapalli and Ananthula implemented an IoT system, to provide better and quality health for everyone. Body sensor networks are inserted into the human body to measure blood pressure, pulse, and temperature easily. This is very helpful for people who have busy schedules and reduces the risk of medical staff [34]. In 2019, Dhanvijay and Patil [35] proposed an

IoT-based health architecture. By using architecture, patients can check their medical records easily and access them remotely. The main advantage is to provide better healthcare to everyone [35]. In 2016, a Wireless Body Area Network was proposed for remote health checking. Here, various sensors were used to capture information like temperature sensor, humidity, barometer, and pressure sensor. Easily, a patient can check his or her medical record from anywhere anytime. It reduces the risk to healthcare workers and caregivers [36].

23.3.4 SMART WEATHER MONITORING

IoT weather-monitoring systems use various parameters such as temperature, humidity, speed of wind, intensity of light, moisture, and level of carbon monoxide in the atmosphere. To monitor weather in real time, sensors can be deployed in the environment for sensing data. The sensed data can be used for further analysis. Whenever there is a change in the environment, it automatically measures the climate conditions. The advantages of smart weather-monitoring systems include higher crop yields and prevention of over- or under-watering. The disadvantages include the lack of compatibility and privacy/security. The applications of smart weather-monitoring systems include measurement of temperature and humidity, automatic control of climate, and monitoring of the environment [37]. In 2016, an IoT-based wireless sensor network was developed to monitor environmental changes by using various sensors such as temperature, humidity, and CO_2. Arduino Uno board with a Wi-Fi module was used to monitor and control environmental changes by sensing and storing data in the cloud. By using an IoT-sensor network weather information can be accessed remotely through a smartphone [38].

In 2018, an IoT-based smart weather-monitoring system was proposed by Kulkarni and Mukhopadhyay [39], to monitor weather and climate changes based on temperature, humidity, and oxygen concentration. It is easy to carry and flexible to use. In 2019, a wireless remote weather-monitoring system based on micro-electro-mechanical systems (MEMS) was proposed by Ma et al. [40]. To monitor weather, it measures the speed of wind, wind direction, and temperature. It is a low-cost weather-monitoring system, wherein the weather changes are accessed from any location from cloud database management [40]. In 2019, an IoT-based weather-monitoring system called smart weather-monitoring system was introduced by Tsao et al. [41]. It is based on the Message Queue Telemetry Transport (MQTT) protocol and measures the rainfall, wind speed, temperature, and light intensity of the sensor. It gives higher crop yields and prevents over- and under-watering. Data attributes and flow-oriented MQTT were proposed that include prototypes like Arduino MKR1000, and Raspberry Pi [41]. The main advantage is to protect public health from pollution by continuous monitoring of the environment.

23.3.5 SMART CRIME ANALYSIS

In recent years, the economy of smart cities has increased. Due to fast commercial growth in smart cities, there is an increase in the number of crimes. Timely detection and prediction of crimes is the major focus of any police system.

Figure 23.6 describes the IoT long-term evaluation network for crime detection, crime visualizations, and so forth. It can be monitored flexibly. A person's emotions who wears a wearable sensing device can be captured and CCTV cameras are used for detecting the movement of a person. For effective management of crime analysis, emotions and movements were recorded. This data is helpful for future processing of crime detection, crime prediction, crime visualizations, and so forth. But it does not give good accuracy. In 2020, Multi UAV allocation framework was proposed to detect crime prediction. A network of UAVs is deployed in major crime areas. However, even in a multi-UAV framework, the reliability of data should be increased [42]. In 2019, an edge-based architectural model was proposed to track and detect criminals. It speeds up the process of computation and the safety of the public is improved [43].

In 2019, a model of spatio-temporal crime forecasting was introduced for detecting high-risk crime regions. It measures the exact time and area of crime. The cities are exposed to high crime rates so it gives better solutions to police forces and citizens [44]. In 2021, crime forecasting was done based on time, date, and year. The highest crime rate can be predicted using miming algorithms [45].

In 2022, to detect crimes automatically, an intelligent forensic analysis mechanism was proposed with an IoT device based on a machine-to-machine framework, which combines various ML algorithms to detect the attacks. The performance of the model was evaluated using metrics such as accuracy, precession, recall, and F1 score [46]. In 2023, to derive the artifacts of criminal analysis, a forensic-based IoT wearable device was proposed. The forensic model was derived based on the ecosystem of wearable devices. Smart watches and smart bands generate personal information which is helpful in criminal investigations [47]. There is a correlation between socioeconomic and crime levels. Crime affects the quality of human life and the reputation of a nation. Crime analysis plays a significant role in predicting and solving crimes. Once the crime is analyzed and predicted then it is useful to police departments, to give a better and safer society to citizens.

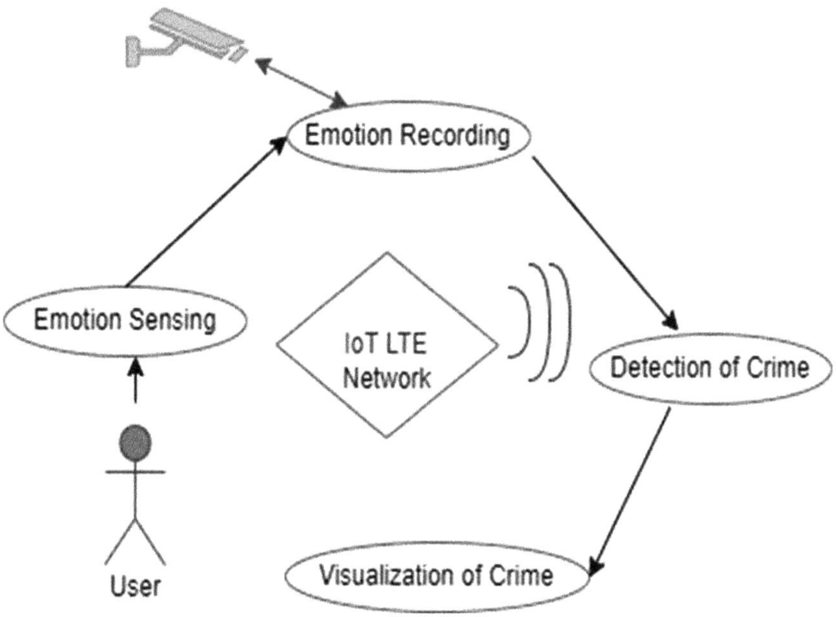

FIGURE 23.6 Smart crime analysis.

23.4 SUMMARY OF VARIOUS SMART APPLICATIONS

In this section, various applications of IoT are discussed with their objectives, proposed methodology, the metrics used for results analysis, and their advantages. Table 23.1 describes the summary of smart cities with various methods, smart farming frameworks, smart health care systems, smart weather-monitoring systems, and smart crime analysis.

Table 23.2 outlines multiple applications, including smart cities, smart farming, smart healthcare, smart weather monitoring, and smart crime analysis, which exemplifies the variety of advantages IoT has in altering many industries. IoT technologies allow smart city deployments that optimize infrastructure, manage resources effectively, and improve urban livability. IoT is used in "smart farming" to give farmers access to real-time data, enhancing farming methods, and boosting crop yields. IoT-enabled devices in smart healthcare provide remote monitoring and individual care, improving patient health outcomes. IoT offers precise data for weather monitoring, assisting disaster preparedness and climate-sensitive decision-making. Last, but not the least, smart crime analysis uses IoT to improve public safety and enable law enforcement to react quickly to illegal activity. Collectively, these IoT applications demonstrate the technology's enormous potential for revolutionizing sectors, encouraging sustainable lifestyles, and enhancing people's quality of life.

23.5 CHALLENGES AND FUTURE DIRECTIONS

The integration of technologies like cloud computing, big data, edge computing, and ML can make a significant impact in various fields of IoT.

23.6 CONCLUSION

This chapter has described the basic details of IoT, enabling technologies used by IoT. Also, various IoT-related application domains such as smart cities, smart farming, smart healthcare-monitoring systems, smart weather management, and smart crime analysis are explained. Further, this chapter summarizes research works done in the above-mentioned smart application areas. It is clear that IoT reduces the dependence on manpower and also improves functionalities in every application area leading to improvements in socioeconomic activities. However, IoT is still not much used in all these fields, and a lot of improvisation is required for better quality of human life.

TABLE 23.2 Summary of Smart Applications.

Year and Reference	Objective	Technology/Algorithm/ Method Used/ Proposed	Metrics Used	Advantages
2019 [44]	To address the difficulties of city problems in efficient and sustainable ways	ICT, content-based recommenders, collaborative recommenders	Ranking quality metrics-precision and recall, rating prediction errors-MAE, RMSE	Provides personalization for customers, managing environment
2017 [21]	To propose a framework for smart cities in practical implications by government professionals	Smart City Initiative Framework	IT infrastructure, security, privacy, operational cost	Increases the sustainability and better management of natural resources
2015 [20]	Implementation of IoT system in smart city context	Smart City IoT System using cloud	Smart buildings, energy, water, transportation, healthcare, and education	Better transport services, infrastructure improvement, new economic development opportunities
2020 [42]	To propose an open-source intelligent IoT platform for crime analysis	Intelligent UAV (unmanned aerial vehicle) platform	Data, connectivity, device, service	It solves issues such as scalability, integration, and interoperability
2019 [29]	RFID-based authentication system for farming applications	Farm Management Information Systems (FMISs)	Usability, security, robustness	Greatly improves security and enables traceability in agricultural practice
2016 [26]	To propose IoT in agriculture by using various sensors and Wi-Fi	Proposed IoT-based polyhouse	Temperature, humidity, soil, and water	Agriculture can be enhanced and simplified and crop production can be optimized
2020 [27]	To propose phase-based IoT system to monitor agriculture	Monitors temperature and humidity in agriculture fields based on a phase-phase-based IoT system	Temperature and humidity	Crop can be protected from animals and birds

TABLE 23.2 *(Continued)*

Year and Reference	Objective	Technology/Algorithm/ Method Used/ Proposed	Metrics Used	Advantages
2018 [28]	To implement an agriculture-monitoring system and crop growth	IoT with wireless sensor networks	Leaf area index (LAI)	Computational power and power autonomy are increased by using the MQTT protocol
2017 [23]	To collect agriculture data in real time and easily access	Proposed remote monitoring system (RMS)	Temperature, humidity, light intensity, and pH levels	It allows the farmers to receive information on crop
2012 [31]	IoT sensors are embedded in hospitals to enhance the traditional medical system	Conducted survey questionnaire for patients	Questionnaire for patients (temperature sensor, ECG sensor, humidity sensor, barometric pressure sensor)	Body sensor networks are used for remote tracking
2019 [35]	To monitor the patient remotely using IoT	Proposed IoT-based health architecture	Temperature and pulse	Patients can easily track their medical records so that better healthcare is provided to everyone
2017 [34]	Design and implementation of wearable sensors to monitor patient's health	Body sensor network using IoT	Temperature, blood pressure, and pulse	Monitors huge numbers of patients easily
2016 [36]	To implement a WBAN-based IoT healthcare system	IoT-based Wireless Body Area Network framework for healthcare applications	Computational speed	Real-time remote monitoring of the health care system, reduced computation overhead
2016 [37]	To monitor and control the environmental conditions	For sensing and storing the data in the cloud, Arduino UNO with a Wi-Fi module is used as an embedded device	Temperature, humidity, light intensity, sound intensity, and CO levels	Efficient and low-cost resolution for nonstop monitoring of the environment

TABLE 23.2 *(Continued)*

Year and Reference	Objective	Technology/Algorithm/ Method Used/ Proposed	Metrics Used	Advantages
2018 [39]	To monitor weather and climate changes	Smart weather-monitoring system	Temperature, humidity, and oxygen concentration	It can be easily carried to remote locations for installation
2011 [40]	To get weather conditions of any location, and it allows to access the up-to-date data of any station	Wireless remote weather-monitoring system based on micro-electro-mechanical systems (MEMS).	Wind speed, wind direction, and temperature	Weather-monitoring system with the lowest cost
2019 [41]	To implement IoT-based weather-monitoring system	Message Queue Telemetry Transport (MQTT)	Rainfall, wind speed, temperature, and light intensity of the sensor	It can help in producing higher crop yields and prevent over- or under-watering
2018 [38]	To monitor environmental changes using Wireless Sensor Network	IoT-enabled sensing and monitoring system	Temperature, humidity, and CO_2	Data can be accessed remotely by using a smartphone
2019 [43]	To track and detect criminals	Edge-based architectural model for criminal tracking	Computational speed of tracking point	The safety of the public is improved
2020 [42]	To detect crime prediction	Multi Unmanned Aerial Vehicles (UAV) allocation framework	Allocation of unmanned aerial vehicles in major crime areas	Robustness is improved
2019 [44]	To detect high-risk crime region	The spatiotemporal crime-forecasting model	Time and area	Protects public lives from high-risk crime regions
2021 [45]	To reduce the crime rate	IoT-based smart crime detection model	Accuracy of emotions	Useful tool for both police agencies to determine crime and citizens to be on the safe side of the places they live in

TABLE 23.2 *(Continued)*

Year and Reference	Objective	Technology/Algorithm/ Method Used/ Proposed	Metrics Used	Advantages
2022 [46]	To detect crimes automatically	Random forest, decision tree, Naïve bayes	Accuracy, precision-recall, F1 score	The machine automates the attacking process of crime and the detection of crimes
2020 [48]	To collect personal information by wearing wearable IoT-based devices	IoT-based wearable devices Samsung smartwatch	Interface and memory chip of the Printed Circuit Board	To collect the information by wearing IoT-based wearable devices

TABLE 23.3 Challenges and Future Directions of IoT.

Challenges	Possible Solution	Future Directions
Privacy risk	Patient's identity in a virtual manner by implementing application layer encryption	Adopting federated learning techniques for enhanced privacy of patients' sensitive data
Smart cities	One of the challenges tackled in this is enhancing the run-time and quality performance of spacetime series clustering	Storage of data: Due to open research issues, a study on protecting the data-collecting process, its sharing protocols, and storage management systems are essential for the smart city systems' expanding amount of data. The main factors to take into consideration for future research initiatives are data confidentiality, non-repudiation, security, and privacy
Smart crimes	AI and drones for public safety	Drone systems powered by AI can serve as CCTV cameras to protect the public
Smart farming	UAV-based farm monitoring using fuzzy logic and remote sensing	Agricultural UAVs using fuzzy logic-based ideas are developing into instruments for giving farmers a wealth of knowledge about their crops and quickly accomplishing certain farming missions
		Real-time and prescriptive temperature, moisture content, heating rate, and power usage walkthroughs are provided by intelligent motion detectors and customized software. For farms that depend largely on seasonal variations, this is quite advantageous

KEYWORDS

- **Internet of Things**
- **smart cities**
- **smart crime analysis**
- **smart farming**
- **smart healthcare**
- **smart weather**

REFERENCES

1. Fox, S. E., Silva, R. M. L., & Rosner, D. K. Beyond the prototype: Maintenance, collective responsibility, and public IoT. In *Proceedings of the 2018 Designing Interactive Systems Conference* (pp. 21–32).
2. Farooqi, N., Gutub, A., & Khozium, M. O. (2019). Smart community challenges: Enabling IoT/M2M technology case study. *Life Science Journal*, *16*(7), 11–17.
3. Gokhale, P., Bhat, O., & Bhat, S. (2018). Introduction to IOT. *International Advanced Research Journal in Science, Engineering and Technology, 5*(1), 41–44.
4. Pourghebleh, B., Hayyolalam, V., & Aghaei Anvigh, A. Service discovery in the internet of things: Review of current trends and research challenges. *Wireless Networks, 26*(7), 5371–5391.
5. Jaidka, H., Sharma, N., & Singh, R. (2020). Evolution of IoT to IIoT: Applications & challenges. *SSRN Electron Journal,* 1–6.
6. Moraes, E., & et al. (2023). Integration of Industry 4.0 technologies with education 4.0: Advantages for improvements in learning. *Interactive Technology and Smart Education, 20*(2), 271–287.
7. Kassab, W. A., & Darabkh, K. A. (2020). A–Z survey of internet of things: Architectures, protocols, applications, recent advances, future directions, and recommendations. *Journal of Network and Computer Applications, 163*, 102663.
8. Hui, H., et al. (2020). A novel secure data transmission scheme in industrial internet of things. *China Communications, 17*(1), 73–88.
9. Yanes, A. R., Martinez, P., & Ahmad, R. (2020). Towards automated aquaponics: A review on monitoring, IoT, and smart systems. *Journal of Cleaner Production, 263*, 121571.
10. Ande, R., Adebisi, B., Hammoudeh, M., & Saleem, J. (2020). Internet of things: Evolution and technologies from a security perspective. *Sustainable Cities and Society, 54*, 101728.
11. Hammi, B., Khatoun, R., Zeadally, S., Fayad, A., & Khoukhi, L. (2018). IoT technologies for smart cities. *IET Networks, 7*(1), 1–13.

12. Xu, L. D., He, W., & Li, S. (2014). Internet of things in industries: A survey. *IEEE Transactions on Industrial Informatics, 10*(4), 2233–2243.
13. Khare, S., & Totaro, M. (2019). Big data in IoT. In *2019 10th International Conference on Computing, Communication and Networking Technologies (ICCCNT)*. IEEE.
14. Al-Fuqaha, A., Guizani, M., Mohammadi, M., Aledhari, M., & Ayyash, M. (2015). Internet of things: A survey on enabling technologies, protocols, and applications. *IEEE Communications Surveys and Tutorials*, *17*(4).
15. Lee, I., & Lee, K. (2015). The internet of things (IoT): Applications, investments, and challenges for enterprises. *Business Horizons*, *58*(4).
16. Ray, P. P., Dash, D.,& De, D. (2018). Edge computing for internet of things: A survey, e-healthcare case study, and future direction. *Journal of Network and Computer Applications*, *140*, 1–22.
17. Dang, L. M., Piran, M. J., Han, D., Min, K., & Moon, H. (2019). A survey on Internet of things and cloud computing for healthcare. *Electronics*, *8*(7), 1–49.
18. Ghosh, D., Chun, S. A., Shafiq, B., & Adam, N. R. (2016). Big Data-based Smart City Platform (pp. 58–66).
19. Mahdavinejad, M. S., & et al. (2018). Machine learning for Internet of Things data analysis: A survey. *Digital Communications and Networks*, *4*(3), 161–175.
20. Harmon, R. R., Castro-Leon, E. G., & Bhide, S. (12015). Smart cities and the Internet of Things. In *Portland International Conference on Management of Engineering and Technology* (pp. 485–494).
21. Pribadi, A., Kurniawan, F., Hariadi, M., & Nugroho, S. M. S. (2017). Urban distribution CCTV for Smart City using decision tree methods. In *International Seminar on Intelligent Technology and It's Application* (pp. 21–24).
22. Kirimtat, A., Krejcar, O., Kertesz, A., & Tasgetiren, M. F. (2020). Future trends and current state of smart city concepts: A survey. *IEEE Access, 8*, 86448–86467.
23. Prathibha, S. R., Hongal, A., & Jyothi, M. P. (2017). IOT based monitoring system in smart agriculture. In *Proceedings of the 2017 International Conference on Recent Advances in Electronics and Communication Technology (ICRAECT)* (pp. 81–84).
24. Lipper, L., & et al. (2014). Climate-smart agriculture for food security. *Nature Climate Change*, *4*(12), 1068–1072.
25. Zamora-Izquierdo, M. A., Santa, J., Martínez, J. A., Martínez, V., & Skarmeta, A. F. (2019). Smart farming IoT platform based on edge and cloud computing. *Biosystems Engineering, 177*, 4–17.
26. Kirimtat, A., & et al. (2020). IoT in agriculture: Smart farming. In *International Conference on Signal Processing, Communication, Power Embedded System (SCOPES 2016)* (Vol. *1*(1), pp. 848–854).
27. Deepa, R., Moorthy, V., Venkataraman, R., & Kundu, S. S. (2020). Smart farming implementation using phase based IOT system. In *2020 IEEE International Conference on Communication and Signal Processing (ICCSP)* (pp. 930–934).
28. Bauer, J., & Aschenbruck, N. (2018). Design and implementation of an agricultural monitoring system for smart farming. In *2018 IoT Vertical and Topical Summit on Agriculture - Tuscany (IOT Tuscany)* (pp. 1–6).
29. Bothe, A., Bauer, J., & Aschenbruck, N. (2019). RFID-assisted continuous user authentication for IoT-based smart farming. In *2019 IEEE International Conference on RFID Technology and Applications (RFID-TA)* (pp. 505–510).

30. Darianian, M., & Michael, M. P. (2008). Smart home mobile RFID-based internet-of-things systems and services. In *Proceedings of the 2008 International Conference on Advanced Computer Theory and Engineering (ICACTE)* (pp. 116–120).
31. Kumar, P.,& Lee, H. J. (2012). Security issues in healthcare applications using wireless medical sensor networks: A survey. *Sensors, 12*(1), 55–91.
32. Saha, G.., Singh, R., & Saini, S. (2019). A survey paper on the impact of 'Internet of Things' in Healthcare. In *Proceedings of the 3rd International Conference of Electronics, Communication and Aerospace Technology* (ICECA) (pp. 331–334).
33. Islam, S. M. R., Kwak, D., Kabir, M. H., Hossain, M., & Kwak, K. S. (2015). The internet of things for health care: A comprehensive survey. *IEEE Access, 3*, 678–708.
34. Vippalapalli, V., & Ananthula, S. (2017). Internet of things (IoT) based smart health care system. In *International Conference on Signal Processing, Communication, Power and Embedded System (SCOPES)* (pp. 1229–1233).
35. Dhanvijay, M. M., & Patil, S. C. (2019). Internet of things: A survey of enabling technologies in healthcare and its applications. *Computer Networks, 153*, 113–131.
36. Mohammad, G., & et al. (2016). A survey on wireless body area networks for healthcare systems in residential environments. *Sensors, 16*(6), 831.
37. Shah, J., & Mishra, B. (2016). IoT enabled environmental monitoring system for smart cities. In *2016 International Conference on Internet of Things and Applications (IOTA)* (pp. 383–388).
38. Rahut, Y., Afreen, R., & Kamini, D. (2018). Smart weather monitoring and real-time alert system using IoT. *International Research Journal of Engineering and Technology, 5*(10), 848–854.
39. Kulkarni, A., & Mukhopadhyay, D. Internet of things-based weather forecast monitoring system. *Indonesian Journal of Electrical Engineering and Computer Science, 9*(3), 555–557.
40. Ma, R. H., Wang, Y. H., & Lee, C. Y. (2011). Wireless remote weather monitoring system based on MEMS technologies. *Sensors, 11*(3), 2715–2727.
41. Tsao, Y. C., Te Tsai, Y., Kuo, Y. W., & Hwang, C. (2019). An implementation of IoT-based weather monitoring system. In *Proceedings of the 2019 IEEE International Conferences on Ubiquitous Computing and Communications and Data Science and Computational Intelligence and Smart ComputIng, Networking and Services* (pp. 648–652).
42. Miyano, K., Shinkuma, R., Shiode, N., Shiode, S., Sato, T., & Oki, E. (2020). Multi-UAV allocation framework for predictive crime deterrence and data acquisition. *Internet of Things, 11*, 100205.
43. Adesola, F., Misra, S., Omoregbe, N., Damasevicius, R., & Maskeliunas, R. (2019). An IOT-based architecture for crime management in Nigeria. In R. K. Shukla, J. Agrawal, S. Sharma, & G. Singh Tomer (Eds.), *Data, Engineering and Applications*. Springer.
44. Catlett, C., Cesario, E., Talia, D., & Vinci, A. (2019). Spatio-temporal crime predictions in smart cities: A data-driven approach and experiments. *Pervasive and Mobile Computing, 53*, 62–74.
45. Yadav, D., & Krishan, K. (2021). A review on resolving crime with prediction. *International Journal for Research in Applied Science & Engineering Technology (IJRASET), 9*(2), 197–201.

46. Mazhar, M.S., Saleem, Y., Almogren, A., Arshad, J., Jeffery, M. H., Rehman, A. U., Shafiq, M., & Hamam, H. (2022). Forensic analysis on internet of things (IoT) device using machine-to-machine (M2M) framework. *Electronics, 11*, 1126. https://doi.org/10.3390/ electronics11071126
47. Kim, M., Shin, Y., Jo, W., et al. (2023). Digital forensic analysis of intelligent and smart IoT devices. *Journal of Supercomputing, 79*(1), 973–997. https://doi.org/10.1007/s11227-022-04639-5
48. Tundis, A., Kaleem, H., & Mühlhäuser, M. (2020). Detecting and tracking criminals in the real world through an IoT-based system. *Sensors, 20*(13), 3795.
49. Senthilkumar, P., & Rajesh, K. (2021). Design of a model-based engineering deep learning scheduler in a cloud computing environment using Industrial Internet of Things (IIOT). *Journal of Ambient Intelligence and Humanized Computing.* https://doi.org/10.1007/s12652-020-02862-7.
50. Wu, J., Zhang, G., Nie, J., Peng, Y., & Zhang, Y. (2021). Deep reinforcement learning for scheduling in an edge computing-based industrial internet of things. *Wireless Communication Mobile Computing, 202*1. https://doi.org/10.1155/2021/8017334.
51. Ren, S., Kim, J. S., Cho, W. S., Soeng, S., Kong, S., & Lee, K. H. (2021). Big data platform for intelligence industrial IoT sensor monitoring system based on edge computing and AI. In *3rd International Conference Artificial Intelligence Information Communication ICAIIC* (Vol. 2021). https://doi.org/10.1109/ICAIIC51459.2021.9415189.
52. Ghabak, V., & Seetharaman, A. (2023). Integration of machine learning in agile supply chain management. In *2023 15th International Conference on Computer and Automation Engineering (ICCAE)*. IEEE.

INDEX

I

K

L

M

N

O

P

R

S

For Product Safety Concerns and Information please contact our EU
representative GPSR@taylorandfrancis.com
Taylor & Francis Verlag GmbH, Kaufingerstraße 24, 80331 München, Germany

www.ingramcontent.com/pod-product-compliance
Ingram Content Group UK Ltd.
Pitfield, Milton Keynes, MK11 3LW, UK
UKHW021512250926
475702UK00043B/248
* 9 7 8 1 7 7 9 6 4 3 4 1 4 *